# Room Acoustics
## Fifth Edition

# Room Acoustics
## Fifth Edition

# Heinrich Kuttruff

Institut für Technische Akustik, Technische
Hochschule Aachen, Aachen, Germany

CRC Press
Taylor & Francis Group
Boca Raton London New York

CRC Press is an imprint of the
Taylor & Francis Group, an **informa** business

First published 1973
by Elsevier Science Publishers Ltd
Second edition 1979
by Elsevier Science Publishers Ltd
Third edition 1991
by Elsevier Science Publishers Ltd
Fourth edition published 2000
by Spon Press

This edition published 2009
by Spon Press
2 Park Square, Milton Park, Abingdon, Oxon OX14 4RN

Simultaneously published in the USA and Canada
by Taylor & Francis
270 Madison Avenue, New York, NY 10016, USA

*Spon Press is an imprint of the Taylor & Francis Group,
an informa business*

© 1973, 1979, 1991 Elsevier Science Publishers;
1999, 2000, 2009 Heinrich Kuttruff

Typeset in Sabon by Keyword Group Ltd

CPI Antony Rowe, Chippenham, Wiltshire

*British Library Cataloguing in Publication Data*
A catalogue record for this book is available
from the British Library

*Library of Congress Cataloging in Publication Data*
Kuttruff, Heinrich.
  Room acoustics/Heinrich Kuttruff.—5th ed.
    p. cm.
  Includes bibliographical references and index.
  1. Architectural acoustics. I. Title.
  NA2800.K87 2009

729'.29—dc22                                          2008051863

ISBN10: 0-415-48021-3 (hbk)
ISBN10: 0-203-87637-7 (ebk)

ISBN13: 978-0-415-48021-5 (hbk)
ISBN13: 978-0-203-87637-4 (ebk)

# Contents

# Preface to the fifth edition

I am delighted to present a new edition of my book *Room Acoustics*. Preparing a new edition enabled me to take into consideration some new publications in the field of room acoustics. Furthermore, several more conventional subjects which were not incorporated into the earlier editions, although being of great interest with regard to room acoustics, have been included, along with some new figures. The chapter on measuring techniques has been restructured because of the rapid progress in signal processing. Finally, the new edition has provided me with the opportunity to eliminate a number of incomplete, inaccurate or even misleading formulations or expressions and to replace them with more adequate ones.

Despite all these modifications and additions, the original translation by Professor Peter Lord, to whom I owe my sincere gratitude, is still the basis of this text. Once more, I would like to thank the publisher for giving me the opportunity to present a revised version of my work, as well as for their excellent cooperation.

Heinrich Kuttruff
*Aachen*

# Preface to the fourth edition

Almost a decade has elapsed since the third edition and during this period many new ideas and methods have been introduced into room acoustics. I therefore welcome the opportunity to prepare a new edition of this book and to include the more important of those developments, while also introducing new topics which were not dealt with in earlier editions.

In room acoustics, as in many other technical fields, the digital computer has continued its triumphant progress; nowadays hardly any acoustical measurements are carried out without using a computer, allowing previously inconceivable improvements in accuracy and rapidity. Therefore, an update of the chapter on measuring techniques (Chapter 8) was essential. Furthermore, the increased availability of computers has opened new ways for the computation and simulation of sound fields in enclosures. These have led to better and more reliable methods in the practical design of halls; indeed, due to its flexibility and low cost, sound field simulation will probably replace the conventional scale model in the near future. Moreover, by simulation it can be demonstrated what a new theatre or concert hall which is still on the drawing board will sound like when completed ('auralisation'). These developments are described in Chapter 9, which contains a separate section on auralisation.

Also included in the new edition are sections on sound scattering and diffuse reflection, on sound reflection from curved walls, on sound absorption by several special arrangements (freely hanging porous material, Schroeder diffusers) and on the measurement of diffuse reflections from walls.

The preparation of a new edition offered the chance to present some subjects in a more comprehensive and logical way, to improve numerous text passages and formulae and to correct errors and mistakes that inevitably crept into the previous editions. I appreciate the suggestions of many critical readers, who drew my attention to weak or misleading material in the book. Most text passages, however, have been adopted from the previous editions without any changes. Therefore I want to express again my most sincere thanks to Professor Peter Lord of the University of Salford for his competent and sensitive translation. Finally, I want to thank the publishers for their cooperation in preparing this new edition.

Heinrich Kuttruff
*Aachen*

# Preface to the first edition

This book is intended to present the fundamentals of room acoustics in a systematic and comprehensive way so that the information thus provided may be used for the acoustical design of rooms and as a guide to the techniques of associated measurement.

These fundamentals are twofold in nature: the generation and propagation of sound in an enclosure, which are physical processes which can be described without ambiguity in the language of the physicist and engineer; and the physiological and psychological factors, of prime importance but not capable of exact description even within our present state of knowledge. It is the interdependence and the equality of importance of both these aspects of acoustics which are characteristic of room acoustics, whether we are discussing questions of measuring techniques, acoustical design, or the installation of a public address system.

In the earlier part of the book ample space is devoted to the objective description of sound fields in enclosures, but, even at this stage, taking into account, as far as possible, the limitations imposed by the properties of our hearing. Equal weight is given to both the wave and geometrical description of sound fields, the former serving to provide a more basic understanding, the latter lending itself to practical application. In both instances, full use is made of statistical methods; therefore, a separate treatment of what is generally known as 'statistical room acoustics' has been dispensed with.

The treatment of absorption mechanisms is based upon the concept that a thorough understanding of the various absorbers is indispensable for the acoustician. However, in designing a room he will not, in all probability, attempt to calculate the absorptivity of a particular arrangement but instead will rely on collected measurements and data based on experience. It is for this reason that in the chapter on measuring techniques the methods of determining absorption are discussed in some detail.

Some difficulties were encountered in attempting to describe the factors which are important in the perception of sound in rooms, primarily because of the fragmentary nature of the present state of knowledge, which seems to consist of results of isolated experiments which are strongly influenced by the conditions under which they were performed.

We have refrained from giving examples of completed rooms to illustrate how the techniques of room acoustics can be applied. Instead, we have chosen to show how one can progress in designing a room and which parameters need to be considered. Furthermore, because model investigations have proved helpful these are described in detail.

Finally, there is a whole chapter devoted to the design of loudspeaker installations in rooms. This is to take account of the fact that nowadays electroacoustic installations are more than a mere crutch in that they frequently present, even in the most acoustically faultless room, the only means of transmitting the spoken word in an intelligible way. Actually, the installations and their performance play a more important role in determining the acoustical quality of what is heard than certain design details of the room itself.

The book should be understood in its entirety by readers with a reasonable mathematical background and some elementary knowledge of wave propagation. Certain hypotheses may be omitted without detriment by readers with more limited mathematical training.

The literature on room acoustics is so extensive that the author has made no attempt to provide an exhaustive list of references. References have only been given in those cases where the work has been directly mentioned in the text or in order to satisfy possible demand for more detailed information.

The author is greatly indebted to Professor Peter Lord of the University of Salford and Mrs Evelyn Robinson of Prestbury, Cheshire, for their painstaking translation of the German manuscript, and for their efforts to present some ideas expressed in my native language into colloquial English. Furthermore, the author wishes to express his appreciation to the publishers for this carefully prepared edition. Last, but not least, he wishes to thank his wife most sincerely for her patience in the face of numerous evenings and weekends which he has devoted to his manuscript.

Heinrich Kuttruff
*Aachen*

# Introduction

We all know that a concert hall, theatre, lecture room or a church may have good or poor 'acoustics'. As far as speech in these rooms is concerned, it is relatively simple to make some sort of judgement on their quality by rating the ease with which the spoken word is understood. However, judging the acoustics of a concert hall or an opera house is generally more difficult, since it requires considerable experience, the opportunity for comparisons and a critical ear. Even so, the inexperienced cannot fail to learn about the acoustical reputation of a certain concert hall should they so desire, for instance by listening to the comments of others, or by reading the critical reviews of concerts in the press.

An everyday experience (although most people are not consciously aware of it) is that living rooms, offices, restaurants and all kinds of rooms for work can be acoustically satisfactory or unsatisfactory. Even rooms which are generally considered insignificant or spaces such as staircases, factories, passenger concourses in railway stations and airports may exhibit different acoustical properties; they may be especially noisy or exceptionally quiet, or they may differ in the ease with which announcements over the public address system can be understood. That is to say, even these spaces have 'acoustics' which may be satisfactory or less than satisfactory.

Despite the fact that people are subconsciously aware of the acoustics to which they are daily subjected, there are only a few who can explain what they really mean by 'good or poor acoustics' and who understand factors which influence or give rise to certain acoustical properties. Even fewer people know that the acoustics of a room is governed by principles which are amenable to scientific treatment. It is frequently thought that the acoustical design of a room is a matter of chance, and that good acoustics cannot be designed into a room with the same precision as a nuclear reactor or space vehicle is designed. This idea is supported by the fact that opinions on the acoustics of a certain room or hall frequently differ as widely as the opinions on the literary qualities of a new book or on the architectural design of a new building. Furthermore, it is well known that sensational failures in this field do occur from time to time. These and similar anomalies add even more weight to the general belief that the acoustics of a room is beyond the scope of calculation or prediction, at least with any reliability, and hence the study of room acoustics is an art rather than an exact science.

In order to shed more light on the nature of room acoustics, let us first compare it to a related field: the design and construction of musical instruments. This comparison is not as senseless as it may appear at first sight, since a concert hall too may be regarded as a large musical instrument, the shape and material of which determine to a considerable extent what the listener will hear. Musical instruments—string instruments for instance—are, as is well known, not designed or built by scientifically trained acousticians but, fortunately, by people who have acquired the necessary experience through long and systematic practical training. Designing or building musical instruments is therefore not a technical or scientific discipline but a sort of craft, or an 'art' in the classical meaning of this word.

Nevertheless, there is no doubt that the way in which a musical instrument functions, i.e. the mechanism of sound generation, the determining of the pitch of the tones generated and their timbre through certain resonances, as well as their radiation into the surrounding air, are all purely physical processes and can therefore be understood rationally, at least in principle. Similarly, there is no mystery in the choice of materials; their mechanical and acoustical properties can be defined by measurements to any required degree of accuracy. (How well these properties can be reproduced is another problem.) Thus, there is nothing intangible nor is there any magic in the construction of a musical instrument: many particular problems which are still unsolved will be understood in the not too distant future. Then one will doubtless be in a position to design a musical instrument according to scientific methods, i.e. not only to predict its timbre but also to give, with scientific accuracy, details for its construction, all of which are necessary to obtain prescribed or desired acoustical qualities.

Room acoustics is in a different position from musical instrument acoustics in that the end product is usually more costly by orders of magnitude. Furthermore, rooms are produced in much smaller numbers and have by no means geometrical shapes which remain unmodified through the centuries. On the contrary, every architect, by the very nature of his profession, strives to create something which is entirely new and original. The materials used are also subject to the rapid development of building technology. Therefore, it is impossible to collect in a purely empirical manner sufficient know-how from which reliable rules for the acoustical design of rooms or halls can be distilled. An acoustical consultant is confronted with quite a new situation with each task, each theatre, concert hall or lecture room to be designed, and it is of little value simply to transfer the experience of former cases to the new project if nothing is known about the conditions under which the transfer may be safely made.

This is in contrast to the making of a musical instrument where the use of unconventional materials as well as the application of new shapes is either firmly rejected as an offence against sacred traditions or dismissed as a whim. As a consequence, time has been sufficient to develop well-established empirical rules. And if their application happens to fail in one case or another, the faulty product is abandoned or withdrawn from service—which is not true for large rooms in an analogous situation.

For the above reasons, the acoustician has been compelled to study sound propagation in closed spaces with increasing thoroughness and to develop the knowledge in this field much further than is the case with musical instruments, even though the acoustical behaviour of a large hall is considerably more complex and involved. Thus, room acoustics has become a science during the past century and those who practise it on a purely empirical basis will fail sooner or later, like a bridge builder who waives calculations and relies on experience or empiricism.

On the other hand, the present level of reliable knowledge in room acoustics is not particularly advanced. Many important factors influencing the acoustical qualities of large rooms are understood only incompletely or even not at all. As will be explained below in more detail, this is due to the complexity of sound fields in closed spaces—or, as may be said equally well—to the large number of 'degrees of freedom' which we have to deal with. Another difficulty is that the acoustical quality of a room ultimately has to be proved by subjective judgements.

In order to gain more understanding about the sort of questions which can be answered eventually by scientific room acoustics, let us look over the procedures for designing the acoustics of a large room. If this room is to be newly built, some ideas will exist as to its intended use. It will have been established, for example, whether it is to be used for the showing of ciné films, for sports events, for concerts or as an open-plan office. One of the first tasks of the consultant is to translate these ideas concerning the practical use into the language of objective sound field parameters and to fix values for them which he thinks will best meet the requirements. During this step he has to keep in mind the limitations and peculiarities of our subjective listening abilities. (It does not make sense, for instance, to fix the duration of sound decay with an accuracy of 1% if no one can subjectively distinguish such small differences.) Ideally, the next step would be to determine the shape of the hall, to choose the materials to be used, to plan the arrangement of the audience, of the orchestra and of other sound sources, and to do all this in such a way that the sound field configuration will develop which has previously been found to be the optimum for the intended purpose. In practice, however, the architect will have worked out already a preliminary design, certain features of which he considers imperative. In this case the acoustical consultant has to examine the objective acoustical properties of the design by calculation, by geometric ray considerations, by model investigations or by computer simulation, and he will eventually have to submit proposals for suitable adjustments. As a general rule there will have to be some compromise in order to obtain a reasonable result.

Frequently the problem is refurbishment of an existing hall, either to remove architectural, acoustical or other technical defects or to adapt it to a new purpose which was not intended when the hall was originally planned. In this case an acoustical diagnosis has to be made first on the basis of appropriate measurement. A reliable measuring technique which yields objective quantities, which are subjectively meaningful at the same time, is an indispensable tool of the acoustician. The subsequent therapeutic step is essentially the same as described above: the acoustical consultant has to propose measures which would result in the intended objective changes in the sound field and consequently in the subjective impressions of the listeners.

In any case, the acoustician is faced with a two-fold problem: on the one hand he has to find and to apply the relations between the structural features of a room—such as shape, materials and so on—with the sound field which will occur in it, and on the other hand he has to take into consideration as far as possible the interrelations between the objective and measurable sound field parameters and the specific subjective hearing impressions effected by them. Whereas the first problem lies completely in the realm of technical reasoning, it is the latter problem which makes room acoustics different from many other technical disciplines in that the success or failure of an acoustical design has finally to be decided by the collective judgement of all 'consumers', i.e. by some sort of average, taken over the comments of individuals with widely varying intellectual, educational and aesthetic backgrounds. The measurement of sound field parameters can replace to a certain extent systematic or sporadic questioning of listeners. But, in the final analysis, it is the average opinion of listeners which decides whether the acoustics of a room is favourable or poor. If the majority of the audience (or that part which is vocal) cannot understand what a speaker is saying, or thinks that the sound of an orchestra in a certain hall is too dry, too weak or indistinct, then even though the measured reverberation time is appropriate, or the local or directional distribution of sound is uniform, the listener is always right; the hall does have acoustical deficiencies.

Therefore, acoustical measuring techniques can only be a substitute for the investigation of public opinion on the acoustical qualities of a room and it will serve its purpose better the closer the measured sound field parameters are related to subjective listening categories. Not only must the measuring techniques take into account the hearing response of the listeners but the acoustical theory too will only provide meaningful information if it takes regard of the consumer's particular listening abilities. It should be mentioned at this point that the sound field in a real room is so complicated that it is not open to exact mathematical treatment. The reason for this is the large number of components which make up the sound field in a closed space regardless of whether we describe it in terms of vibrational modes or, if we prefer, in terms of sound rays which have undergone one or more reflections from boundaries. Each of these components depends on the sound source, the shape of the room and on the materials from which it is made; accordingly, the exact computation of the sound field is usually quite involved. Supposing this procedure were possible with reasonable expenditure, the results would be so confusing that such a treatment would not provide a comprehensive survey and hence would not be of any practical use. For this reason, approximations and simplifications are inevitable; the totality of possible sound field data has to be reduced to averages or average functions which are more tractable and condensed to provide a clearer picture. This is why we have to resort so frequently to statistical methods and models in room acoustics, whichever way we attempt to describe sound fields. The problem is to perform these reductions and simplifications once again in accordance with the properties of human hearing, i.e. in such a way that the remaining average parameters correspond as closely as possible to particular subjective sensations.

From this it follows that essential progress in room acoustics depends to a large extent on the advances in psychological acoustics. As long as the physiological and psychological processes which are involved in hearing are not completely understood, the relevant relations between objective stimuli and subjective sensations must be investigated empirically—and should be taken into account when designing the acoustics of a room.

Many interesting relations of this kind have been detected and successfully investigated during the past few decades. But other questions which are no less important for room acoustics are unanswered so far, and much work remains to be carried out in this field.

It is, of course, the purpose of all efforts in room acoustics to avoid acoustical deficiencies and mistakes. It should be mentioned, on the other hand, that it is neither desirable nor possible to create the 'ideal acoustical environment' for concerts and theatres. It is a fact that the enjoyment when listening to music is a matter not only of the measurable sound waves hitting the ear but also of the listener's personal attitude and his individual taste, and these vary from one person to another. For this reason there will always be varying shades of opinion concerning the acoustics of even the most marvellous concert hall. For the same reason, one can easily imagine a wide variety of concert halls with excellent, but nevertheless different, acoustics. It is this 'lack of uniformity' which is characteristic of the subject of room acoustics, and which is responsible for many of its difficulties, but it also accounts for the continuous power of attraction it exerts on many acousticians.

# Chapter 1

# Some facts on sound waves, sources and hearing

In principle, any complex sound field can be considered as a superposition of numerous simple sound waves, e.g. plane waves. This is especially true of the very involved sound fields which we have to deal with in room acoustics. So it is useful to describe first the properties of a simple plane or a spherical sound wave or, more basically, the general features of sound propagation. We can, however, restrict our attention to sound propagation in gases, because in room acoustics we are only concerned with air as the medium.

In this chapter we assume the sound propagation to be free of losses and ignore the effect of any obstacles such as walls, i.e. we suppose the medium to be unbounded in all directions. Furthermore, we assume our medium to be homogeneous and at rest. In this case the velocity of sound is constant with reference to space and time. For air, its magnitude is

$$c = (331.4 + 0.6\,\theta) \quad \text{m/s} \tag{1.1}$$

where $\theta$ is the temperature in degrees centigrade.

In large halls, variations of temperature and hence of the sound velocity with time and position cannot be entirely avoided. Likewise, because of temperature differences and air conditioning, the air is not completely at rest, and so our assumptions are not fully realised. But the effects which are caused by these inhomogeneities are so small that they can be neglected.

## 1.1 Basic relations, the wave equation

In any sound wave, the particles of the medium undergo vibrations about their mean positions. Therefore, a wave can be described completely by indicating the instantaneous displacements of these particles. It is more customary, however, to consider the velocity of particle displacement as a basic acoustical quantity rather than the displacement itself.

The vibrations in a sound wave do not take place at all points with the same phase. We can, in fact, find points in a sound field where the particles vibrate in opposite phase. This means that in certain regions the particles are pushed together or compressed and in other regions they are pulled apart or rarefied. Therefore, under the influence of a sound wave, variations of gas density and pressure occur, both of which are functions of time and position. The difference between the instantaneous pressure and the static pressure is called the sound pressure.

The changes of gas pressure caused by a sound wave in general occur so rapidly that heat cannot be exchanged between adjacent volume elements. Consequently, a sound wave causes adiabatic variations of the temperature, and so the temperature too can be considered as a quantity characterising a sound wave.

The various acoustical quantities are connected by some basic laws which enable us to set up a general differential equation governing sound propagation. Firstly, conservation of momentum is expressed by the relation

$$\text{grad } p = -\rho_0 \frac{\partial \mathbf{v}}{\partial t} \tag{1.2}$$

where $p$ denotes the sound pressure, $\mathbf{v}$ the vector particle velocity, $t$ the time and $\rho_0$ the static value of the gas density. The unit of sound pressure is the pascal: 1 pascal (Pa) $= 1 \text{ N/m}^2 = 1 \text{ kg m}^{-1} \text{ s}^{-2}$.

Furthermore, conservation of mass leads to

$$\rho_0 \text{ div } \mathbf{v} = -\frac{\partial \rho}{\partial t} \tag{1.3}$$

$\rho$ being the total density including its variable part, $\rho = \rho_0 + \delta\rho$. In these equations, it is assumed that the changes of $p$ and $\rho$ are small compared with the static values $p_0$ and $\rho_0$ of these quantities; furthermore, the absolute value of the particle velocity $\mathbf{v}$ should be much smaller than the sound velocity $c$.

Under the further supposition that we are dealing with an ideal gas, the following relations hold between the sound pressure, the density variations and the temperature changes $\delta\theta$:

$$\frac{p}{p_0} = \kappa \frac{\delta\rho}{\rho_0} = \frac{\kappa}{\kappa - 1} \cdot \frac{\delta\theta}{\theta + 273} \tag{1.4}$$

Here $\kappa$ is the adiabatic or isentropic exponent (for air $\kappa = 1.4$).

The particle velocity $\mathbf{v}$ and the variable part $\delta\rho$ of the density can be eliminated from eqns (1.2) to (1.4). This yields the differential equation

$$c^2 \Delta p = \frac{\partial^2 p}{\partial t^2} \tag{1.5}$$

where

$$c^2 = \kappa \frac{p_0}{\rho_0} \tag{1.6}$$

$\Delta(= \nabla^2 = \text{div grad})$ is the Laplacian operator. This differential equation governs the propagation of sound waves in any lossless fluid and is therefore of central importance for almost all acoustical phenomena. We shall refer to it as the 'wave equation'. It holds not only for sound pressure but also for density and temperature variations.

## 1.2 Plane waves and spherical waves

Now we assume that the acoustical quantities depend only on the time and on one single direction, which may be chosen as the $x$-direction of a cartesian coordinate system. Then eqn (1.5) reads

$$c^2 \frac{\partial^2 p}{\partial x^2} = \frac{\partial^2 p}{\partial t^2} \tag{1.7}$$

The general solution of this differential equation is

$$p(x,t) = F(ct - x) + G(ct + x) \tag{1.8a}$$

where $F$ and $G$ are arbitrary functions, the second derivatives of which exist. The first term on the right represents a pressure wave travelling in the positive $x$-direction with a velocity $c$, because the value of $F$ remains unaltered if a time increase $\delta t$ is associated with an increase in the coordinate $\delta x = c\delta t$. For the same reason the second term describes a pressure wave propagated in the negative $x$-direction. Therefore the constant $c$ is the sound velocity.

Each term of eqn (1.8a) represents a progressive 'plane wave': As shown in Fig. 1.1a, the sound pressure $p$ is constant in any plane perpendicular to the $x$-axis. These planes of constant sound pressure are called 'wavefronts', and any line perpendicular to them is a 'wave normal'.

According to eqn (1.2), the particle velocity has only one non-vanishing component, which is parallel to the gradient of the sound pressure, i.e. to the $x$-axis. This means sound waves in fluids are longitudinal waves. The particle velocity may be obtained from applying eqn (1.2) to eqn (1.8a):

$$v(x,t) = \frac{1}{\rho_0 c} [F(ct - x) - G(ct + x)] \tag{1.8b}$$

As may be seen from eqns (1.8a) and (1.8b) the ratio of sound pressure and particle velocity in a plane wave propagated in the positive direction ($G = 0$)

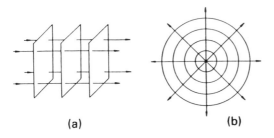

*Figure 1.1* Simple types of waves: (a) plane wave; (b) spherical wave.

is frequency independent:

$$\frac{p}{v} = \rho_0 c \tag{1.9}$$

This ratio is called the 'characteristic impedance' of the medium. For air at 20° centigrade its value is

$$\rho_0 c = 416 \frac{\mathrm{Pa}}{\mathrm{m/s}} = 416 \ \mathrm{kg \ m^{-2} \ s^{-1}} \tag{1.10}$$

If the wave is travelling in the negative $x$-direction, the ratio of sound pressure and particle velocity is negative.

Of particular importance are harmonic waves in which the time and space dependence of the acoustical quantities, for instance of the sound pressure, follows a sine or cosine function. If we set $G = 0$ and specify $F$ as a cosine function, we obtain an expression for a plane, progressive harmonic wave:

$$p(x, t) = \hat{p} \cos\left[k\left(ct - x\right)\right] = \hat{p} \cos\left(\omega t - kx\right) \tag{1.11}$$

with the arbitrary constants $\hat{p}$ and $k$. Here the angular frequency

$$\omega = kc \tag{1.12}$$

was introduced which is related to the temporal period

$$T = \frac{2\pi}{\omega} \tag{1.13a}$$

of the harmonic vibration represented by eqn (1.11). At the same time this equation describes a spatial harmonic vibration with the period

$$\lambda = \frac{2\pi}{k} \tag{1.13b}$$

This is the 'wavelength' of the harmonic wave. It denotes the distance in the $x$-direction where equal values of the sound pressure (or any other field quantity) occur. According to eqn (1.12) it is related to the angular frequency by

$$\lambda = \frac{2\pi c}{\omega} = \frac{c}{f} \tag{1.14}$$

where $f = \omega/2\pi = 1/T$ is the frequency of the vibration. It has the dimension second$^{-1}$; its units are hertz (Hz), kilohertz (1 kHz $= 10^3$ Hz), megahertz (1 MHz $= 10^6$ Hz), etc. The quantity $k = \omega/c$ is the propagation constant or the (angular) wave number of the wave, and $\hat{p}$ is its amplitude.

A very useful and efficient representation of harmonic oscillations and waves is obtained by applying the relation (with $i = \sqrt{-1}$):

$$\exp(ix) = \cos x + i \sin x \tag{1.15}$$

This is the complex or symbolic notation of harmonic vibrations and will be employed quite frequently in what follows. Using this relation, eqn (1.11) can be written in the form

$$p(x,t) = \text{Re}\left\{\hat{p} \exp\left[i\left(\omega t - kx\right)\right]\right\}$$

or, omitting the sign Re:

$$p(x,t) = \hat{p} \exp\left[i\left(\omega t - kx\right)\right] \tag{1.16}$$

The complex notation has several advantages over the real representation of eqn (1.11). Differentiation or integration with respect to time is equivalent to multiplication or division by i. Furthermore, only the complex notation allows a clear-cut definition of impedances and admittances (see Section 2.1). It fails, however, in all cases where vibrational quantities are to be multiplied or squared. If doubts arise concerning the physical meaning of an expression it is advisable to recall the origin of this notation, i.e. to take the real part of the expression.

As with any complex quantity the complex sound pressure in a plane wave may be represented in a rectangular coordinate system with the horizontal and the vertical axis corresponding to the real and the imaginary part of the pressure, respectively. The quantity is depicted as an arrow, often called 'phasor', pointing from the origin to the point which corresponds to the value of the pressure (see Fig. 1.2). The length of this arrow corresponds to the magnitude of the complex quantity while the angle it includes with the real axis is its phase angle or 'argument' (abbreviated arg $p$). In the present case the magnitude of the phasor equals the amplitude $\hat{p}$ of the oscillation, the phase angle depends on time $t$ and position $x$:

$$\arg p = \omega t - kx$$

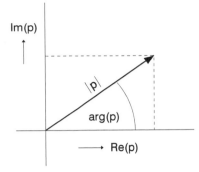

Im(p)

|p|

arg(p)

Re(p)

*Figure 1.2* Phasor representation of a complex quantity p.

This means, that for a fixed position the arrow or phasor rotates around the origin with an angular velocity $\omega$, which explains the expression 'angular frequency'.

So far it has been assumed that the wave medium is free of losses. If this is not the case, the pressure amplitude does not remain constant in the course of wave propagation but decreases according to an exponential law. Then eqn (1.16) is modified in the following way:

$$p(x,t) = \hat{p} \exp(-mx/2) \exp\left[i\left(\omega t - kx\right)\right] \tag{1.16a}$$

We can even use the representation of eqn (1.16) if we conceive the wave number $k$ as a complex quantity containing half the attenuation constant $m$ as its imaginary part:

$$k = \frac{\omega}{c} - i\frac{m}{2} \tag{1.17}$$

Another simple wave type is the spherical wave in which the surfaces of constant pressure, i.e. the wave fronts, are concentric spheres (see Fig. 1.1b). In their common centre we have to imagine some very small source which introduces or withdraws fluid. Such a source is called a 'point source'. The appropriate coordinates for this geometry are polar coordinates with the distance $r$ from the centre as the relevant space coordinate. Transformed into this system, the differential equation (1.5) reads:

$$\frac{\partial^2 p}{\partial r^2} + \frac{2}{r}\frac{\partial p}{\partial r} = \frac{1}{c^2}\frac{\partial^2 p}{\partial t^2} \tag{1.18}$$

A simple solution of this equation is

$$p(r,t) = \frac{\rho_0}{4\pi r}\dot{Q}\left(t - \frac{r}{c}\right) \tag{1.19}$$

It represents a spherical wave produced by a point source at $r = 0$ with the 'volume velocity' $Q$, which is the rate (in m³/s) at which fluid is expelled by the source. The overdot means partial differentiation with respect to time. Again, the argument $t - r/c$ indicates that any disturbance created by the sound source is propagated outward with velocity $c$, its strength decreasing as $1/r$. Reversing the sign in the argument of $\dot{Q}$ would result in the unrealistic case of an in-going wave.

The only non-vanishing component of the particle velocity is the radial one, which is calculated by applying eqn (1.2) to eqn (1.19):

$$v_r = \frac{1}{4\pi r^2}\left[Q\left(t - \frac{r}{c}\right) + \frac{r}{c}\dot{Q}\left(t - \frac{r}{c}\right)\right] \tag{1.20}$$

If the volume velocity of the source varies according to $Q(t) = \hat{Q}\exp(i\omega t)$, we obtain from eqn (1.19)

$$p(r, t) = \frac{i\omega\rho_0}{4\pi r}\hat{Q}\cdot\exp\left[i\left(\omega t - kr\right)\right] \tag{1.21}$$

This expression represents—with $k = \omega/c$—the sound pressure in a harmonic spherical wave. The particle velocity after eqn (1.20) is

$$v_r = \frac{p}{\rho_0 c}\left(1 + \frac{1}{ikr}\right) \tag{1.22}$$

This formula indicates that the ratio of sound pressure and particle velocity in a spherical sound wave depends on the distance $r$ and the frequency $\omega = kc$. Furthermore, it is complex, i.e. between both quantities there is a phase difference. For $kr \gg 1$, i.e. for distances which are large compared with the wavelength, the ratio $p/v_r$ tends asymptotically to $\rho_0 c$, the characteristic impedance of the medium.

A plane wave is an idealised wave type which does not exist in the real world, at least not in its pure form. However, sound waves travelling in a rigid tube can come very close to a plane wave. Furthermore, a limited region of a spherical wave may also be considered as a good approximation to a plane wave provided the distance $r$ from the centre is large compared with all wavelengths involved, i.e. $kr \gg 1$, see eqn (1.22).

On the other hand, a point source producing a spherical wave can be approximated by any sound source which is small compared with the wavelength and which expels fluid, for instance by a small pulsating sphere or a loudspeaker mounted into one side of an airtight box. Most sound sources, however, do not behave as point sources. In these cases, the sound pressure depends not only on the distance $r$ but also on the direction, which can be characterised by a polar angle $\vartheta$ and an azimuth angle $\varphi$. For distances exceeding a characteristic range, which depends on the sort of sound source and the frequency, the sound pressure is given by

$$p(r, \vartheta, \varphi, t) = \frac{A}{r} \Gamma(\vartheta, \varphi) \exp\left[\mathrm{i}(\omega t - kr)\right] \tag{1.23}$$

where the 'directional factor' $\Gamma(\vartheta, \varphi)$ is normalised so as to make $\Gamma = 1$ for its absolute maximum. $A$ is a constant.

## 1.3 Energy density and intensity, radiation

If a sound source, for instance a musical instrument, is to generate a sound wave it has to deliver some energy to a fluid. This energy is carried away by the sound wave. Accordingly we can characterise the amount of energy contained in one unit volume of the wave by the energy density. As with any kind of mechanical energy one has to distinguish between potential and kinetic energy density:

$$w_{\mathrm{pot}} = \frac{p^2}{2\rho_0 c^2}, \qquad w_{\mathrm{kin}} = \frac{\rho_0 |\mathbf{v}|^2}{2} \tag{1.24}$$

where $|\mathbf{v}|$ is the magnitude of the vector $\mathbf{v}$. The total energy density is

$$w = w_{\mathrm{pot}} + w_{\mathrm{kin}} \tag{1.25}$$

Another important quantity is sound intensity, which is a measure of the energy transported in a sound wave. Imagine a window of 1 m² perpendicular to the direction of sound propagation. Then the intensity is the energy per second passing this window. Generally the intensity is a vector parallel to the vector $\mathbf{v}$ of the particle velocity and is given by

$$\mathbf{I} = p\mathbf{v} \tag{1.26}$$

The principle of energy conservation requires

$$\frac{\partial w}{\partial t} + \mathrm{div}\,\mathbf{I} = 0 \tag{1.27}$$

It should be noted that—in contrast to the sound pressure and particle velocity—these energetic quantities do not simply add if two waves are superimposed on each other.

In a plane wave the sound pressure and the longitudinal component of the particle velocity are related by $p = \rho_0 c v$, and the same holds for a spherical wave at a large distance from the centre ($kr \gg 1$, see eqn (1.22)). Hence we can express the particle velocity in terms of the sound pressure. Then the energy density and the intensity are

$$w = \frac{p^2}{\rho_0 c^2} \quad \text{and} \quad I = \frac{p^2}{\rho_0 c} \tag{1.28}$$

They are related by

$$I = cw \tag{1.29}$$

Stationary signals which are not limited in time may be characterised by time averages over a sufficiently long time. We introduce the root-mean-square of the sound pressure by

$$p_{rms} = \left( \frac{1}{t_a} \int_0^{t_a} p^2 \, dt \right)^{1/2} = \left( \overline{p^2} \right)^{1/2} \tag{1.30}$$

where the overbar is a shorthand notation indicating time averaging. The eqns (1.28) yield

$$\overline{w} = \frac{p_{rms}^2}{\rho_0 c^2} \quad \text{and} \quad I = \frac{p_{rms}^2}{\rho_0 c} \tag{1.28a}$$

Finally, for a harmonic sound wave with the sound pressure amplitude $\hat{p}$, $p_{rms}$ equals $\hat{p}/\sqrt{2}$, which leads to

$$\overline{w} = \frac{\hat{p}^2}{2\rho_0 c^2} \quad \text{and} \quad I = \frac{\hat{p}^2}{2\rho_0 c} \tag{1.28b}$$

The last equation can be used to express the total power output of a point source $P = 4\pi r^2 I(r)$ by its volume velocity. According to eqn (1.21), the sound pressure amplitude in a spherical wave is $\rho_0 \omega \hat{Q}/4\pi r$ and therefore

$$P = \frac{\rho_0}{8\pi c} \hat{Q}^2 \omega^2 \tag{1.31}$$

If, on the other hand, the power output of a point source is given, the root-mean-square of the sound pressure at distance $r$ from the source is

$$p_{rms} = \frac{1}{r} \sqrt{\frac{\rho_0 c P}{4\pi}} \tag{1.32}$$

## 1.4 Signals and systems

Any acoustical signal can be unambiguously described by its time function $s(t)$ where $s$ denotes a sound pressure, a component of particle velocity, or the instantaneous density of air, for instance. If this function is a sine or cosine function—or an exponential with imaginary argument—we speak of a harmonic signal, which is closely related to the harmonic waves as introduced in Section 1.2. Harmonic signals and waves play a key role in acoustics although real sound signals are almost never harmonic but show a much more complicated time dependence. The reason for this apparent contradiction is the fact that virtually all signals can be considered as superposition of harmonic signals. This is the fundamental statement of the famous Fourier theorem.

The Fourier theorem can be formulated as follows: let $s(t)$ be a real, non-periodic time function describing, for example, the time dependence of the sound pressure or the volume velocity, this function being sufficiently steady (a requirement which is fulfilled in all practical cases), and the integral $\int_{-\infty}^{\infty} [s(t)]^2 dt$ having a finite value. Then

$$s(t) = \int_{-\infty}^{\infty} S(f) \exp(2\pi i f t) \, df \tag{1.33a}$$

and

$$S(f) = \int_{-\infty}^{\infty} s(t) \exp(-2\pi i f t) \, dt \tag{1.33b}$$

Because of the symmetry of these formulae $S(f)$ is not only the Fourier transform of $s(t)$ but $s(t)$ is the (inverse) Fourier transform of $S(f)$ as well. The complex function $S(f)$ is called the 'spectral function' or the 'complex amplitude spectrum', or simply the 'spectrum' of the signal $s(t)$. It can easily be shown that $S(-f) = S^*(f)$, where the asterisk denotes the transition to the complex conjugate function. $S(f)$ and $s(t)$ are different but equivalent representations of a signal.

According to eqn (1.33a), the signal $s(t)$ is composed of harmonic oscillations with continuously varying frequencies $f$. The absolute value of the spectral function, which can be written as

$$S(f) = |S(f)| \exp[i\psi(f)] \tag{1.34}$$

is the complex amplitude of the harmonic oscillation; $\psi(f)$ is its phase angle. The functions $|S(f)|$ and $\psi(f)$ are called the amplitude and the phase spectrum of the signal $s(t)$. A few examples of time function and their amplitude spectra are shown in Fig. 1.8.

The Fourier theorem assumes a slightly different form if $s(t)$ is a periodic function with period $T$, i.e. if $s(t) = s(t+T)$. Then the integral in eqn (1.33$a$) has to be replaced by a series:

$$s(t) = \sum_{n=-\infty}^{\infty} S_n \exp\left(\frac{2\pi i n t}{T}\right) \tag{1.35a}$$

with the coefficients

$$S_n = \int_{-\infty}^{\infty} s(t) \exp\left(\frac{-2\pi i n t}{T}\right) dt \tag{1.35b}$$

where $n$ is an integer. The steady spectral function has changed now into a set of discrete 'Fourier coefficients', for which $S_{-n} = S_n^*$ as before. Hence a periodic signal consists of discrete harmonic vibrations, the frequencies of which are multiples of a fundamental frequency $1/T$. These components are called 'partial oscillations' or 'harmonics', the first harmonic being identical with the fundamental oscillation.

If the signal is not continuous but consists of a sequence of $N$ discrete, periodically repeated numbers

$$s_0, s_1, s_2, \ldots, s_{N-2}, s_{N-1}, s_0, s_1, \ldots$$

these coefficients are given by

$$S_m = \sum_{n=0}^{N-1} s_n \exp\left(-\frac{2\pi i n m}{N}\right) \qquad (m = 0, 1, 2 \ldots N-1) \tag{1.36a}$$

The sequence of these coefficients, which is also periodic with the period $N$, is called the discrete fourier transform (DTF) of $s_n$. The inverse transformation reads

$$s_n = \sum_{n=0}^{N-1} S_n \exp\left(\frac{2\pi i n m}{N}\right) \qquad (m = 0, 1, 2 \ldots N-1) \tag{1.36b}$$

Of course all the formulae above can be written with the angular frequency $\omega = 2\pi f$ instead of the frequency $f$. A real notation of the formulae above is also possible. For this purpose one just has to separate the real parts from the imaginary parts in eqns (1.33), (1.35) and (1.36), respectively.

Equations (1.33) cannot be applied in this form to stationary non-periodic signals, i.e. to signals which are not limited in time and the statistical properties of which are not time dependent. This holds, for instance, for any kind of random noise. In this case the integrals would not converge. Therefore, firstly a 'window' of width $T_0$ is cut out of the signal. For this section the spectral function $S_{T_0}(f)$ is well defined and can be evaluated numerically or experimentally. The 'power spectrum' of the whole signal is then given by

$$W(f) = \lim_{T_0 \to \infty} \left[ \frac{1}{T_0} S_{T_0}(f) S^*_{T_0}(f) \right] \tag{1.37}$$

The determination of the complex spectrum $S(f)$ or of the power spectrum $W(f)$, known as 'spectral analysis', is of great theoretical and practical importance. Nowadays it is most conveniently achieved by using digital computers. A particularly efficient procedure for computing spectral functions is the 'fast Fourier transform' (FFT) algorithm. To give at least an idea of it we suppose that $N$ is an even number. We note that the sum in eqn (1.36a) can be decomposed into one sum containing terms of even order $m$ and a second one comprising the terms with odd $m$:

$$S_m = \sum_{j=0}^{(N/2)-1} s_{2j} \exp\left(-2\pi i \frac{jm}{N/2}\right)$$

$$+ \exp\left(-2\pi i \frac{m}{N}\right) \sum_{j=0}^{(N/2)-1} s_{2j+1} \exp\left(-2\pi i \frac{2jm}{N/2}\right)$$

Each of these sums represents a discrete Fourier transform of $N/2$ elements. But while the calculation according to eqn (1.36a) requires $N^2$ multiplications, the number of multiplications occurring in the above equation is only $2(N/2)^2 + N$. If $N$ is an integral power of 2, the decomposition of sums can be repeated again and again until the final sums consist just of one term each. It can be shown that the final number of required multiplications is only $(N/2) \log_2(N)$. Thus, the saving of computing time provided by FFT is considerable, particularly if $N$ is large. For a more detailed description of the fast Fourier transform, the reader is referred to the extensive literature on this subject (e.g. Ref. 1). A rough, experimental spectral analysis can also be carried out by applying the signal to a set of bandpass filters.

The power spectrum, which is an even function of the frequency, does not contain all the information on the original signal $s(t)$, because it is based on the absolute value of the spectral function only, whereas all phase information has been eliminated. Inserted into eqn (1.33a), it does not restore the original function $s(t)$ but instead yields another important time function, called the 'autocorrelation function' of $s(t)$:

$$\phi_{ss}(\tau) = \int_{-\infty}^{\infty} W(f) \cdot \exp(2\pi i f \tau)df = 2\int_{0}^{\infty} W(f) \cdot \cos(2\pi f \tau)df \qquad (1.38a)$$

The time variable has been denoted by $\tau$ in order to indicate that it is not identical with the real time. In the usual definition of the autocorrelation function it occurs as a time shift:

$$\phi_{ss}(\tau) = \lim_{T_0 \to \infty} \frac{1}{T_0} \int_{-T_0/2}^{T_0/2} s(t)s(t+\tau)\,dt = \overline{s(t)s(t+\tau)} \qquad (1.38b)$$

The autocorrelation function characterizes the similarity of a signal at time $t$ and the same signal at a different time $t+\tau$.

Since $\phi_{ss}$ is the Fourier transform of the power spectrum, the latter is also obtained by inverse Fourier transformation of the autocorrelation function:

$$W(f) = \int_{-\infty}^{\infty} \phi_{ss}(\tau) \cdot \exp(-2\pi i f \tau)d\tau = 2\int_{0}^{\infty} \phi_{ss}(\tau) \cdot \cos(2\pi f \tau)d\tau \qquad (1.39)$$

Equations (1.38a) and (1.39) are the mathematical expressions of the theorem of Wiener and Khinchine: power spectrum and autocorrelation function are Fourier transforms of each other.

If $s(t+\tau)$ in eqn (1.38b) is replaced with $s'(t+\tau)$, where $s'$ denotes a time function different from $s$, one obtains the 'cross-correlation function' of the two signals $s(t)$ and $s'(t)$:

$$\phi_{ss'}(\tau) = \lim_{T_0 \to \infty} \frac{1}{T_0} \int_{-T_0/2}^{T_0/2} s(t)s'(t+\tau)dt = \overline{s(t)s'(t+\tau)} \qquad (1.40)$$

The cross-correlation function provides a measure of the statistical similarity of two functions $s$ and $s'$. It is closely related to the correlation coefficient

$$\Psi = \frac{\overline{s(t) \cdot s'(t)}}{\sqrt{\overline{s(t)^2} \cdot \overline{s'(t)^2}}} \qquad (1.41)$$

which may vary between $+1$ and $-1$. If $\Psi = 0$, the functions $s(t)$ and $s'(t)$ are said to be uncorrelated (or incoherent). This is the case when these functions are random or nearly random. As an example, consider the signals

$$s(t) = \sum_n a_n \cos(\omega_n t + \varphi_n) \qquad \text{and} \qquad s'(t) = \sum_n b_n \cos(\omega_n t + \varphi'_n)$$

each of them consisting of many sinusoidal components with phase angles which are randomly distributed over the interval from 0 to $2\pi$. First we calculate the averaged square of both sums:

$$s^2 = \sum_n \sum_m a_n a_m \cos(\omega_n t + \varphi_n)\cos(\omega_m t + \varphi_m)$$

$$= \frac{1}{2}\sum_n \sum_m a_n a_m \left[\cos(\omega_n t + \omega_m t + \varphi_n + \varphi_m) + \cos(\omega_n t - \omega_m t + \varphi_n - \varphi_m)\right]$$

Time averaging eliminates all time-dependent terms, i.e. all terms except those with $n = m$, which are reduced to $a_n^2$. Thus we arrive at the important result that under the given conditions the averaged square of the signal $s(t)$ is just

$$\overline{s^2} = \frac{1}{2}\sum_n a_n^2 \tag{1.42}$$

In a similar way it is shown that

$$\overline{s \cdot s'} = \frac{1}{2}\sum_n a_n b_n \cos(\varphi_n - \varphi'_n)$$

which is zero since the cos terms cancel due to the random distribution of phase angles.

It should be noted that $\Psi = 0$ is a necessary but not a sufficient condition for two signals being uncorrelated.

In a certain sense a sine or cosine signal can be considered as an elementary signal; it is unlimited in time and steady in all its derivatives, and its spectrum consists of a single line. The counterpart of it is Dirac's delta function $\delta(t)$: it has one single line in the time domain, so to speak, whereas its amplitude spectrum is constant for all frequencies, i.e. $S(f) = 1$ for the delta function. This leads to the following representation:

$$\delta(t) = \lim_{f_0 \to \infty} \frac{1}{f_0} \int_{-f_0/2}^{f_0/2} \exp(2\pi ft)\,df \tag{1.43}$$

The delta function has the following fundamental property:

$$s(t) = \int_{-\infty}^{\infty} s(\tau)\delta(t-\tau)\,d\tau \tag{1.44}$$

where $s(t)$ is any function of time. Accordingly, any signal can be considered as a close succession of very short pulses, as indicated in Fig. 1.3. Especially, for $s(t) \equiv 1$ we obtain

$$\int_{-\infty}^{\infty} \delta(\tau)\,d\tau = 1 \tag{1.45}$$

Since the delta function $\delta(t)$ is zero for all $t \neq 0$ it follows from eqn (1.45) that its value at $t = 0$ must be infinite.

Now consider a linear and time-independent but otherwise unspecified transmission system. Examples of acoustical transmission systems are all kinds of ducts (air ducts, mufflers, wind instruments, etc.) and resonators. Likewise, any two points in an enclosure may be considered as the input and output terminal of an acoustic transmission system. Linearity means that multiplying the input signal with a factor results in an output signal which is augmented by the same factor. The properties of such a system are completely characterised by the so-called 'impulse response' $g(t)$, i.e. the output signal which is the response to an impulsive input signal represented by the Dirac function $\delta(t)$ (see Fig. 1.4). Since the response cannot precede the excitation, the impulse response of any causal system must vanish for $t < 0$. If $g(t)$ is known, the output signal $s'(t)$ with respect to any input signal $s(t)$ can be obtained by replacing the Dirac function in eqn (1.44) with its

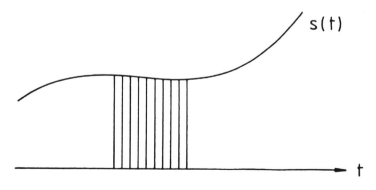

Figure 1.3 Continuous function as the limiting case of a close succession of short impulses.

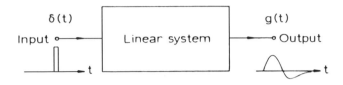

*Figure 1.4* Impulse response of a linear system.

response, i.e. with $g(t)$:

$$s'(t) = \int\limits_{-\infty}^{\infty} s(\tau)g(t-\tau)\,\mathrm{d}\tau = \int\limits_{-\infty}^{\infty} g(\tau)s(t-\tau)\,\mathrm{d}\tau \tag{1.46}$$

This operation is known as the convolution of two functions $s$ and $g$. A common shorthand notation of it is

$$s'(t) = s(t) * g(t) = g(t) * s(t) \tag{1.46a}$$

Equation (1.46) has its analogue in the frequency domain, which looks even simpler. Let $S(f)$ be the complex spectrum of the input signal $s(t)$ of our linear system; then the spectrum of the resulting output signal is

$$S'(f) = G(f) \cdot S(f) \tag{1.47}$$

The complex function $G(f)$ is the 'transmission function' or 'transfer function' of the system; it is related to the impulse response by the Fourier transformation:

$$G(f) = \int\limits_{-\infty}^{\infty} g(t) \exp(-2\pi i f t)\,\mathrm{d}t \tag{1.48a}$$

$$g(t) = \int\limits_{-\infty}^{\infty} G(f) \exp(2\pi i f t)\,\mathrm{d}f \tag{1.48b}$$

with $G(-f) = G^*(f)$ since $g(t)$ is a real function. The transfer function $G(f)$ has also a direct meaning: if a harmonic signal with frequency $f$ is applied to a transmission system, its amplitude will be changed by the factor $|G(f)|$ and its phase will be shifted by the phase angle of $G(f)$.

## 1.5 Sound pressure level and sound power level

In the frequency range in which our hearing is most sensitive (500–5000 Hz) the intensity of the threshold of sensation and of the threshold of pain in hearing differ by about 13 orders of magnitude. For this reason it would be impractical to characterise the strength of a sound signal by its sound pressure or its intensity. Instead, a logarithmic quantity, the so-called 'sound pressure level' is generally used for this purpose, defined by

$$SPL = 20\log_{10}\left(\frac{p_{\text{rms}}}{p_0}\right) \quad \text{decibels} \tag{1.49}$$

In this definition, $p_{\text{rms}}$ denotes the 'root-mean-square' pressure, as introduced in Section 1.3. The quantity $p_0$ is an internationally standardised reference pressure; its value is $2 \times 10^{-5}$ Pa, which corresponds roughly to the normal hearing threshold at 1000 Hz. The 'decibel' (abbreviated dB) is not a unit in a physical sense but is used rather to recall the above level definition. Strictly speaking, the $p_{\text{rms}}$ as well as the SPL are defined only for stationary sound signals since they both imply an averaging process.

According to eqn (1.49), two different sound fields or signals may be compared by their level difference:

$$\Delta SPL = 20\log_{10}\left(\frac{p_{\text{rms}1}}{p_{\text{rms}2}}\right) \quad \text{decibels} \tag{1.49a}$$

It is often convenient to express the sound power delivered by a sound source in terms of the 'sound power level', defined by

$$PL = 10\log_{10}\left(\frac{P}{P_0}\right) \quad \text{decibels} \tag{1.50}$$

where $P_0$ is a reference power of $10^{-12}$ W. Using this quantity, the sound pressure level produced by a point source with power $P$ in the free field can be expressed as follows (see eqn (1.32).

$$SPL = PL - 20\log_{10}\left(\frac{r}{r_0}\right) - 11\,\text{dB} \quad \text{with } r_0 = 1 \text{ m} \tag{1.51}$$

## 1.6 Some properties of human hearing

Since the ultimate consumer of all room acoustics is the listener, it is important to consider at least a few facts relating to aural perception. More information may be found in Ref. 2, for instance.

One of the most obvious facts of human hearing is that the ear is not equally sensitive to sounds of different frequencies. Generally, the loudness at which a sound is perceived depends, of course, on its objective strength, i.e. on its sound pressure level. Furthermore, it depends in a complicated manner on the spectral composition of the sound signal, on its duration and

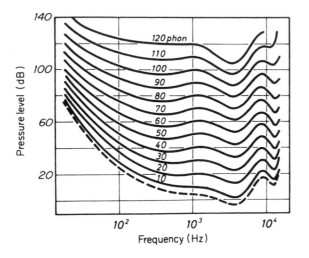

*Figure 1.5* Contours of equal loudness level for frontal sound incidence. The dashed curve corresponds to the average hearing threshold.

on several other factors. The loudness is often characterised by the 'loudness level', which is the sound pressure level of a 1000 Hz tone which appears equally loud as the sound to be characterised. The unit of the loudness level is the 'phon'.

Figure 1.5 presents the contours of equal loudness level for sinusoidal sound signals which are presented to a listener in the form of frontally incident plane waves. The numbers next to the curves indicate the loudness level. The lowest, dashed curve which corresponds to a loudness level of 3 phons, marks the threshold of hearing. According to this diagram, a pure tone with a SPL of 40 decibels has by definition a loudness level of 40 phons if its frequency is 1000 Hz. However, at 100 Hz its loudness level would be only 24 phons whereas at 50 Hz it would be almost inaudible.

Using these curves, the loudness level of any pure tone can be determined from its frequency and its sound pressure level. In order to simplify this somewhat tedious procedure, electrical instruments have been constructed which measure the sound pressure level. Basically, they consist of a calibrated microphone which converts the sound into an electrical signal, amplifiers and, most important, a weighting network, the frequency-dependent attenuation of which approximates the contours of equal loudness. Several weighting functions are in use and have been standardised internationally; the most common of them is the A-weighting curve. Consequently, the result of such a measurement is not the loudness level in phons, but the 'A-weighted sound pressure level' in dB(A).

When such an instrument is applied to a sound signal with more complex spectral structure, the result may deviate considerably from the true loudness level. The reason for such errors is the fact that, in our hearing, weak spectral components are partially or completely masked by stronger ones and that this effect is not modelled in the above-mentioned sound level meters. Apart from masking in the frequency domain, temporal masking may occur in non-stationary signals. In particular, a strong time-variable signal may mask a subsequent weaker signal. This effect is very important in listening in closed spaces as will be described in more detail in Chapter 7.

A more fundamental shortcoming of the above-mentioned measurement of loudness level is its unsatisfactory relation to our subjective perception. In fact, doubling the subjective sensation of loudness does not correspond to twice the loudness level as should be expected. Instead it corresponds only to an increase of about 10 phons. This shortcoming is avoided by the loudness scale with the 'sone' as a unit. The sone scale is defined in such a way that 40 phons correspond to 1 sone and every increase of the loudness level by 10 phons corresponds to doubling the number of sones. Nowadays instruments as well as computer programs are available which are able to measure or to calculate the loudness of almost any type of sound signal, taking into account the above-mentioned masking effect.

Another important property of our hearing is its ability to detect the direction from which a sound wave is arriving, and thus to localise the direction of sound sources. For sound incidence from a lateral direction it is easy to understand how this effect is brought about: an originally plane or spherical wave is distorted by the human head, by the pinnae and—to a minor extent—by the shoulders and the trunk. This distortion depends on sound frequency and the direction of incidence. As a consequence, the sound signals at both ears show characteristic differences in their amplitude and phase spectrum or, to put it more simply, at lateral sound incidence one ear is within the shadow of the head but the other is not. The interaural amplitude and phase differences caused by these effects enable our hearing to reconstruct the direction of sound incidence.

Quantitatively, the changes a sound signal undergoes on its way to the entrance of the ear canal can be described by the so-called 'head-related transfer functions (HRTF)' which characterise the transmission from a very remote point source to the ear canal, for instance its entrance. Such transfer functions have been measured by many researchers.[3] As an example, Fig. 1.6 shows head-related transfer functions for 11 lateral angles of incidence $\varphi$ relative to the sound pressure at frontal sound incidence ($\varphi = 0°$); Fig. 1.6a presents the magnitude expressed in decibels, while Fig. 1.6b plots the group delays, i.e. the functions $(1/2\pi)(d\psi/df)$ where $\psi$ is the phase of the HRTF. By comparing the curves of $\varphi = 90°$ and $\varphi = 270°$, the shadowing effect of the head becomes obvious.

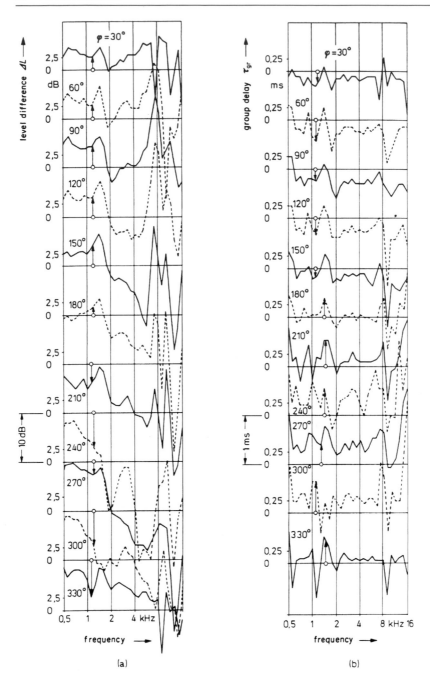

*Figure 1.6* Head-related transfer functions for several directions of sound incidence in the horizontal plane, relative to that for frontal incidence, average over 25 subjects: (a) amplitude (logarithmic scale, in decibels); (b) group delay (after Blauert[3]).

However, if the sound source is situated within the vertical symmetry plane, this explanation fails since then the source produces equal sound signals at both ear canals. But even then the ear transfer functions show characteristic differences for various elevation angles of the source, and it is commonly believed that the way in which they modify a sound signal enables us to distinguish whether a sound source is behind, above or in front of our head.

These considerations are valid only for the localisation of sound sources in a free sound field. In a closed room, however, the sound field is made up of many sound waves propagating in different directions, and accordingly, matters are more complicated. We shall discuss the subjective effects of more complex sound fields as they are encountered in room acoustics in Chapter 7.

## 1.7 Sound sources

In room acoustics we are mainly concerned with three types of sound source: the human voice, musical instruments and technical noise sources. (We do not consider loudspeakers here because they reproduce sound signals originating from other sources.)

It is a common feature of all these sources that the sounds they produce have a more or less complicated spectral structure—apart from some rare exceptions. In fact, it is the spectral content of speech signals (phonemes) which gives them their characteristics. Similarly, the timbre of musical sounds is determined by their spectra.

The signals emitted by most musical instruments, in particular by string and wind instruments, including the organ, are nearly periodic. Therefore, their spectra consist of many discrete Fourier coefficients (see Section 1.4) which may be represented by equally spaced vertical lines. The time signals themselves are mixtures of harmonic vibrations with frequencies which are integral multiples of the fundamental frequency $1/T$ ($T$ = period of the signal). The component with the lowest frequency $1/T$ is the fundamental tone, and the higher-order components are called overtones. It is the fundamental which determines what we perceive as the pitch of a tone. This means that our ear receives many harmonic components of quite different frequencies even if we listen to a single tone. Likewise, the spectra of many speech sounds, in particular vowels and voiced consonants, have a line structure. As an example, Fig. 1.7 presents the time function and the amplitude spectrum of three vowels.[4] There are some characteristic frequency ranges in which the overtones are especially strong. These are called the 'formants' of the vowel.

For normal speech the fundamental frequency lies between 50 and 350 Hz and is identical to the frequency at which the vocal chords vibrate. The total frequency range of conversational speech may be seen from Fig. 1.8,

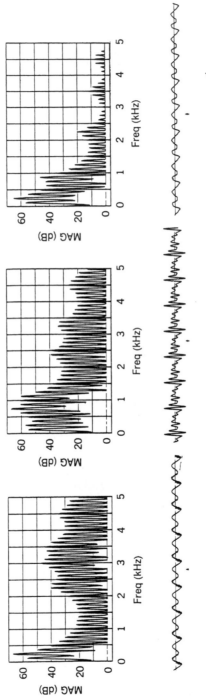

Figure 1.7 Amplitude spectrum and time function (sound pressure) of vowels /i/ (left), /a/ (middle) and /u/ (right) (after Flanagan[4]).

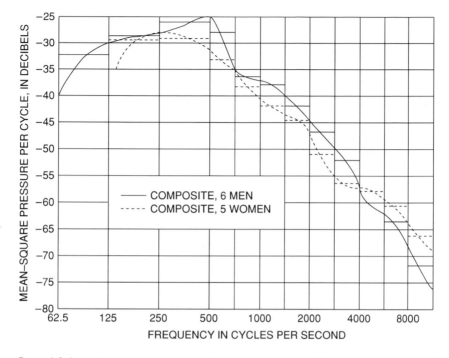

*Figure 1.8* Long-time power spectrum of continuous speech 30 cm from the mouth (after Flanagan[5]).

which plots the long-time power spectrum of continuous speech, both for male and female speakers.[5] The high-frequency energy is mainly due to the consonants, for instance to fricatives such as /s/ or /f/, or to plosives such as /p/ or /t/. Since consonants are of particular importance for the intelligibility of speech, a room or hall intended for speech, as well as a public address system, should transmit the high frequencies with great fidelity. The transmission of the fundamental vibration, on the other hand, is less important since our hearing is able to reconstruct it if the sound signal is rich in higher harmonics.

Among musical instruments, large pipe organs have the widest frequency range, reaching from 16 Hz to about 9 kHz. (There are some instruments, especially percussion instruments, which produce sounds with even higher frequencies.) The piano follows, having a frequency range which is smaller by about three octaves, i.e. by nearly a decade. The frequencies of the remaining instruments lie somewhere within this range. This is true, however, only for the fundamental frequencies. Since almost all instruments produce higher harmonics, the actual range of frequencies occurring in music extends still further, up to about 15 kHz. In music, unlike speech, all frequencies are of almost equal importance, so it is not permissible deliberately to suppress or to neglect certain frequency ranges. On the other hand, the entire frequency range is not the responsibility of the acoustical engineer. At 10 kHz and above the attenuation in air is so dominant that the influence of a room on the propagation of high-frequency sound components can safely be neglected. At frequencies lower than 50 Hz geometrical considerations are almost useless because of the large wavelengths of the sounds; furthermore, at these frequencies it is almost impossible to assess correctly the sound absorption by vibrating panels or walls and hence to control the reverberation. This means that, in this frequency range too, room acoustical design possibilities are very limited. On the whole, it can be stated that the frequency range relevant to room acoustics reaches from 50 to 10 000 Hz, the most important part being between 100 and 5000 Hz.

The acoustical power output of the sound sources as considered here is relatively low by everyday standards. Table 1.1 lists a few typical data. The human voice generates a sound power ranging from 0.001 $\mu$W (whispering) to 1000 $\mu$W (shouting), the power produced in conversational speech is of the order of 10 $\mu$W, corresponding to a sound power level of 70 dB. The power of a single musical instrument may lie in the range from 10 $\mu$W to 100 mW. A full symphony orchestra can easily generate a sound power of 10 W in fortissimo passages. It may be added that the dynamic range of most musical instruments is about 30 dB (woodwinds) to 50 dB (string instruments). A large orchestra can cover a dynamic range of 100 dB.

An important property of the human voice and musical instruments is their directionality, i.e. the fact that they do not emit sound with equal intensity in all directions. In speech this is because of the 'sound shadow' cast by the head. The lower the sound frequency, the less pronounced is the reduction of sound intensity by the head, because with decreasing frequencies the sound waves are increasingly diffracted around the head.

Table 1.1 Sound power and power level of some sound sources (the data of musical instruments are for fortissimo)

| Source or signal | Sound power (mW) | Sound power level (dB) |
| --- | --- | --- |
| Whisper | $10^{-6}$ | 30 |
| Conversational speech | 0.01 | 70 |
| Human voice, maximum | 1 | 90 |
| Violin | 1 | 90 |
| Clarinet, French horn | 50 | 107 |
| Trumpet | 100 | 110 |
| Organ, large orchestra | $10^4$ | 130 |

In Figs 1.9a and 1.9b the distribution of the relative pressure level for different frequency bands is plotted on a horizontal plane and a vertical plane, respectively. These curves are obtained by filtering out the respective frequency bands from natural speech; the direction denoted by 0° is the frontal direction.

Musical instruments usually exhibit a pronounced directionality because of the linear dimensions of their sound-radiating surfaces, which, in the interest of high efficiency, are often large compared with the wavelengths. Unfortunately general statements are almost impossible, since the directional distribution of the radiated sound changes very rapidly, not only from one frequency to the other; it can be quite different for instruments of the same sort but different manufacture. This is true especially for string instruments, the bodies of which exhibit complicated vibration patterns, particularly at higher frequencies. The radiation from a violin takes place in a fairly uniform way only at frequencies lower than about 450 Hz; at higher frequencies, however, matters become quite involved. For wind instruments the directional distributions exhibit more common features, since here the sound is not radiated from a curved anisotropic plate with complicated vibration patterns but from a fixed opening which is very often the end of a horn. The 'directional characteristics of an orchestra' are highly involved, but space is too limited here to discuss this in detail. For the room acoustician, however, it is important to know that strong components, particularly from the strings but likewise from the piano, the woodwinds and, of course, from the tuba, are radiated upwards. For further details we refer to the exhaustive account of J. Meyer.[6]

In a certain sense, the sounds from natural sources can be considered as statistical or stochastic signals, and in this context their autocorrelation function is of interest as it gives some measure of a signal's 'tendency of conservation'. Autocorrelation measurements on speech and music have been performed by several authors.[7,8] Here we are reporting results obtained by Ando, who passed various signals through an A-weighting filter and formed their autocorrelation function according to eqn (1.38b) with a finite integration time of $T_0 = 35$ s. Two of his results are depicted in Fig. 1.10. A useful measure of the effective duration of the autocorrelation function is the delay $\tau_e$, at which its envelope is just one-tenth of its maximum. These values are indicated in Table 1.2 for a few signals. They range from about 10 to more than 100 ms.

The variety of possible noise sources is too large to discuss in any detail. A common kind of noise in a room is sound intruding from adjacent rooms or from outside through walls, doors and widows, due to insufficient sound insulation. A typical noise source in halls is the air conditioning system; some of the noise produced by the machinery propagates in the air ducts and is radiated into the hall through the air outlets.

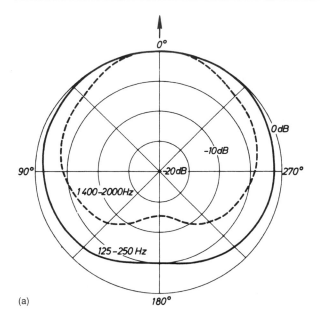

(a)

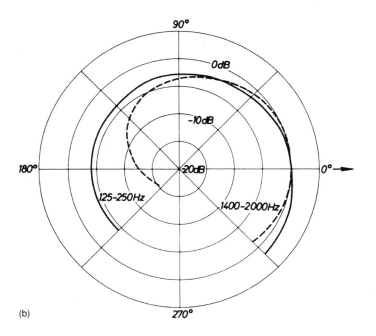

(b)

*Figure 1.9* Directional characteristics of speech sounds for two different frequency bands. The arrow points in the viewing direction: (a) in the horizontal plane; (b) in the vertical plane.

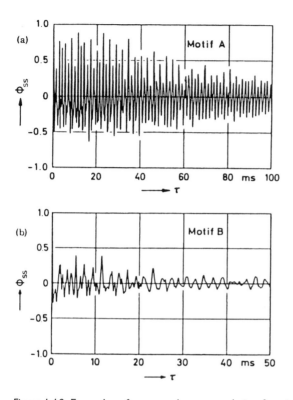

*Figure 1.10* Examples of measured autocorrelation functions: (a) music motif A; (b) music motif B (both from Table 1.2) (after Ando[8]).

*Table 1.2* Duration of autocorrelation functions of various sound signals (after Ando[8])

| Motif | Name of piece | Composer | Duration $\tau_e$, (ms) |
|---|---|---|---|
| A | Royal Pavane | Gibbons | 127 |
| B | Sinfonietta opus 48, 4th movement (Allegro con brio) | Arnold | 43 |
| C | Symphony No. 102 in B flat major, 2nd movement (Adagio) | Haydn | 65 |
| D | Siegfried Idyll; bar 322 | Wagner | 40 |
| E | Symphony KV 551 in C major (Jupiter), 4th movement (Molto allegro) | Mozart | 38 |
| F | Poem read by a female | Kunikita | 10 |

## References

1 Bracewell RN. The Fourier Transform and its Applications. Singapore: McGraw-Hill, 1986.
2 Zwicker E, Fastl H. Psychoacoustics—Facts and Models. Berlin: Springer-Verlag, 1990.
3 Blauert J. Spatial Hearing. Cambridge, Mass: MIT Press, 1997.
4 Flanagan JJ. Speech communication. In: Crocker MJ (ed.), Encyclopedia of Acoustics. New York: John Wiley, 1997.
5 Flanagan JJ. Speech Analysis Synthesis and Perception. Berlin: Springer-Verlag, 1965.
6 Meyer J. Acoustics and the Performance of Music. Berlin: Springer-Verlag, 2008.
7 Furdujev V. Evaluation objective de l'acoustique des salles. Proc Fifth Intern Congr on Acoustics, Liege, 1965, p 41.
8 Ando YJ. Subjective preference in relation to objective parameters of music sound fields with a single echo. Acoust Soc Am 1977; 62:1436. Proc of the Vancouver Symposium (Acoustics and Theatres for the Performing Arts) Canadian Acoust Association, Ottawa, Canada, 1986, p 112.

# Chapter 2

# Reflection and scattering

Up to now we have dealt with sound propagation in a medium which was unbounded in every direction. In contrast to this simple situation, room acoustics is concerned with sound propagation in enclosures where the sound-conducting medium is bounded on all sides by walls, ceiling and floor. These room boundaries usually reflect a certain fraction of the sound energy impinging on them. Another fraction of the energy is 'absorbed', i.e. it is extracted from the sound field inside the room, either by conversion into heat or by being transmitted to the outside by the walls. It is just this combination of the numerous reflected components which is responsible for what is known as 'the acoustics of a room' and also for the complexity of the sound field in a room.

Before we discuss the properties of such involved sound fields we shall consider in this chapter the process which is fundamental for their occurrence: the reflection of a plane sound wave by a single wall or surface. In this context we shall encounter the concepts of wall impedance and absorption coefficient, which are of special importance in room acoustics. The sound absorption by a wall will be dealt with mainly from a formal point of view, whereas the discussion of the physical causes of sound absorption and of the functional principles of various absorbent arrangements will be postponed to Chapter 6.

Strictly speaking, the simple laws of sound reflection to be explained in this chapter hold only for unbounded walls. Any free edge of a reflecting wall or panel will scatter some sound energy in all directions. The same happens when a sound wave hits any other obstacle of limited extent, such as a pillar or a listener's head, or when it arrives at a basically plane wall which has an irregular surface. Since scattering is a common phenomenon in room acoustics, we shall briefly deal with it in this chapter.

Throughout this chapter we shall assume that the incident, undisturbed wave is a plane wave. In reality, however, all waves originate from a sound source and are therefore spherical waves or superpositions of spherical waves. The reflection of a spherical wave from a plane wall is highly complicated unless we can assume that the wall is rigid. More on this matter may be found in the literature (see, for instance, Ref. 1 which presents a comprehensive discussion of the exact theory and its various approximations). For our discussion it may be sufficient to assume that the sound source is not too close to the reflecting wall or to the scattering obstacle so that the curvature of the wave fronts can be neglected without too much error.

## 2.1 Reflection factor, absorption coefficient and wall impedance

If a plane wave strikes a plane and uniform wall of infinite extent, in general a part of the sound energy will be reflected from it in the form of a reflected wave originating from the wall, the amplitude and the phase of which differ from those of the incident wave. Both waves interfere with each other and form a 'standing wave', at least partially.

The changes in amplitude and phase which take place during the reflection of a wave are expressed by the complex reflection factor

$$R = |R| \exp(i\chi)$$

which is a property of the wall. Its absolute value as well as its phase angle depend on the frequency and on the direction of the incident wave.

According to eqn (1.28), the intensity of a plane wave is proportional to the square of the pressure amplitude. Therefore, the intensity of the reflected wave is smaller by a factor $|R|^2$ than that of the incident wave and the fraction $1 - |R|^2$ of the incident energy is lost during reflection. This quantity is called the 'absorption coefficient' of the wall:

$$\alpha = 1 - |R|^2 \tag{2.1}$$

For a wall with zero reflectivity ($R = 0$) the absorption coefficient has its maximum value 1. Then the wall is said to be totally absorbent or 'matched to the sound field'. If $R = 1$ (in-phase reflection, $\chi = 0$), the wall is 'rigid' or 'hard'; in the case of $R = -1$ (phase reversal, $\chi = \pi$), we speak of a 'soft' wall. In both cases there is no sound absorption ($\alpha = 0$). Soft walls, however, are very rarely encountered in room acoustics and only in limited frequency ranges.

The acoustical properties of a wall surface—as far as they are of interest in room acoustics—are completely described by the reflection factor for all angles of incidence and for all frequencies. Another quantity which is even more closely related to the physical behaviour of the wall and to its construction is based on the particle velocity normal to the wall which is generated by a given sound pressure at the surface. It is called the wall impedance and is defined by

$$Z = \left(\frac{p}{v_n}\right)_{\text{surface}} \tag{2.2}$$

where $v_n$ denotes the velocity component normal to the wall. For non-porous walls which are excited into vibration by the sound field, the normal component of the particle velocity is identical to the velocity of the wall vibration. Like the reflection factor, the wall impedance is generally complex and a function of the angle of sound incidence.

Frequently the 'specific acoustic impedance' is used, which is the wall impedance divided by the characteristic impedance of the air:

$$\zeta = \frac{Z}{\rho_0 c} \tag{2.2a}$$

The reciprocal of the wall impedance is the 'wall admittance'; the reciprocal of $\zeta$ is called the 'specific acoustic admittance' of the wall.

As explained in Section 1.2 any complex quantity can be represented in a rectangular coordinate system (see Fig. 1.2). This holds also for the wall impedance. In this case, the length of that arrow corresponds to the magnitude of $Z$ while its inclination angle is the phase angle of the wall impedance:

$$\mu = \arg(Z) = \arctan\left(\frac{\operatorname{Im} Z}{\operatorname{Re} Z}\right) \tag{2.3}$$

If the frequency changes, the impedance will usually change as well and also the length and inclination of the arrow representing it. The curve connecting the tips of all arrows is called the 'locus of the impedance in the complex plane'. A simple example of such a curve is shown in Fig. 2.9a.

## 2.2 Sound reflection at normal incidence

First we assume the wall to be normal to the direction in which the incident wave is travelling, which is chosen as the $x$-axis of a rectangular coordinate system. The wall intersects the $x$-axis at $x = 0$ (Fig. 2.1). The wave is coming from the left and its sound pressure is

$$p_i(x, t) = \hat{p}_0 \exp\left[i(\omega t - kx)\right] \tag{2.4a}$$

The particle velocity in the incident wave is according to eqn (1.9):

$$v_i(x, t) = \frac{\hat{p}_0}{\rho_0 c} \exp\left[i(\omega t - kx)\right] \tag{2.4b}$$

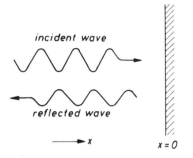

*incident wave*

*reflected wave*

$x = 0$

*Figure 2.1* Reflection of a normally incident sound wave from a plane surface.

The reflected wave has a smaller amplitude and has undergone a phase change; both changes are described by the reflection factor $R$. Furthermore, we must reverse the sign of $k$ because of the reversed direction of travel. The sign of the particle velocity is also changed since $p/v$ has opposite signs for positive and negative travelling waves. So we obtain for the reflected wave:

$$p_{\mathrm{r}}(x,t) = R\hat{p}_0 \exp\left[\mathrm{i}(\omega t + kx)\right] \tag{2.5a}$$

$$v_{\mathrm{r}}(x,t) = -R\frac{\hat{p}_0}{\rho_0 c} \exp\left[\mathrm{i}(\omega t + kx)\right] \tag{2.5b}$$

The total sound pressure and particle velocity in the plane of the wall are obtained simply by adding the above expressions and by setting $x = 0$:

$$p(0,t) = \hat{p}_0(1 + R)\exp(\mathrm{i}\omega t)$$

and

$$v(0,t) = \frac{\hat{p}_0}{\rho_0 c}(1 - R)\exp(\mathrm{i}\omega t)$$

Since the only component of particle velocity is normal to the wall, dividing $p(0,t)$ by $v(0,t)$ gives the wall impedance:

$$Z = \rho_0 c\frac{1 + R}{1 - R} \tag{2.6}$$

and from this

$$R = \frac{Z - \rho_0 c}{Z + \rho_0 c} = \frac{\zeta - 1}{\zeta + 1} \tag{2.7}$$

A rigid wall ($R = 1$) has impedance $Z = \infty$; for a soft wall ($R = -1$) the impedance will vanish. For a completely absorbent wall the impedance equals the characteristic impedance of the medium.

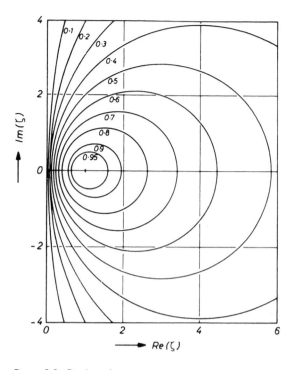

Figure 2.2 Circles of constant absorption coefficient in the complex wall impedance plane. The numbers denote the absorption coefficient.

Inserting eqn (2.7) into the definition (2.1) gives for the absorption coefficient:

$$\alpha = \frac{4\,\mathrm{Re}\,(\zeta)}{|\zeta|^2 + 2\,\mathrm{Re}\,(\zeta) + 1} \tag{2.8}$$

In Fig. 2.2 this relation is represented graphically. The diagram shows the circles of constant absorption coefficient in the complex $\zeta$-plane, i.e. abscissa and ordinate in this figure are the real and imaginary part of the specific wall impedance, respectively. As $\alpha$ increases, the circles contract towards the point $\zeta = 1$, which corresponds to complete matching of the wall to the medium.

The combination of the incident and the reflected wave forms what is called a standing wave. The pressure amplitude in this wave is found by adding eqns (2.4a) and (2.5a), and evaluating the absolute value:

$$|p(x)| = \hat{p}_0 \left[ 1 + |R|^2 + 2\,|R|\cos(2kx + \chi) \right]^{1/2} \tag{2.9a}$$

Similarly, for the amplitude of the particle velocity we find

$$|v(x)| = \frac{\hat{p}_0}{\rho_0 c} \left[ 1 + |R|^2 - 2|R| \cos(2kx + \chi) \right]^{1/2} \tag{2.9b}$$

The time dependence of the pressure and the velocity is taken into account simply by multiplying these expressions by $\exp(i\omega t)$. According to eqns (2.9a) and (2.9b), the pressure amplitude and the velocity amplitude in the standing wave vary periodically between the maximum values

$$p_{max} = \hat{p}_0(1 + |R|) \quad \text{and} \quad v_{max} = \frac{\hat{p}_0}{\rho_0 c}(1 + |R|) \tag{2.10a}$$

and the minimum values

$$p_{min} = \hat{p}_0(1 - |R|) \quad \text{and} \quad v_{min} = \frac{\hat{p}_0}{\rho_0 c}(1 - |R|) \tag{2.10b}$$

but in such a way that each maximum of the pressure amplitude coincides with a minimum of the velocity amplitude and vice versa (see Fig. 2.3). The distance of one maximum to the next is $\pi/k = \lambda/2$. So, by measuring the pressure amplitude as a function of $x$, we can evaluate the wavelength. Furthermore, the absolute value and the phase angle of the reflection factor can also be evaluated. This is the basis of a standard method of measuring the impedance and the absorption coefficient of wall materials (see Section 8.6).

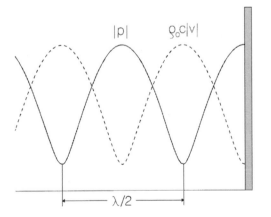

Figure 2.3 Standing sound wave in front of a plane surface with the real reflection factor $R = 0.7$. ——————— magnitude of sound pressure; - - - - - - - - - magnitude of particle velocity.

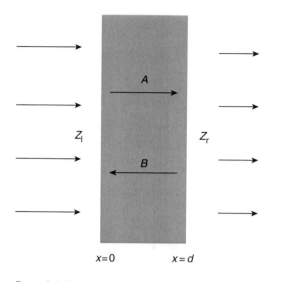

Figure 2.4 Sound transmission through a slab of different material.

Finally we consider the impedance transformation effected by a layer of a homogeneous material which is characterized by the characteristic impedance $Z'_0$ and the angular wave number $k'$. Its thickness is $d$; accordingly, we assume that the layer extends from $x = 0$ to $x = d$ (see Fig. 2.4). At its rear side the layer is loaded with the specific characteristic impedance $Z_r$. A sound wave arriving from the left will excite two plane waves in the layer travelling in the opposite $x$-direction. Hence, the sound pressure and the particle velocity within the layer are given by

$$p(x) = A \exp(-ik'x) + B \exp(ik'x)$$

and

$$Z'_0 v(x) = A \exp(-ik'x) - B \exp(ik'x)$$

Dividing both expressions by each other above leads to

$$Z_1 = Z'_0 \frac{A + B}{A - B} \qquad \text{at } x = 0$$

and

$$
\begin{aligned}
Z_r &= Z'_0 \frac{A \exp(-ik'd) + B \exp(ik'd)}{A \exp(-ik'd) - B \exp(ik'd)} \\
&= Z'_0 \frac{(A + B) \cos(k'd) - i(A - B) \sin(k'd)}{(A - B) \cos(k'd) - i(A + B) \sin(k'd)} \qquad \text{at } x = d
\end{aligned}
$$

By introducing $A + B = Z_r(A - B)/Z_0'$ into the latter fraction, we obtain after a little algebra:

$$Z_l = Z_0' \frac{Z_r + iZ_0' \tan(k'd)}{Z_0' + iZ_r \tan(k'd)} \tag{2.11}$$

This relation tells us that the impedance $Z_r$ with which the layer is loaded at its rear side is transformed into another impedance $Z_l$ appearing at its front side. If, for instance, the layer is backed by a rigid plane ($Z_r \to \infty$) its wall impedance at the front side is $Z_0'/i \tan(k'd) = -iZ_0' \cot(k'd)$. It should be noted that eqn (2.11), which is valid for normal sound incidence holds also for lossy materials where the characteristic impedance $Z_0'$ and wave number $k'$ are complex quantities.

## 2.3 Sound reflection at oblique incidence

In this section we consider the more general case of sound waves whose angles of incidence $\theta$ may be any value between 0 and 90°. Without loss of generality, we can assume that the wall normal as well as the wave normal of the incident wave lie in the $x - y$ plane of a rectangular coordinate system. The new situation is depicted in Fig. 2.5.

Suppose we replace in eqn (2.4a) $x$ by $x'$, the latter belonging to a coordinate system, the axes of which are rotated by an angle $\theta$ with respect to the $x - y$ system. The result is a plane wave propagating in positive $x'$-direction. According to the well-known formulae for coordinate

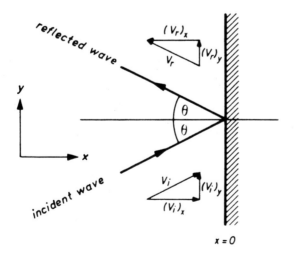

Figure 2.5 Sound reflection at oblique incidence.

transformation, $x'$ and $x$ are related by

$$x' = x \cos \theta + y \sin \theta$$

Inserting this into eqn (2.4$a$), we obtain the pressure in the incident plane wave

$$p_i(x, t) = \hat{p}_0 \exp\left[-ik(x \cos \theta + y \sin \theta)\right] \tag{2.12a}$$

(In this and the following expressions we omit, for the sake of simplicity, the factor exp $(i\omega t)$, which is common to all pressures and particle velocities.) For the calculation of the wall impedance we require the velocity component normal to the wall, i.e. the $x$-component. It is obtained from eqn (1.2), which reads in the present case:

$$v_x = -\frac{1}{i\omega\rho_0}\frac{\partial p}{\partial x}$$

Applied to eqn (2.12$a$) this yields

$$(v_i)_x = \frac{\hat{p}_0}{\rho_0 c}\cos \theta \cdot \exp\left[-ik(x \cos \theta + y \sin \theta)\right] \tag{2.12b}$$

When the wave is reflected, as for normal incidence, the sign of $x$ in the exponent of eqns (2.12$a$) and (2.12$b$) is reversed, since the direction is altered with reference to this coordinate. Furthermore, the pressure and the velocity are multiplied by the reflection factor $R$ and $-R$, respectively:

$$p_r(x, t) = R\hat{p}_0 \exp\left[-ik(-x \cos \theta + y \sin \theta)\right] \tag{2.13a}$$

$$(v_r)_x = \frac{-R\hat{p}_0}{\rho_0 c}\cos \theta \cdot \exp\left[-ik(-x \cos \theta + y \sin \theta)\right] \tag{2.13b}$$

Again, the direction of the reflected wave includes the angle $\theta$ with the wall normal. This is the important reflection law, which is also well-known in optics.

By setting $x = 0$ in eqns (2.12$a$) to (2.13$b$) and by dividing $p_i + p_r$ by $(v_i)_x + (v_r)_x$ we obtain

$$Z = \frac{\rho_0 c}{\cos \theta}\frac{1 + R}{1 - R} \tag{2.14}$$

or, after solving for $R$:

$$R = \frac{Z \cos \theta - \rho_0 c}{Z \cos \theta + \rho_0 c} = \frac{\zeta \cos \theta - 1}{\zeta \cos \theta + 1} \tag{2.15}$$

The pressure amplitude in front of the wall resulting from adding eqns (2.12a) and (2.13a) is

$$|p(x)| = \hat{p}_0 \left[ 1 + |R|^2 + 2|R| \cos(2kx \cos\theta + \chi) \right]^{1/2} \tag{2.16}$$

This pressure distribution again corresponds to a standing wave, the maxima of which are separated by a distance $(\lambda/2) \cos\theta$. The common factor $\exp(-iky \sin\theta)$ in eqns (2.12a) and (2.13a) tells us that this pressure distribution moves parallel to the wall with a velocity

$$c_y = \frac{\omega}{k_y} = \frac{\omega}{k \sin\theta} = \frac{c}{\sin\theta}$$

Of special interest are surfaces, the impedance of which is independent of the direction of incident sound. This applies if the normal component of the particle velocity at any wall element depends only on the sound pressure at that element and not on the pressure at neighbouring elements. Walls or surfaces with this property are referred to as 'locally reacting'.

In practice, surfaces with local reaction are rather the exception than the rule. They are encountered whenever the wall itself or the space behind it is unable to propagate waves or vibrations in a direction parallel to its surface. Obviously this is not true for a panel whose neighbouring elements are coupled together by bending stiffness. Moreover, this does not apply to a porous layer with an air space between it and a rigid rear wall. In the latter case, however, local reaction of the various surface elements of the arrangement can be brought about by rigid partitions which obstruct the air space in any lateral direction and prevent sound propagation parallel to the surface.

Using eqn (2.15) the absorption coefficient is given by

$$\alpha = \frac{4 \operatorname{Re}(\zeta) \cos\theta}{|\zeta|^2 \cos^2\theta + 2 \operatorname{Re}(\zeta) \cos\theta + 1} \tag{2.17}$$

Its dependence on the angle of incidence is plotted in Fig. 2.6 for various values of $\zeta$.

## 2.4 A few examples

In this section we consider as examples two types of surface which are of some practical importance as linings to room walls.

The first arrangement consists of a thin porous layer of fabric, cloth or something similar which is stretched or hung in front of a rigid wall at a distance $d$ from it and parallel to it. The $x$-axis is normal to the layer and the wall, the former having the coordinate $x = 0$. Hence, the wall is located

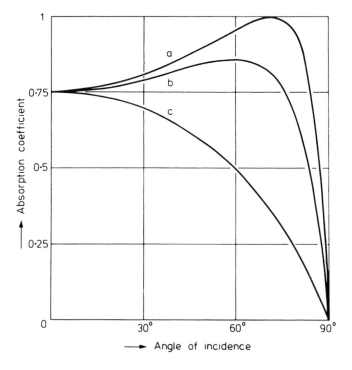

*Figure 2.6* Absorption coefficient of walls with specific impedances: (a) $\zeta = 3$; (b) $\zeta = 1.5 + 1.323i$; (c) $\zeta = 1/3$.

at $x = d$ (see Fig. 2.7). At first, we assume that the porous layer is so heavy that it does not vibrate under the influence of an incident sound wave. Any pressure difference between the two sides of the layer forces an air stream through the pores with an air velocity $v_s$. The latter is related to the pressures $p$ in front of and $p'$ behind the layer by

$$r_s = \frac{p - p'}{v_s} \tag{2.18}$$

where $r_s$ is the flow resistance of the porous layer. We assume that this relation is valid for a steady flow of air as well as for alternating flow. The unit of flow resistance is 1 Pa s/m = 1 kg m$^{-2}$ s$^{-1}$. Another commonly used unit is the Rayl (1 Rayl = 10 kg m$^{-2}$ s$^{-1}$).

Because of the conservation of matter, the particle velocities at both sides of the layer must be equal to each other and to the flow velocity through the layer:

$$v(0) = v'(0) = v_s \tag{2.19}$$

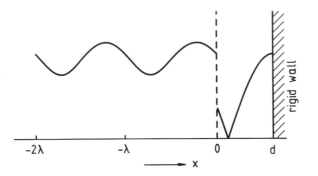

*Figure 2.7* Sound reflection from a thin porous layer at a distance *d* from a rigid wall. The plotted curve is the pressure amplitude for $r = \rho_0 c$ and $d/\lambda = 5/16$.

Then it follows from eqn (2.18) that the wall impedance of the arrangement in Fig. 2.7 is

$$Z = \frac{p}{v_s} = r_s + \frac{p'}{v_s} \qquad (2.20)$$

The second term on the right hand side of the equation represents the impedance $Z'$ of the air cushion between the fabric and the rigid wall, which is given by

$$Z' = \frac{p'}{v_s} = -i\rho_0 c \cdot \cot(kd) \qquad (2.21)$$

This formula follows from eqn (2.11) with $k' = k$ and $Z_r = \infty$ and by identifying $Z'_0$ with $\rho_0 c$, the characteristic impedance of air. Since the only non-zero component of the particle velocity is perpendicular to the rigid wall, $Z'$ represents also the input impedance of a hard-walled tube with the length $d$ and a rigid termination, provided its lateral dimensions are small compared with the wavelength $\lambda$.

If the thickness $d$ of the air space is much smaller than the wavelength, i.e. if $kd \ll 1$, then we have approximately

$$Z' \approx \frac{\rho_0 c}{ikd} = \frac{\rho_0 c^2}{i\omega d} \qquad (2.22)$$

Introducing eqn (2.21) into eqn (2.20) yields

$$Z = r_s - i\rho_0 c \cdot \cot(kd) \qquad (2.23)$$

Hence the impedance of the air space behind the porous layer is simply added to the flow resistance to give the impedance of the complete arrangement.

In the complex plane of Fig. 2.2, this wall impedance would be represented by a vertical line at a distance $r_s/\rho_0 c$ from the imaginary axis. Increasing the wave number or the frequency is equivalent to going repeatedly from $-i\infty$ to $+i\infty$ on that line. As can be seen from the circles of constant absorption coefficient, the latter has a maximum whenever $Z$ is real, i.e. whenever the depth $d$ of the air space is an odd multiple of $\lambda/4$. Introducing $\zeta = Z/\rho_0 c$ from eqn (2.23) into eqn (2.8) yields the following formula for the absorption coefficient of a porous layer in front of a rigid wall:

$$\alpha(f) = \frac{4r'_s}{(r'_s + 1)^2 + \cot^2(2\pi fd/c)} \tag{2.24}$$

with $r'_s = r_s/\rho_0 c$.

In Fig. 2.8a the absorption coefficient of this arrangement is plotted as a function of the frequency for $r_s = 0.25\rho_0$, $r_s = \rho_0 c$ and $r_s = 4\rho_0 c$. Beginning from very low values, the absorption coefficient assumes alternate maximum and minimum values. Minimum absorption occurs for all such frequencies at which the distance $d$ between the porous layer and the rigid rear wall is an integral multiple of half the wavelength. This can be easily understood since, at these distances or frequencies, the standing wave behind the porous layer has a zero of particle velocity in the plane of the layer, but energy losses can take place only if the air is moving in the pores of the layer.

In reality a porous layer will not remain at rest in a sound field as assumed so far but will vibrate as a whole, due to its finite mass. Then the total velocity in the plane $x = 0$ of Fig. 2.7 consists of two components. One of them, $v_m$, is the velocity of the layer which is set into motion by the pressure difference between its faces:

$$p - p' = M'\frac{\partial v_m}{\partial t} = i\omega M' v_m \tag{2.25}$$

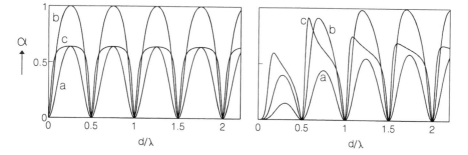

Figure 2.8 Absorption coefficient of a porous layer according to Fig. 2.7. (a) $r_s = 0.25\rho_0 c$, (b) $r_s = \rho_0 c$, (c) $r_s = 4\rho_0 c$. Left: $r_s d/M'c = 0$ (i.e. the layer is kept at rest); right: $r_s d/M'c = 4$.

where $M'$ is the specific mass of the layer, i.e. its mass per unit area. The second velocity component $v_s$ is due to the air flow forced through its pores according to eqn (2.18). Hence, the ratio of the pressure difference between both sides of the layer and the total velocity $v_m + v_s$ is no longer $r_s$ but

$$Z_r = \left( \frac{1}{r_s} + \frac{1}{i\omega M'} \right)^{-1} = \frac{r_s}{1 - i\omega_s/\omega} \tag{2.26}$$

where the characteristic frequency $\omega_s = r_s/M'$ has been introduced. Accordingly, $r_s$ in eqn (2.23) – but not in eqn (2.24) – must be replaced by $Z_r$. It is left to the reader to work out a modified formula corresponding to eqn (2.24). Fig. 2.8b demonstrates the influence of the finite mass on the absorption coefficient of the arrangement.

In practical applications it may be advisable to provide for a varying distance between the porous fabric and the rigid wall in order to smooth out the irregularities of the absorption coefficient. This can be achieved by hanging or stretching the fabric in pleats as is usually done with draperies.

In the next example we consider an arrangement similar to that shown in Fig. 2.7, however with an additional non-porous layer of specific mass $M'$ placed immediately at the left side of the porous one without touching it. For the sake of simplicity we assume that the porous layer cannot vibrate as a whole, i.e. that $\omega M' \gg r_s$ in eqn (2.26). Again we suppose that the direction of the incident sound wave is perpendicular to both layers. In contrast to the situation discussed before, both layers vibrate with the same velocity, $v_m = v_s$, but the pressure differences $\delta p$ generating these motions are different. For the porous layer we have $(\delta p)_s = r_s v_s$ after eqn (2.18). The motion of the non-porous layer is controlled by its mass; according to eqn (2.25), the pressure difference is $(\delta p)_m = i\omega M' v_s$. The total pressure difference, divided by the velocity $v_s$, is

$$\frac{(\delta p)_s + (\delta p)_m}{v_s} = r_s + i\omega M'$$

To obtain the wall impedance of the whole arrangement the impedance $Z'$ of the air cushion must be added to this expression, as in the first example. Using eqn (2.22) we finally arrive at

$$Z = r_s + i\left( \omega M' - \frac{\rho_0 c^2}{\omega d} \right) \tag{2.27}$$

In the complex plane this impedance is represented as a vertical with distance $r_s$ from the imaginary axis (see Fig. 2.9a). But now the locus moves only once from $-i\infty$ to $+i\infty$ if the frequency is varied from zero to infinity. When it

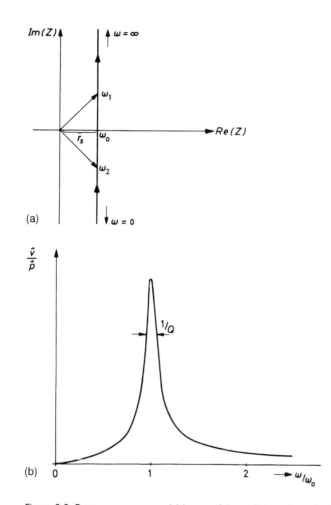

Figure 2.9 Resonance system: (a) locus of the wall impedance in the complex plane; (b) ratio of velocity to pressure amplitude as a function of the driving frequency.

crosses the real axis, the absolute value of the wall impedance reaches its minimum. Since $Z = p/v_s$, a given sound pressure will then cause a particularly high velocity of the impervious sheet. According to eqn (2.27), this 'resonance' occurs at the angular frequency:

$$\omega_0 = \left(\frac{\rho_0 c^2}{M'd}\right)^{1/2}$$

(2.28)

$f_0 = \omega_0/2\pi$ is the 'resonance frequency' of the system.

In Fig. 2.9b a typical resonance curve is depicted, i.e. the ratio of the velocity amplitude and the pressure amplitude as a function of the sound frequency. This ratio is equal to the magnitude of the wall admittance, $1/|Z|$:

$$\left|\frac{v}{p}\right| = \frac{\omega}{M'\left[(\omega^2 - \omega_0^2)^2 + 4\delta^2\omega^2\right]^{1/2}} \tag{2.29}$$

where the damping constant

$$\delta = \frac{r_s}{2M'} \tag{2.30}$$

has been introduced. Assuming that $\delta$ is small compared with $\omega_0$,

$$\omega_1 = \omega_0 + \delta \quad \text{and} \quad \omega_2 = \omega_0 - \delta$$

are the angular frequencies for which the phase angle of the wall impedance becomes $\pm45°$. Furthermore, at these frequencies the velocity amplitude is below its maximum $|p|/2M'\delta$ by a factor $\sqrt{2}$. The difference $\Delta\omega = \omega_1 - \omega_2$ is the 'half-width' or, divided by the resonance angular frequency $\omega_0$, the 'relative half-width' of the resonance, which is the reciprocal of the 'quality factor' or '$Q$-factor' $Q$ of the system:

$$\frac{\Delta\omega}{\omega_0} = \frac{1}{Q} = \frac{2\delta}{\omega_0} = \frac{r_s}{\omega_0 M'} \tag{2.31}$$

All these quantities may be used to characterise the width of a resonance curve.

Using eqns (2.8) and (2.27), the absorption coefficient of the arrangement is obtained as

$$\alpha = \frac{4r_s\rho_0 c}{(r_s + \rho_0 c)^2 + (M'/\omega)^2 (\omega^2 - \omega_0^2)^2}$$

$$= \frac{\alpha_{max}}{1 + Q_a^2 (\omega/\omega_0 - \omega_0/\omega)^2} \tag{2.32}$$

with the abbreviations

$$\alpha_{max} = \frac{4r_s\rho_0 c}{(r_s + \rho_0 c)^2} \tag{2.33}$$

and

$$\frac{1}{Q_a} = \frac{1}{Q} + \frac{\rho_0 c}{\omega_0 M'} \tag{2.34}$$

Equation (2.32) shows clearly that the combination of a mass layer with an air cushion acts as an acoustic resonance absorber. The physical meaning of $Q_a$ is similar to that of $Q$; it is related to the frequencies at which the absorption coefficient has fallen to half its maximum value. It takes into account not only the losses caused by the porous sheet but also the loss of vibrational energy due to re-radiation (reflection) of sound from the surface of the whole arrangement.

Practical resonance absorbers as applied in room acoustical design will be discussed in Section 6.4.

As a last example we consider the 'open window', i.e. an imaginary, laterally bounded surface behind which free space extends. This concept played an important role as an absorption standard in the early days of modern room acoustics, since by definition its absorption coefficient is 1.

When a plane sound wave strikes an open window at an angle $\theta$ to its normal, the component of the particle velocity normal to the 'wall' is $v_n = v \cos \theta$, where $v$ is the particle velocity in the direction of propagation. Hence the 'wall impedance' is given by

$$Z = \frac{p}{v \cos \theta} = \frac{\rho_0 c}{\cos \theta} \tag{2.35}$$

Thus the 'open window' does not react locally; its 'impedance' depends on the angle of incidence. By inserting this into eqns (2.15) and (2.17) it is easily shown that—as expected—$R = 0$ and $\alpha = 1$ for all angles of incidence.

We can enforce local reaction by filling the opening of the window with a large number of parallel tubes with thin and rigid walls whose axes are perpendicular to the plane of the window. The entrances of these tubes are flush with the window opening; the tubes are either infinitely long or their opposite ends are sealed by a perfect absorber. In each of these tubes and hence on the front face of this 'wall' we can apply

$$\frac{p}{v_n} = \rho_0 c = Z$$

and with eqn (2.15):

$$R(\theta) = \frac{\cos \theta - 1}{\cos \theta + 1} \tag{2.36a}$$

$$\alpha(\theta) = \frac{4 \cos \theta}{(\cos \theta + 1)^2} \tag{2.36b}$$

The absorption of this modified window is perfect only at normal incidence; at grazing incidence its value approaches zero. If the absorption coefficient is averaged over all directions of incidence (see Section 2.5), a value of 0.91 is obtained, whereas if the lateral subdivision of the window opening is absent, the mean absorption coefficient is of course 1.

## 2.5 Random sound incidence

In a closed room, the typical sound field does not consist of a single plane wave but is composed of many such waves, each with its own particular amplitude, phase and direction. To find the effect of a wall on such a complicated sound field we ought to consider the reflection of each wave separately and then to add all sound pressures.

With certain assumptions we can resort to some simplifications which allow general statements on the effect of a reflecting wall. If the phases of the waves incident on a wall are randomly distributed one can neglect all phase relations and the interference effects caused by them. Then the components are called incoherent. According to eqn (1.42), the total energy at some point can be calculated just by adding the energies of the components, which are proportional to the squares of the sound pressures:

$$p_{rms}^2 = \sum_n (p_{rms})_n^2 \qquad \text{or} \qquad I = \sum_n I_n \tag{2.37}$$

e.g. the total intensity is just the sum of all component intensities

Furthermore, we assume that the intensities of the incident sound waves are uniformly distributed over all possible directions; hence, each solid angle element carries the same energy per second to the wall. In this case we speak of 'random sound incidence', and the sound field associated with it is referred to as isotropic or 'diffuse'.

In room acoustics, the diffuse sound field plays the role of a standard field, and the reader will frequently encounter it in this book.

In the following it is convenient to use a spherical polar coordinate system as depicted in Fig. 2.10. Its origin is the centre of a wall element $dS$; the wall normal is its polar axis. We consider an element of solid angle $d\Omega$ around a direction which is determined by the polar angle $\theta$ and the azimuth angle $\phi$. Expressed in these angular coordinates, the solid angle element is $d\Omega = \sin\theta\, d\theta\, d\phi$.

First we calculate how the square of the sound pressure amplitude depends on the distance from the wall which, for the moment, is assumed to be perfectly rigid ($R = 1$). A wave hitting this wall under an angle $\theta$ gives rise to a standing wave with the squared pressure amplitude, according to eqn (2.16):

$$|p|^2 = 2\hat{p}_0^2 [1 + \cos(2kx \cos\theta)]$$

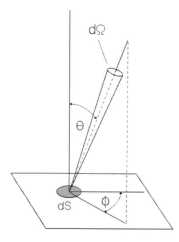

*Figure 2.10* Spherical polar coordinates.

By averaging this expression over all directions of incidence we obtain

$$|p|^2_{av} = 2\hat{p}_0^2 \cdot \frac{1}{2\pi} \int_0^{2\pi} d\phi \int_0^{\pi/2} [1 + \cos(2kx\cos\theta)] \sin\theta \, d\theta$$

$$= 2\hat{p}_0^2 \left[1 + \frac{\sin(2kx)}{2kx}\right] \tag{2.38}$$

This quantity, divided by $|p_\infty|^2 = 2\hat{p}_0^2$, is plotted in Fig. 2.11 (solid curve) as a function of the distance $x$ from the wall. Next to the wall the square pressure fluctuates, as it does in every standing wave. With increasing distance, however, these fluctuations fade out and the square pressure approaches a constant limiting value which is half of that immediately on the wall. Accordingly, the sound pressure level close to the wall would surpass that measured far away by 3 dB. For the same reason, the sound absorption of an absorbent surface adjacent and perpendicular to a rigid wall is higher near the edge than at a distance of several wavelengths from the wall.

When the sinusoidal excitation signal is replaced with random noise of limited bandwidth the pressure distribution is obtained by averaging eqn (2.38) over the frequency band. As an example, the dashed curve of Fig. 2.11 plots the result of averaging $|p|^2_{av}$ over an octave band, i.e. a frequency band with $f_2 = 2f_1$ where $f_1$ and $f_2$ denote the lower and the upper limiting frequency of the band. Here, the wavelength $\lambda$ corresponds to the frequency $\sqrt{f_1 f_2} = f_1 \sqrt{2}$. Now the standing wave has virtually levelled out for all

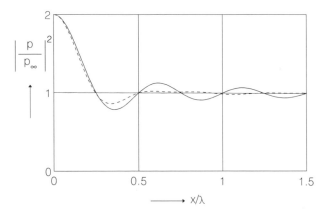

*Figure 2.11* Squared sound pressure amplitude in front of a rigid wall for random sound incidence: ——— sine tone, - - - - - - - - - - random noise of octave bandwidth.

distances exceeding $x > 0.5\lambda$, but there is still a pronounced increase of sound pressure if the wall is approached. This increase of $|p|^2_{av}$ to twice its far distance value is obviously caused by a certain relation between the phases of all impinging and reflected waves which is enforced by the wall. In any case we can conclude that in a diffuse sound field phase effects are limited to a relatively small range next to the walls which is of the order of half a wavelength.

Now we again consider a wall element with area $dS$. Its projection in the direction $\phi$, $\theta$ is $dS \cos\theta$ (see Fig. 2.10). If $I$ denotes the intensity arriving from that direction, $I \cos\theta \, dS \, d\Omega$ is the sound energy arriving per second on $dS$ from the solid angle $d\Omega$ around the considered direction. By integrating this over all solid angle elements, assuming $I$ independent of $\phi$ and $\theta$, we obtain the total energy per second arriving at $dS$:

$$E_i = I \, dS \int_0^{2\pi} d\phi \int_0^{\pi/2} \cos\theta \sin\theta \, d\theta = \pi I \, dS \tag{2.39}$$

From the energy $I \cos\theta \, dS \, d\Omega$ the fraction $\alpha(\theta)$ is absorbed; thus the totally absorbed energy per second is

$$E_a = I \, dS \int_0^{2\pi} d\phi \int_0^{\pi/2} \alpha(\theta) \cos\theta \sin\theta \, d\theta = 2\pi I \, dS \int_0^{\pi/2} \alpha(\theta) \cos\theta \sin\theta \, d\theta \tag{2.40}$$

By dividing both expressions we get the absorption coefficient for random or uniformly distributed incidence:

$$\alpha_{\text{uni}} = \frac{E_a}{E_i} = 2 \int\limits_0^{\pi/2} \alpha(\theta) \cos\theta \sin\theta \, d\theta = \int\limits_0^{\pi/2} \alpha(\theta) \sin(2\theta) \, d\theta \qquad (2.41)$$

This is occasionally referred to as the 'Paris' formula' in the literature.

For locally reacting surfaces we can express the angular dependence of the absorption coefficient by eqn (2.17). If this is done and the integration is performed, we obtain

$$\alpha_{\text{uni}} = \frac{8}{|\zeta|^2} \cos\mu \left[ |\zeta| + \frac{\cos(2\mu)}{\sin\mu} \arctan\left( \frac{|\zeta| \sin\mu}{1 + |\zeta| \cos\mu} \right) \right.$$
$$\left. - \cos\mu \ln\left( 1 + 2|\zeta| \cos\mu + |\zeta|^2 \right) \right] \qquad (2.42)$$

Here the wall impedance is characterised by its absolute value $|\zeta|$ and the phase angle $\mu$ of the specific impedance

$$\mu = \arctan\left( \frac{\text{Im}\,\zeta}{\text{Re}\,\zeta} \right)$$

The content of eqn (2.42) is represented in Fig. 2.12 in the form of curves of constant absorption coefficient $\alpha_{\text{uni}}$ in a coordinate system, the abscissa and the ordinate of which are the phase angle and the absolute value of the specific impedance, respectively. The absorption coefficient has its absolute maximum 0.951 for the real impedance $\zeta = 1.567$. Thus, in a diffuse sound field, a locally reacting wall can never be totally absorbent.

It should be mentioned that the validity of the Paris formula has been called into question by Makita and Hidaka.[2] These authors recommend replacing the factor $\cos\theta$ in eqns (2.39) and (2.40) by a somewhat more complicated weighting function. This, of course, would also modify eqn (2.42).

## 2.6 Reflection from finite-sized plane surfaces

So far we have considered sound reflection from walls of infinite extension. If a reflecting wall has finite dimensions with a free boundary the latter will become the origin of an additional sound wave when it is irradiated with sound. This additional wave is brought about by diffraction and hence may be referred to as a 'diffraction wave'. It spreads more or less in all directions.

The simplest example is diffraction by a semi-infinite wall, i.e. a rigid plane with one straight edge as depicted in Fig. 2.13. If this wall is exposed

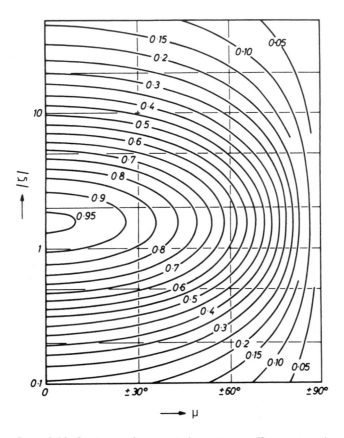

*Figure 2.12* Contours of constant absorption coefficient at random sound incidence. The ordinate is the magnitude, and the abscissa is the phase angle of the specific impedance ζ.

to a plane sound wave at normal incidence one might expect that it reflects some sound into a region A, while another region B, the 'shadow zone', remains completely free of sound. This would indeed be true if the acoustical wavelength were vanishingly small. In reality, however, the diffraction wave originating from the edge of the wall modifies this picture. Behind the wall, i.e. in region B, there is still some sound intruding into the shadow zone. And in region C the plane wave is disturbed by interferences with the diffraction wave. On the whole, there is a steady transition from the undisturbed, i.e. the primary sound wave, to the field in the shadow zone. This is shown in Fig. 2.13 where the squared sound pressure is depicted in a plane parallel to the half-plane. Of course, the extension of this transition depends on the angular wave number $k$ and the distance $d$. A similar effect occurs at the upper boundary of region A.

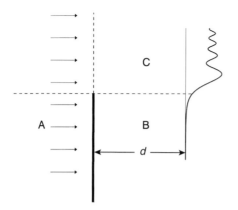

*Figure 2.13* Diffraction of a plane wave from a rigid half-plane. The diagram shows the squared sound pressure amplitude across the boundary B–C ($kd = 100$) (after Ref. 3).

If the 'wall' is a bounded reflector of limited extension, for example a freely suspended panel, the line source which the diffraction wave originates from is wound around the edge of the reflector, so to speak. As an example, Fig. 2.14a shows a rigid circular disc with radius $a$, irradiated from a point source S. Consider the sound pressure at point P. Both P and S are situated on the middle axis of the disc at distances $R_1$ and $R_2$ from its centre, respectively. Figure 2.14b shows the squared sound pressure of the reflected wave in P as a function of the disc radius as calculated from the approximation

$$|p|^2 = p_{max}^2 \sin^2\left(\frac{ka^2}{2\check{R}}\right) \quad \text{with} \quad \frac{1}{\check{R}} = \frac{1}{2}\left(\frac{1}{R_1} + \frac{1}{R_2}\right) \quad (2.43)$$

which is valid for distances that are large compared to the disc radius $a$. For very small discs—or for very low frequencies—the reflected sound is negligibly weak since the primary sound wave is nearly completely diffracted around the disc, and the obstacle is virtually not present. With increasing disc radius or frequency, the pressure in P grows rapidly; for higher values of the frequency parameter $x = a(k/2\check{R})^{1/2}$ it shows strong fluctuations. The latter are caused by interferences between the sound reflected specularly from the disc and the diffraction wave originating from its rim.

Another way to explain these fluctuations is by drawing a number of concentric circles on the disc. The radii $\rho_n$ of these circles are chosen in such a way that the length of the path connecting S with P (see Fig. 2.14a) via a circle point exceeds $R_1 + R_2$ by an integral multiple of half the wavelength.

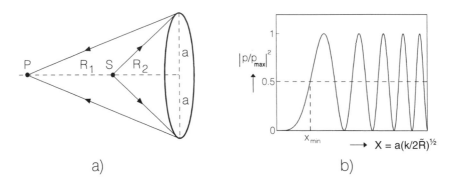

*Figure 2.14* Sound reflection from a circular disc: (a) arrangement (S = point source, P = observation point); (b) squared sound pressure amplitude of reflected wave.

Expressed in formulae:

$$\sqrt{R_1^2 + \rho_n^2} + \sqrt{R_2^2 + \rho_n^2} - R_1 - R_2 = n\frac{\lambda}{2}$$

or, since

$$\sqrt{R_1^2 + \rho_n^2} \approx R_1 + \frac{\rho_n^2}{2R_1}, \text{etc.}$$

$$\rho_n^2 = n\lambda \left( \frac{1}{R_1} + \frac{1}{R_2} \right)^{-1} = n\lambda \breve{R}$$

Each of these circles separates two zones on the plane called 'Fresnel zones', which contribute with opposite signs to the sound pressure in P. As long as the disc radius $a$ is smaller than the radius of the first zone, i.e. $a < (\lambda \breve{R})^{1/2}$, the contributions of all disc points have the same sign. With increasing disc radius (or decreasing wavelength), additional Fresnel zones will enter the disc from its rim, each of them lessening the effect of the preceding one.

We consider the reflection from the disc as fully developed if $|p|^2$ equals its average value, which is half its maximum value. This is the case if the argument of the sine in eqn (2.43) is $\pi/4$. This condition defines a minimum frequency $f_{min}$ above which the disc can be considered as an efficient reflector:

$$f_{min} = \frac{c\breve{R}}{4a^2} \approx 85\frac{\breve{R}}{a^2} \quad \text{Hz} \tag{2.44}$$

(In the second version $\breve{R}$ and $a$ are expressed in metres.) A circular panel with a diameter of 1 m, for instance, viewed from a distance of 5 m ($R_1 = R_2 = 5$ m) reflects an incident sound wave at frequencies above 1700 Hz. For lower frequencies its effect is much smaller.

Similar considerations applied to a rigid strip with the width $h$, yield for the minimum frequency of geometrical reflections

$$f_{min} \approx 185 \frac{\check{R}}{h^2} \quad \text{Hz} \tag{2.45}$$

($\check{R}$ and $h$ in metres) with the same meaning of $\check{R}$ as in eqn (2.43). If the reflector is tilted by an angle $\theta$ against the primary sound wave $a$ in eqn (2.44) and $h$ in eqn (2.45) must be multiplied with a factor $\cos \theta$.

Generally, any body or surface of limited extension distorts a primary sound wave by diffraction unless its dimensions are very small compared to the wavelength. One part of the diffracted sound is scattered more or less in all directions. For this reason this process is also referred to as 'sound scattering'. (The role of sound scattering by the human head in hearing has already been mentioned in Section 1.6.)

The scattering efficiency of a body is often characterised by its 'scattering cross-section', defined as the ratio of the total power scattered $P_s$ and the intensity $I_0$ of the incident wave:

$$Q_s = \frac{P_s}{I_0} \tag{2.46}$$

If the dimensions of the scattering body are small compared to the wavelength, $P_s$ and hence $Q_s$ is very small. In the opposite case of short wavelengths, the scattering cross-section approaches twice its visual cross-section, i.e. $2\pi a^2$ for a sphere or a circular disc with radius $a$. Then one half of the scattered power is concentrated into a narrow beam behind the obstacle and forms its shadow by interference with the primary wave while the other half is deflected from its original direction.

## 2.7 Scattering by wall irregularities

Very often a wall is not completely smooth but contains regular or irregular coffers, bumps or other projections. If these are very small compared with the wavelength, they do not disturb the wall's 'specular' reflection (see Fig. 2.15a). In the opposite case, i.e. if they are large compared with the wavelength, each of their faces may be treated as a plane or curved wall section, reflecting the incident sound specularly as shown in Fig. 2.15c. There is an intermediate range of wavelengths, however, in which each projection adds a scattered component wave to the reflected sound field (see Fig. 2.15b). If the wall has an irregular surface structure, a noticeable fraction of the incident sound energy will be scattered in all directions. In this case we speak of a

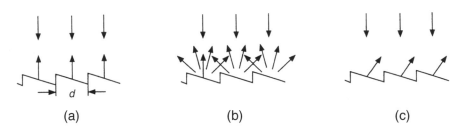

*Figure 2.15* Scattering by wall irregularities: (a) $d \ll \lambda$, (b) $d \approx \lambda$, (c) $d \gg \lambda$.

'diffusely reflecting wall'. In Section 8.5 methods for measuring the scattering efficiency of acoustically rough surfaces will be described.

As an example of a sound scattering boundary, we consider the ceiling of a particular concert hall.[4] It is covered with many bodies made of gypsum such as pyramids and spherical segments; their depth is about 30 cm on the average. Figure 2.16 shows the directional distribution of the sound reflected from that ceiling, measured at a frequency of 1000 Hz at normal incidence of the primary sound wave; the plotted quantity is the sound pressure amplitude. (This measurement has been carried out on a model ceiling.) The pronounced maximum at 0° corresponds to the specular component which is still of considerable strength.

The occurrence of diffuse or partially diffuse reflections is not restricted to walls with a geometrically structured surface; they may also be produced by walls with non-uniform impedance. To understand this we return to Fig. 2.13 and imagine that the dotted vertical line marks a totally absorbing wall. This would not change the structure of the sound field at the left side of the wall, i.e. the disturbance caused by the diffraction wave. Therefore we can conclude that any change of wall impedance creates a diffraction wave.

A practical example of this kind are walls lined with relatively thin panels which are mounted on a rigid framework. At the points where the lining is fixed the panel is very stiff and cannot react to the incident sound field. Between these points, however, the lining is more compliant because it can perform bending vibrations, particularly if the frequency of the exciting sound is close to the resonance frequency of the lining (see eqn (2.28)). Scattering will be even stronger if adjacent partitions are tuned to different resonance frequencies due to variations of the panel masses or the depths of the air space behind them.

In Fig. 2.17 we consider a plane wall subdivided into strips with equal width $d$ and with different reflection factors $R_n = |R_n| \exp(i\chi_n)$. We assume that $d$ is noticeably smaller than the wavelength. If a plane wave arrives normally at the wall, it will excite all strips with equal amplitude and phase, and each of them will react to it by emitting a secondary wave or wavelet.

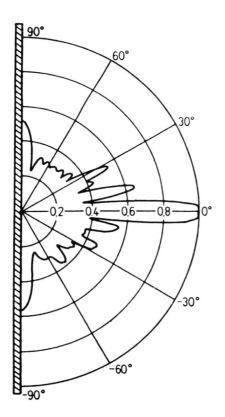

Figure 2.16 Polar diagram of the sound (pressure amplitude), scattered from a highly irregular ceiling.

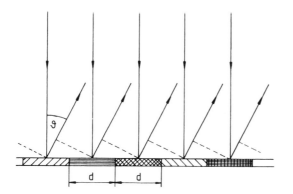

Figure 2.17 Sound reflection from strips with equal widths but different reflection factors, arranged in a plane.

We consider those wave portions which are re-emitted (i.e. reflected) from corresponding points of the strips at some angle $\vartheta$. The phase difference between the contributions of two adjacent wall elements are due to the different phase shifts $\chi_n$ and to the path differences $d \sin\vartheta$. The sound pressure far from the wall is obtained by summation over all contributions:

$$p(\vartheta) \propto \sum_n |R_n| \exp\left[i(\chi_n - nkd \sin \vartheta)\right] \tag{2.47}$$

If all strips had the same reflection factor, all contributions would have the same phase angles for $\vartheta = 0$ and add to a particularly high pressure amplitude, corresponding to the specular reflection. It is evident that by varying $|R_n|$ and $\chi_n$ the specular reflection can be destroyed more or less and that its energy will be scattered into non-specular directions instead.

In the following we assume that all reflection factors in eqn (2.47) have the magnitude 1. Our goal is to achieve maximum diffusion of the reflected sound. In principle, this could be effected by distributing the phase angles $\chi_n$ randomly within the interval from 0 to $2\pi$ since randomising the phase angles is equivalent to randomising the directions in eqn (2.47). Complete randomness, however, would require a very large number of elements.

A similar effect can be reached with so-called pseudorandom sequences of phase angles. If these sequences are periodic, arrangements of this kind act as phase gratings, with the grating constant $Nd$ if $N$ denotes the number of elements within one period. As with optical gratings, constructive interference of the wavelets reflected from corresponding elements will occur if the condition

$$\sin \vartheta_m = m\frac{2\pi}{Nkd} = m\frac{\lambda}{Nd} \qquad \left(m < \frac{Nd}{\lambda}\right) \tag{2.48}$$

is fulfilled. The integer $m$ with $|m| \leq m_{\max}$ is the 'diffraction order' where $m_{\max}$ is the highest integer not exceeding $Nd/\lambda$. Introducing $\sin \vartheta_m$ from eqn (2.48) into eqn (2.47) yields the far-field sound pressure in the $m$th diffraction order:

$$p_{\mathrm{m}} = \sum_{n=0}^{N-1} \exp(i\chi_n) \exp\left(-\frac{2\pi imn}{N}\right) \tag{2.49}$$

$p_{\mathrm{m}}$ is the discrete Fourier transform of the sequence $\exp(i\chi_n)$ as may be seen by comparing this expression with eqn (1.36a). Hence, a uniform distribution of the reflected energy over all diffraction orders can be achieved by finding phase shifts $\chi_n$ for which the power spectrum of $\exp(i\chi_n)$ is flat.

It was Schroeder's idea[5] to select the phase angles according to certain periodic number sequences which are known to have the required spectral properties, and to realise them in the form of properly corrugated surfaces. Nowadays such scattering devices are known as 'Schroeder diffusers'. They are believed to improve the acoustics of concert halls and recording studios by creating 'lateral sound waves' which are known to be relevant for good music acoustics (see Section 9.2). From a physical standpoint they are reflection phase gratings.

Probably the most popular kind of them is based on quadratic residues. Suppose that the period $N$ is a prime number. Then a set of phase $N$ angles $\chi_n$ which guarantees a flat spectrum of $\exp(i\chi_n)$ is given by

$$\chi_n = \frac{2\pi}{N} n^2 \bmod N \qquad (n = 0, 1, \ldots, N-1) \tag{2.50}$$

i.e. the phase shifts are integral multiples of $2\pi/N$. The modulus function accounts for the fact that all relevant phase angles are in the range from 0 to $2\pi$. Other useful sequences exploit the properties of primitive roots of prime numbers or the index function. Diffusers of the latter kind suppress the specular reflection completely.[6]

The phase shifts $\chi_n$ are generated by a series of equally spaced wells which have different depths $h_n$ and are separated by thin and rigid partitions. A sound wave hitting such an arrangement will excite secondary waves in the wells. Each of these waves travels towards the rigid bottom of the well. When the reflected wave reappears at the opening of the well it will have gained a phase shift

$$\chi_n = 2kh_n = 4\pi\left(h_n/\lambda\right)$$

hence the required depths are

$$h_n = \frac{\chi_n}{2k} = \frac{\lambda_d}{4\pi}\chi_n \tag{2.51}$$

$\lambda_d$ is a free design parameter of the diffuser, the 'design wavelength'. The diffuser works optimally for the 'design frequency' $f_d = c/\lambda_d$ and integral multiples of it. On the other hand, there are critical frequencies at which no scattering takes place at all. This occurs when all depths are integral multiples of the acoustical wavelength. It goes without saying that such a diffuser works not only for normal incident sound but also for waves arriving from oblique directions.

A second design parameter is the width $d$ of the wells. If it is too small, the number of allowed diffraction orders according to eqn (2.48) may become very small or even zero at low frequencies, i.e. there will be no effective redistribution of the reflected sound. Furthermore, narrow wells show increased losses due to the viscous and thermal boundary layer on their walls. If, on the other hand, the wells are too wide, not only the fundamental wave mode

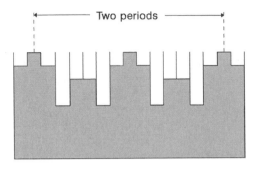

*Figure 2.18* Quadratic residue diffuser (QRD) for $N = 7$.

but also higher-order modes may be excited inside the wells, resulting in a more involved sound field. Practical diffusers have well widths of a few centimetres. Figure 2.18 shows a section of a quadratic residue diffuser (QRD) with $N = 7$.

Real diffusers have finite extension, of course. As a consequence the scattered sound energy is not concentrated in discrete directions but in lobes of finite widths around the angles $\vartheta_m$ of eqn (2.48). Figure 2.19 presents polar scattering diagrams, i.e. plots of the scattered sound pressure amplitude calculated with eqn (2.47) (with $|R_n| \equiv 1$) for a QRD consisting of three periods with $N = 7$; its design frequency is 285 Hz.

The concept of Schroeder diffusers is easily applied to two-dimensional structures which scatter the incident sound into all diffraction orders within the solid angle. They consist of periodic arrays of parallel and rigid-walled channels. For instance, a two-dimensional quadratic residue diffuser can be constructed by choosing the phase angles of the reflection factors according to

$$\chi_{n_1 n_2} = \frac{2\pi}{N} \cdot (n_1^2 + n_2^2) \bmod N \qquad (n_{1,2} = 0, 1, \ldots N - 1) \qquad (2.52)$$

the array is periodic in both directions with the prime number $N$.

The explanation of pseudorandom diffusers as presented here is only qualitative since it neglects all losses and assumes all wells behave independently. In reality the orifices of the wells or channels are coupled to each other by air flows that tend to equalize local pressure differences. A more rigorous treatment of a one-dimensional diffuser starts from the spatial Fourier expansion of the scattered sound field:

$$p_r(x, y) = \sum_{m=-\infty}^{\infty} C_m \exp(-ik_m y) \exp\left(-i\frac{2\pi m x}{Nd}\right) \qquad (2.53)$$

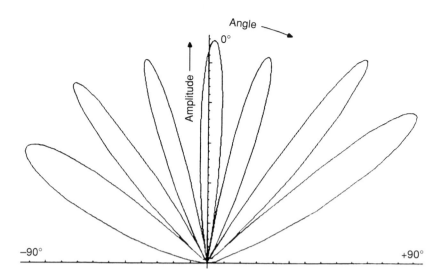

*Figure 2.19* Scattering diagram of a QRD according to Fig. 2.18: two periods, the spacing is λ/2 (see Schroeder[7]).

Here $x$ is the coordinate running parallel to the scattering surface while the $y$-axis points into the surface. The $x$-dependent factors in the sum terms of the equation above reflect the periodicity of the surface. Each term of the sum represents a plane wave travelling within the $x - y$-plane. The angle $\vartheta_m$ between the wave direction and the $y$-axis is given by eqn (2.48), accordingly

$$k_m - k \cos \vartheta_m = k\sqrt{1 - \left(m \frac{\lambda}{Nd}\right)^2}$$

The relevant terms are those with $m \le Nd/\lambda$. Their amplitudes $C_m$ are determined from the boundary condition, which reads

$$\frac{p_i + p_r}{v_y} = Z(x) \qquad \text{for } y = 0 \qquad (n = 0, 1, \ldots N) \tag{2.54}$$

$Z(x) = Z(x + Nd)$ is the wall impedance of the scattering surface, $p_i$ denotes the pressure of the incident wave, and $v_y$ is the $y$-component of the total sound field:

$$v_y = \frac{i}{\rho_0 c k} \frac{\partial}{\partial y} (p_i + p_r)$$

In the present case the impedance assumes $N$ values $Z_n = -i\rho_0 c \cot(kh_n)$ within one period (see eqn (2.21). This leads us to $N$ linear inhomogeneous equations for the amplitudes $C_m$. For a more detailed description of this method the reader is referred to Ref. 8 and Ref. 9.

The phase grating diffusers invented by Schroeder are certainly based on an ingenious concept. Nevertheless, they suffer from the fact that the scattered energy is concentrated in a number of grating lobes which are separated by large minima. This is caused by the periodic repetition of a base element. One way to overcome this disadvantage is to use aperiodic number sequences. Another method is to combine two different base schemes—for instance QR diffusers of different lengths or different 'polarity'—according to an aperiodic binary sequence, for instance a Barker code. These and other possibilities are discussed in Ref. 9.

Another peculiarity of this type of diffusers is their relatively high sound absorption. More on the cause of this absorption will be said in Section 6.8.

## References

1 Mechel F, Schallabsorber S. Stuttgart: Hirzel Verlag, 1989/95/98.
2 Makita Y, Hidaka T. Revision of the $\cos\theta$ law of oblique incident sound energy and modification of the fundamental formulations in geometrical acoustics in accordance with the revised law. Acustica 1987; 63:163. Comparison between reverberant and random incident sound absorption coefficients of a homogeneous and isotropic sound absorbing porous material—experimental, Acustica 1988; 67:214.
3 Morse PM, Ingard KU. Theoretical Acoustics. New York: McGraw-Hill, 1968.
4 Meyer E, Kuttruff H. Zur Akustik der neuerbauten Beethovenhalle in Bonn. Acustica 1959; 9:465.
5 Schroeder MR. Binaural dissimilarity and optimum ceilings for concert halls: more lateral sound diffusion. J Acoust Soc Am, 1979; 65:958.
6 Schroeder MR. Phase gratings with suppressed specular reflection. Acustica 1985; 81:364
7 Schroeder M.R. Number Theory in Science and Communication, 2nd edn. Berlin: Springer-Verlag, 1986.
8 Mechel FP. The wide-angle diffuser—a wide-angle absorber? Acustica 1985; 81:379
9 Cox TJ, D'Antonio P. Acoustic Absorbers and Diffusers. London: Spon Press, 2004.

# Chapter 3

# Sound waves in a room

In the preceding chapter we saw the laws which a plane sound wave obeys upon reflection from a single plane wall and how this reflected wave is superimposed on the incident one. Now we shall try to obtain some insight into the complicated distribution of sound pressure or sound energy in a room which is enclosed on all sides by walls.

We could try to describe the resulting sound field by means of a detailed calculation of all the reflected sound components and by finally adding them together; that is to say, by a manifold application of the reflection laws which we dealt with in the previous chapter. Since each wave which has been reflected from a wall A will be reflected from walls B, C, D, etc., and will arrive eventually once more at wall A, this procedure leads only asymptotically to a final result, not to mention the calculations which grow like an avalanche. Nevertheless, this method is highly descriptive and therefore it is frequently applied in a much simplified form in geometrical room acoustics. We shall return to it in the next chapter.

In this chapter we shall choose a different way of tackling our problem which will lead to a solution in closed form—at least a formal one. This advantage is paid for by a higher degree of abstraction, however. Characteristic of this approach are certain boundary conditions which have to be set up along the room boundaries and which describe mathematically the acoustical properties of the walls, the ceiling and the other surfaces. Then solutions of the wave equations are sought which satisfy these boundary conditions. This method is the basis of what is frequently called 'the wave theory of room acoustics'.

It will turn out that this method in its exact form too can only be applied to highly idealised cases with reasonable effort. The rooms with which we are concerned in our daily life, however, are more or less irregular in shape, partly because of the furniture, which forms part of the room boundary. Rooms such as concert halls, theatres or churches deviate from their basic shape because of the presence of balconies, galleries, pillars, columns and other wall irregularities. Then even the formulation of boundary conditions may turn out to be quite involved, and the solution of a given problem requires extensive numerical calculations. For this reason the immediate application of wave theory to practical problems in room acoustics is very limited. Nevertheless, wave theory is the most reliable and appropriate theory from a physical point of view, and therefore it is essential for a more than superficial understanding of sound propagation in enclosures. For the same reason we should keep in mind the results of wave theory when we are applying more simplified methods, in order to keep our ideas in perspective.

## 3.1 Formal solution of the wave equation

The starting point for a wave theory representation of the sound field in a room is again the wave equation (1.5), which will be used here in a time-independent form. That is to say, we assume, as earlier, a harmonic time law for the pressure, the particle velocity, etc., with an angular frequency $\omega$. Then the equation, known as the Helmholtz equation, reads

$$\Delta p + k^2 p = 0 \quad \text{with} \quad k = \frac{\omega}{c} \tag{3.1}$$

Furthermore, we assume that the room under consideration has locally reacting walls and ceiling, the acoustical properties of which are completely characterised by a wall impedance depending on the coordinates and the frequency but not on the angle of sound incidence.

According to eqn (1.2), the velocity component normal to any wall or boundary is

$$v_{\mathrm{n}} = -\frac{1}{\mathrm{i}\omega\rho_0} (\mathrm{grad}\ p)_{\mathrm{n}} = \frac{\mathrm{i}}{\omega\rho_0} \frac{\partial p}{\partial n} \tag{3.2}$$

The symbol $\partial/\partial n$ denotes partial differentiation in the direction of the outward normal to the wall. We replace $v_n$ by $p/Z$ (see eqn (2.2)) and obtain

$$Z\frac{\partial p}{\partial n} + \mathrm{i}\omega\rho_0 p = 0 \tag{3.2a}$$

or, using the specific impedance $\zeta = Z/\rho_0 c$,

$$\zeta\frac{\partial p}{\partial n} + \mathrm{i}kp = 0 \tag{3.2b}$$

Now it can be shown that the wave equation yields non-zero solutions fulfilling the boundary condition (3.2a) or (3.2b) only for particular discrete values of $k$, called 'eigenvalues'.[1,2] In the following material, we shall frequently distinguish these quantities from each other by a single index number

$n$ or $m$, though it would be more adequate to use a trio of subscripts because of the three-dimensional nature of the problem.

Each eigenvalue $k_n$ is associated with a solution $p_n(\mathbf{r})$, which is known as an 'eigenfunction' or 'characteristic function'. Here $\mathbf{r}$ is used as an abbreviation for the three spatial coordinates. It represents a three-dimensional standing wave, a so-called 'normal mode' of the room. Whenever the boundary or a part of it has non-zero absorption both the eigenfunctions and the eigenvalues are complex. Hence the explicit evaluation of the eigenvalues and eigenfunctions is generally quite difficult and requires the application of numerical methods such as the finite element method (FEM).[3] There are only a few room shapes combined with simple boundary conditions for which the eigenfunctions can be presented in closed form. An important example will be given in the next section.

At this point we need to comment on the wave number $k$ in the boundary condition (3.2a) or (3.2b). Implicitly, it is also contained in $\beta$, since the specific wall admittance depends in general on the frequency $\omega = kc$. If the room contains a sound source operating at a certain frequency, $k$ is given by this frequency. In this case the eigenfunctions as well as the eigenvalues are frequency-dependent. Otherwise, one can identify $k$ with $k_n$, the eigenvalue to be evaluated by solving the boundary problem. The eigenfunctions are mutually orthogonal, which means that

$$\iiint_V p_n(\mathbf{r}) p_m(\mathbf{r})\,\mathrm{d}V = \begin{cases} K_n & \text{for } n = m \\ 0 & \text{for } n \neq m \end{cases} \tag{3.3}$$

where the integration has to be extended over the whole volume $V$ enclosed by the walls. Here $K_n$ is a constant with the dimension $\mathrm{Pa}^2\mathrm{m}^3 = \mathrm{N}^2/\mathrm{m}$.

If all the eigenvalues and eigenfunctions were known, we could—at least in principle—evaluate any desired acoustical property of the room, for instance, its steady state response to arbitrary sound sources. Suppose the sound sources are distributed continuously over the room according to a density function $q(\mathbf{r})$, where $q(\mathbf{r})\mathrm{d}V$ is the volume velocity of a volume element $\mathrm{d}V$ at $\mathbf{r}$. Furthermore, we assume a common driving frequency $\omega$. By adding $\rho_0 q(\mathbf{r})$ to the right-hand side of eqn (1.3) it is easily seen that the Helmholtz equation (3.1) has to be modified into

$$\Delta p + k^2 p = -\mathrm{i}\omega \rho_0 q(\mathbf{r}) \tag{3.4}$$

Since the eigenfunctions form a complete and orthogonal set of functions, we can expand the source function in a series of $p_n$:

$$q(\mathbf{r}) = \sum_n C_n p_n(\mathbf{r}) \quad \text{with} \quad C_n = \frac{1}{K_n} \iiint_V p_n(\mathbf{r}) q(\mathbf{r})\,\mathrm{d}V \tag{3.5}$$

where the summation is extended over all possible combinations of subscripts. In the same way the solution $p_\omega(\mathbf{r})$, which we are looking for, can be expanded in eigenfunctions:

$$p_\omega(\mathbf{r}) = \sum_n D_n p_n(\mathbf{r}) \tag{3.6}$$

Our problem is solved if the unknown coefficients $D_n$ are expressed by the known coefficients $C_n$. For this purpose we insert both series into eqn (3.4):

$$\sum_n D_n(\Delta p_n + k^2 p_n) = -i\omega\rho_0 \sum_n C_n p_n$$

Now $\Delta p_n = -k_n^2 p_n$. Using this relation and equating term by term in the equation above, we obtain:

$$D_n = i\omega\rho_0 \frac{C_n}{k_n^2 - k^2} \tag{3.7}$$

The final solution assumes a particularly simple form if the sound source is a point source with the volume velocity $Q$, located at the arbitrary point $\mathbf{r}_0$. Then the source function is represented mathematically by a delta function:

$$q(\mathbf{r}) = Q\delta(\mathbf{r} - \mathbf{r}_0)$$

Because of eqn (1.44) the coefficients $C_n$ in eqn (3.5) are then given by

$$C_n = \frac{1}{K_n} Q p_n(\mathbf{r}_0)$$

Using this relation and eqns (3.7) and (3.6), we finally find for the sound pressure in a room which is excited by a point source emitting a sine signal with the angular frequency $\omega$:

$$p_\omega(\mathbf{r}) = i\omega Q\rho_0 \sum_n \frac{p_n(\mathbf{r})p_n(\mathbf{r}_0)}{K_n(k_n^2 - k^2)} \tag{3.8}$$

This function is also called the 'Green's function' of the room under consideration. It is interesting to note that it is symmetric in the coordinates of the sound source and of the point of observation. If we put the sound source at $\mathbf{r}$, we observe at point $\mathbf{r}_0$ the same sound pressure as we did before at $\mathbf{r}$, when the sound source was at $\mathbf{r}_0$. Thus eqn (3.8) is the mathematical expression of the famous reciprocity theorem which can be applied sometimes with advantage to measurements in room acoustics.

As mentioned before, the eigenvalues are in general complex quantities. Putting

$$k_n = \frac{\omega_n}{c} + i\frac{\delta_n}{c} \qquad (3.9)$$

and assuming that $\delta_n \ll \omega_n$, we obtain from eqn (3.8)

$$p_\omega(\mathbf{r}) = \rho_0 c^2 \omega Q \sum_n \frac{p_n(\mathbf{r})p_n(\mathbf{r}_0)}{\left[2\delta_n\omega_n + i\left(\omega^2 - \omega_n^2\right)\right]K_n} \qquad (3.10)$$

Considered as a function of the frequency, this expression is the transfer function of the room between the two points $\mathbf{r}$ and $\mathbf{r}_0$. Each term of this sum represents a resonance of the room, since it assumes a maximum when the driving frequency $\omega$ is close to $\omega_n$. Therefore the corresponding frequencies $f_n = \omega_n/2\pi$ are often called the 'resonance frequencies' of the room, or, another commonly used name, 'eigenfrequencies'. The $\delta_n$ will turn out to be 'damping constants'.

If the sound source is not emitting a sinusoidal signal but instead a signal which is composed of several spectral components, then $Q = Q(\omega)$ can be considered as its spectral function and we can represent the source signal as a Fourier integral (see Section 1.4):

$$s(t) = \frac{1}{2\pi} \int_{-\infty}^{+\infty} Q(\omega) \exp(i\omega t)\, d\omega$$

where $\omega = 2\pi f$ has been introduced as the integration variable. Since the response to any spectral component with angular frequency $\omega$ is just given by eqns (3.8) or (3.10), the sound pressure at the point $\mathbf{r}$ as a function of time is

$$p(\mathbf{r}, t) = \frac{1}{2\pi} \int_{-\infty}^{+\infty} p_\omega(\mathbf{r}) \exp(i\omega t)\, d\omega$$

where $Q$ in the formulae for $p_\omega$ has to be replaced by $Q(\omega)$.

## 3.2 Normal modes in rectangular rooms with rigid boundaries

In order to put some life into the abstract formalism outlined in the preceding section, we consider a room with parallel pairs of walls, the pairs being perpendicular to each other. It will be referred to in the following as a 'rectangular room'. In practice, rooms with exactly this shape do not exist. On the other hand, many concert halls or other halls, churches, lecture rooms and so on are much closer in shape to the rectangular room than to any other of simple geometry, and so the results obtained for strictly rectangular rooms can be applied at least qualitatively to many rooms encountered in practice. Therefore our example is not only intended for the elucidation of the theory discussed above but also has some practical bearing.

Our room is assumed to extend from $x = 0$ to $x = L_x$ in the $x$-direction, and similarly from $y = 0$ to $y = L_y$ in the $y$-direction and from $z = 0$ to $z = L_z$ in the $z$-direction (see Fig. 3.1). As far as the properties of the wall are concerned, we start with the simplest case, namely that of all the walls being rigid. That is to say, that at the surface of the walls the normal components of the particle velocity must vanish.

In cartesian coordinates the Helmholtz equation (3.1) may be written

$$\frac{\partial^2 p}{\partial x^2} + \frac{\partial^2 p}{\partial y^2} + \frac{\partial^2 p}{\partial z^2} = 0$$

The variables can be separated, which means that we can compose the solution of three factors:

$$p(x, y, z) = p_x(x) \cdot p_y(y) \cdot p_z(z)$$

each of them depending only on one of the space variables. If this product is inserted into the Helmholtz equation, the latter splits up into three ordinary

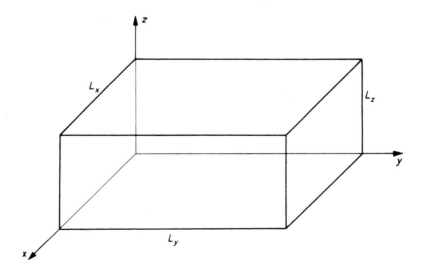

Figure 3.1 Rectangular room.

differential equations. The same is true for the boundary conditions. For instance, $p_x$ must satisfy the equation

$$\frac{d^2 p_x}{dx^2} + k_x^2 p_x = 0 \tag{3.11}$$

together with the boundary condition

$$\frac{dp_x}{dx} = 0 \qquad \text{for } x = 0 \text{ and } x = L_x \tag{3.11a}$$

Analogous equations hold for $p_y(y)$ and $p_z(z)$; the newly introduced constants are related by

$$k_x^2 + k_y^2 + k_z^2 = k^2 \tag{3.12}$$

Equation (3.11) has the general solution

$$p_x(x) = A_1 \cos(k_x x) + B_1 \sin(k_x x) \tag{3.13}$$

The constants $A_1$ and $B_1$ are used for adapting this solution to the boundary conditions (3.11a). So it is seen immediately that we must put $B_1 = 0$, since only the cosine function possesses at $x = 0$ the horizontal tangent required by eqn (3.11a). To obtain a horizontal tangent too at $x = L_x$, we must require $\cos(k_x L_x) = \pm 1$; thus, $k_x L_x$ must be an integral multiple of $\pi$. The constant $k_x$ must therefore assume one of the values

$$k_x = \frac{n_x \pi}{L_x} \tag{3.14a}$$

$n_x$ being a non-negative integer. Similarly, we obtain for the allowed values of $k_y$ and $k_z$

$$k_y = \frac{n_y \pi}{L_y} \tag{3.14b}$$

$$k_z = \frac{n_z \pi}{L_z} \tag{3.14c}$$

By inserting these values into eqn (3.12) one arrives at the eigenvalues of the wave equation:

$$k_{n_x n_y n_z} = \pi \left[ \left( \frac{n_x}{L_x} \right)^2 + \left( \frac{n_y}{L_y} \right)^2 + \left( \frac{n_z}{L_z} \right)^2 \right]^{1/2} \tag{3.15}$$

The eigenfunctions associated with these eigenvalues are simply obtained by multiplication of the three cosines, each of which describes the dependence of the pressure on one coordinate:

$$p_{n_x n_y n_z}(x, y, z) = C \cdot \cos\left(\frac{n_x \pi x}{L_x}\right) \cdot \cos\left(\frac{n_y \pi y}{L_y}\right) \cdot \cos\left(\frac{n_z \pi z}{L_z}\right) \qquad (3.16)$$

where $C$ is an arbitrary constant. This formula represents a 'normal mode' of the room, corresponding to a three-dimensional standing wave: (of course, it is incomplete without the factor exp ($i\omega t$) describing the time dependence of the sound pressure). The pressure amplitude is zero at all points at which at least one of the cosines becomes zero. This occurs for all values of $x$ which are odd integers of $L_x/2n_x$, and for the analogous values of $y$ and $z$. So these points of vanishing sound pressure form three sets of equidistant planes, called 'nodal planes', which are mutually orthogonal. The numbers $n_x$, $n_y$ and $n_z$ indicate the numbers of nodal planes perpendicular to the $x$-axis, the $y$-axis and the $z$-axis, respectively. (For non-rectangular rooms the surfaces of vanishing sound pressure are generally no planes. They are referred to as 'nodal surfaces'.)

In Fig. 3.2 the sound pressure distribution in the plane $z = 0$ is depicted for $n_x = 3$ and $n_y = 2$. The loops are curves of constant pressure amplitude, namely for $|p/p_{max}| = 0.25, 0.5$ and $0.75$. The intersections of vertical nodal planes with the plane $z = 0$ are indicated by dotted lines. On either side of a nodal line the sound pressures have opposite signs.

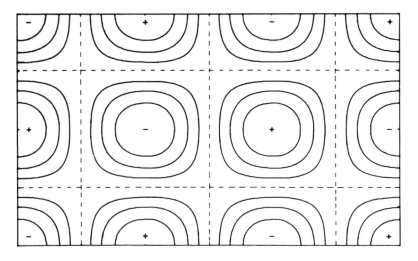

*Figure 3.2* Sound pressure distribution in the plane $z = 0$ of a rectangular room for $n_x = 3$ and $n_y = 2$.

Table 3.1 Eigenfrequencies of a rectangular room with dimensions 4.7 × 4.1 × 3.1 m$^3$ (in hertz)

| $f_n$ | $n_x$ | $n_y$ | $n_z$ | $f_n$ | $n_x$ | $n_y$ | $n_z$ |
|---|---|---|---|---|---|---|---|
| 36.17 | 1 | 0 | 0 | 90.47 | 1 | 2 | 0 |
| 41.46 | 0 | 1 | 0 | 90.78 | 2 | 0 | 1 |
| 54.84 | 0 | 0 | 1 | 99.42 | 0 | 2 | 1 |
| 55.02 | 1 | 1 | 0 | 99.80 | 2 | 1 | 1 |
| 65.69 | 1 | 0 | 1 | 105.79 | 1 | 2 | 1 |
| 68.75 | 0 | 1 | 1 | 108.51 | 3 | 0 | 0 |
| 72.34 | 2 | 0 | 0 | 109.68 | 0 | 0 | 2 |
| 77.68 | 1 | 1 | 1 | 110.05 | 2 | 2 | 0 |
| 82.93 | 0 | 2 | 0 | 115.49 | 1 | 0 | 2 |
| 83.38 | 2 | 1 | 0 | 116.16 | 3 | 1 | 0 |

The eigenfrequencies corresponding to the eigenvalues of eqn (3.15), which are real because of the particular boundary condition (3.11a), are given by

$$f_{n_x n_y n_z} = \frac{c}{2\pi} k_{n_x n_y n_z} \tag{3.17}$$

In Table 3.1 the lowest 20 eigenfrequencies (in Hz) of a rectangular room with dimensions $L_x = 4.7$ m, $L_y = 4.1$ m and $L_z = 3.1$ m are listed for $c = 340$ m/s, together with the corresponding combinations of subscripts, which indicate the structure of the mode which belongs to the respective eigenfrequency.

By employing the relation $\cos x = (e^{ix} + e^{-ix})/2$ (see eqn (1.15), eqn (3.16) can be written in the following form:

$$p_{n_x n_y n_z} = \frac{C}{8} \sum \exp\left[\pi i \left(\pm \frac{n_x}{L_x} x \pm \frac{n_x}{L_x} x \pm \frac{n_x}{L_x} x\right)\right] \tag{3.18}$$

wherein the summation has to be extended over the eight possible combinations of signs in the exponent. Each of these eight terms—multiplied by the usual time factor exp (iωt)—represents a plane travelling wave, whose direction of propagation is defined by the angles $\beta_x$, $\beta_y$ and $\beta_z$, which it makes with the coordinate axes, where

$$\cos \beta_x : \cos \beta_y : \cos \beta_z = \left(\pm \frac{n_x}{L_x}\right) : \left(\pm \frac{n_y}{L_y}\right) : \left(\pm \frac{n_z}{L_z}\right) \tag{3.19}$$

If one of the three characteristic integers $n$, for instance $n_z$, equals zero, then the corresponding angle ($\beta_z$ in this example) is 90°; the propagation

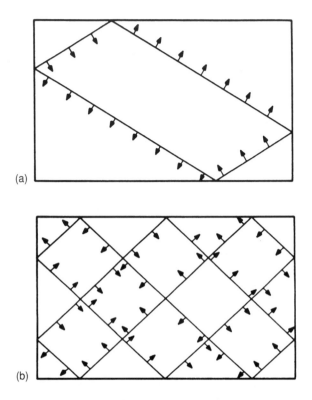

(a)

(b)

*Figure 3.3* Plane wavefronts creating standing waves in a rectangular room: (a) $n_x : n_y = 1:1$; (b) $n_x : n_y = 3:2$.

takes place perpendicularly to the respective axis, i.e. parallel to all planes which are perpendicular to that axis. The corresponding vibration pattern is frequently referred to as a 'tangential mode'. If there is only one non-zero integer $n$, the propagation is parallel to one of the coordinate axes, i.e. parallel to one of the room edges. Then we are speaking of an 'axial mode'. Modes with all integers different from zero are called 'oblique modes'. In Fig. 3.3, for the two-dimensional case, two combinations of plane waves are shown which correspond to two different eigenfunctions.

We can get an illustrative survey on the arrangement, the types and the number of the eigenvalues by the following geometrical representation. We interpret $k_x$, $k_y$ and $k_z$ as cartesian coordinates in a $k$-space. Each of the allowed values of $k_x$, given by eqn (3.14$a$), corresponds to a plane which is perpendicular to the $k_x$-axis. The same statement holds for the values of $k_y$ and $k_z$, given by eqns (3.14$b$) and (3.14$c$). These three equations therefore represent three sets of equidistant, mutually orthogonal planes in the $k$-space.

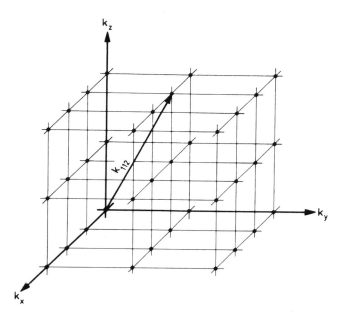

*Figure 3.4* Eigenvalue lattice in the *k*-space for a rectangular room. The arrow pointing from the origin to an eigenvalue point indicates the direction of one of the eight wave planes which the corresponding mode consists of (see Fig. 3.3); its length is proportional to the eigenvalue.

Since for one eigenvalue these equations have to be satisfied simultaneously, each intersection of three mutually orthogonal planes corresponds to a certain eigenvalue. These intersections in their totality form a rectangular point lattice in the first octant of our *k*-space (see Fig. 3.4). (Negative values obviously do not yield additional eigenvalues, since eqn (3.15) is not sensitive to the signs of the characteristic integers!) The lattice points corresponding to tangential and to axial modes are situated on the coordinate planes and on the axes, respectively. The straight line connecting the origin of the coordinate system to a certain lattice point has—according to eqn (3.19)—the same direction as one of the plane waves of which the associated mode is made up (see eqn (3.18)).

This representation allows a simple estimate of the number of eigenfrequencies which are located between the frequency 0 and some other given frequency $f$. Considered geometrically, eqn (3.12) represents a spherical surface in the *k*-space with radius $k$, enclosing a 'volume' $4\pi k^3/3$. Of this volume, however, only the portion situated in the first octant is of interest, i.e. the volume $\pi k^3/6$. On the other hand, the distances between one certain lattice point and its nearest neighbours in the three coordinate directions are $\pi/L_x$, $\pi/L_y$ and $\pi/L_z$. The *k*-'volume' per lattice point is therefore

$\pi^3/L_x L_y L_z = \pi^3/V$, where $V$ is the geometrical volume of the room under consideration. Now we are ready to write down the number of lattice points inside the first octant up to radius $k$, which is equivalent to the number of eigenfrequencies from 0 to an upper limit $f = kc/2\pi$:

$$N_f = \frac{\pi k^3/6}{\pi^3/V} = \frac{Vk^3}{6\pi^2} = \frac{4\pi}{3} V \left(\frac{f}{c}\right)^3 \tag{3.20}$$

The average density of eigenfrequencies on the frequency axis, i.e. the number of eigenfrequencies per Hz at the frequency $f$, is

$$\frac{dN_f}{df} = 4\pi V \frac{f^2}{c^3} \tag{3.21}$$

A more rigorous derivation of $N_f$ must account for the fact that the lattice points situated on the coordinate planes, representing tangential modes, are shared by two adjacent octants. Hence only half of them have been accounted for so far. Similarly, all lattice points on the coordinate axes related to axial modes are common to four octants; therefore only one fourth of them are contained in eqn (3.20) and their contribution to the total number of modes is too small by a factor 4.

The number of all lattice point corresponding to tangential modes can be calculated in about the same way which led us to eqn (3.20). The result is

$$N_{\tan} = \frac{1}{4\pi} k^2 (L_x L_y + L_y L_z + L_z L_x) = \frac{k^2 S}{8\pi} = \frac{\pi S}{2} \left(\frac{f}{c}\right)^2$$

where we introduced the total area of all walls, $S = 2(L_x L_y + L_y L_z + L_z L_x)$. Thus, one correction term to eqn (3.20) reads $(\pi S/4)(f/c)^2$. Likewise, the total number of lattice points situated on the coordinate axes within the first octant is $kL/4\pi = Lf/2c$, with $L = 4(L_x + L_y + L_z)$ denoting the sum of all edge lengths of the room. However, we should keep in mind that half of these points are already contained in the expression above while another quarter of them have been, as mentioned, counted in eqn (3.20). Hence, the correction term due to the axial modes is $Lf/8c$ and the corrected expression of the number of modes with eigenfrequencies up to a frequency $f$ is

$$N_f = \frac{4\pi}{3} V \left(\frac{f}{c}\right)^3 + \frac{\pi}{4} S \left(\frac{f}{c}\right)^2 + \frac{L}{8} \frac{f}{c} \tag{3.20a}$$

It can be shown that in the limiting case $f \to \infty$ eqn (3.20) is valid not only for rectangular rooms but also for rooms of arbitrary shape. This is not too surprising since any enclosure can be conceived as being composed of many (or even infinitely many) rectangular rooms. For each of them eqn (3.20) yields the number $N_i$ of eigenfrequencies. Since this equation is linear in $V$, the total number of eigenfrequencies is just the sum of all $N_i$.

We bring this section to a close by applying eqns (3.20) and (3.21) to two simple examples. The rectangular room for which the eigenfrequencies listed in Table 3.1 have been calculated has the volume 59.7 m³. For an upper frequency limit of 116 Hz, eqn (3.20) indicates—with $c = 343$ m/s—10 eigenfrequencies as compared with the 20 we have listed in the table. Using the more accurate formula (3.20a) we obtain 20 eigenfrequencies. That means that we must not neglect the corrections due to tangential and axial modes when dealing with such small rooms at low frequencies.

Now we consider a rectangular room with dimensions 50 m × 24 m × 14 m whose volume is 16 800 m³. (This might be a large concert hall, for instance.) In the frequency range from 0 to 10 000 Hz there are, according to eqn (3.20), about $1.7 \times 10^9$ eigenfrequencies. At 1000 Hz the number of eigenfrequencies per hertz is about 5200; thus, the average distance of two eigenfrequencies on the frequency axis is less than 0.0002 Hz. These figures underline the impossibility of evaluating the sound field in a room by calculating normal modes.

## 3.3 Non-rigid walls

In this section we are still dealing with rectangular rooms. But now we consider a room the walls of which are not completely rigid. This means, the normal components of particle velocity may have non-vanishing values along the boundary. Accordingly, we have to replace the boundary condition (3.11a) by the more general condition of eqn (3.2a) or (3.2b). As in the preceding section the solution of the wave equation consists of three factors $p_x$, $p_y$ and $p_z$, each of which depends on one spatial coordinate only. If the specific wall impedance is constant over each wall, the boundary condition for $p_x$ reads

$$\zeta_x \frac{\mathrm{d}p_x}{\mathrm{d}x} = \mathrm{i}kp_x \qquad \text{for } x = 0$$

$$\zeta_x \frac{\mathrm{d}p_x}{\mathrm{d}x} = -\mathrm{i}kp_x \qquad \text{for } x = L_x$$

$$\tag{3.22}$$

Analogous conditions are set up for the walls perpendicular to the $y$-axis and the $z$-axis, respectively.

For the present purpose it is more useful to write the general solution for $p_x$ in the complex form:

$$p_x(x) = C_1 \exp(-\mathrm{i}k_x x) + D_1 \exp(\mathrm{i}k_x x) \tag{3.23}$$

which is equivalent to eqn (3.13). By inserting $p_x$ into the boundary conditions (3.22) we obtain two linear and homogeneous equations for the constants $C_1$ and $D_1$:

$$C_1(k_x\zeta_x + k) - D_1(k_x\zeta_x - k) = 0$$
$$C_1(k_x\zeta_x - k)\exp(-ik_xL_x) - D_1(k_x\zeta_x + k)\exp(ik_xL_x) = 0 \tag{3.24}$$

which have a non-vanishing solution only if the determinant of their coefficients is zero. This leads to the following condition:

$$\exp(ik_xL_x) = \pm\frac{k_x\zeta_x - k}{k_x\zeta_x + k} \tag{3.25}$$

Solving this equation for $k_x\zeta_x$ (separately for both signs) and using the relation (1.15) yields two transcendent equations

$$\tan u = i\frac{kL_x}{2u\zeta_x} \quad \text{and} \quad \tan u = i\frac{2u\zeta_x}{kL_x} \tag{3.25a}$$

with

$$u = \frac{1}{2}k_xL_x$$

In general, eqn (3.25a) must be solved numerically. Once the allowed values of $k_x$ have been determined, the ratio of the two constants $C_1$ and $D_1$ can be evaluated from eqns (3.24); for instance, from the first of them:

$$\frac{C_1}{D_1} = \frac{k_x\zeta_x - k}{k_x\zeta_x + k} = \pm\exp(ik_xL_x) \tag{3.26}$$

the latter equality resulting from eqn (3.25). Thus the x-dependent factor of the eigenfunction reads, apart from a constant factor:

$$p_x(x) = \begin{cases} \cos[k_x(x - L_x/2)] & \text{(even solution)} \\ \sin[k_x(x - L_x/2)] & \text{(odd solution)} \end{cases} \tag{3.27}$$

Evidently, these functions are symmetric or antisymmetric, depending on the sign of the exponential in eqn (3.26). Of course, this is a consequence of the symmetry of wall properties. As mentioned before, the complete eigenfunction is made up of three such factors.

From now on we restrict the discussion to enclosures with a nearly rigid boundary, i.e. to the limiting case $|\zeta| \gg 1$. Then we expect that the eigenvalues and eigenfunctions are not very different from those of the rigid-walled room.

An approximate solution of eqn (3.25) can be found by employing the series expansion of the exponential function and to truncate it after the second term:

$$\exp(\pm z) = 1 \pm z + \frac{z^2}{2!} \pm \ldots \approx 1 \pm z \qquad \text{for } z \ll 1$$

Since $k \ll k_x \zeta_x$ this relation can be applied to right-hand side of eqn (3.25), with the result:

$$\exp(ik_x L_x) \approx \pm \exp\left(-\frac{2k}{k_x \zeta_x}\right)$$

One might be tempted to solve this equation by equating the exponents. However, before doing so we should remember that the exponential function with imaginary argument is a periodic function with the period $2\pi$, $\exp(iz) = \exp(iz + i2\pi)$. Thus, to obtain the complete solution of the last equation its right-hand side must be multiplied by the factor $\exp(i\pi n_x)$, which also includes both signs. Here $n_x$ is an arbitrary integer. This leads us to the result:

$$k_x L_x \approx n_x \pi + i\frac{2k}{k_x \zeta_x}$$

According to our assumption the second term on the right is much smaller than the first one; hence we can replace $k_x$ in the denominator with $n_x \pi / L_x$:

$$k_x \approx \frac{n_x \pi}{L_x} + i\frac{2k}{n_x \pi \zeta_x} \tag{3.28}$$

Comparing this result with eqn (3.14$a$) confirms our expectation that the allowed values of $k_x$ are not much different from those of the hard-walled room. With increasing 'order' $n_x$ of the mode the difference becomes even smaller.

Suppose that the wall is reactive, i.e. that it is free of absorption. Then its specific impedance is purely imaginary and the correction term is real. If Im $\zeta$ is positive, which indicates that the motion of the wall is mass-controlled, then the allowed value is higher than in the rigid-walled room. Conversely, a compliance wall, i.e. a wall with the impedance of a spring, will lower the allowed value $k_x$.

Whenever the specific impedance has a non-vanishing real part indicating wall losses $k_x$ is complex; the imaginary part of $k_x$ is related to the damping constant (see also eqn (3.9)). If the allowed values of $k_x$ are denoted by $k_{xn_x}$, the eigenvalues of the original differential equation are given as earlier by (see eqn (3.12)):

$$k_{n_x n_y n_z} = \left( k^2_{x n_x} + k^2_{y n_y} + k^2_{z n_z} \right)^{1/2} \tag{3.29}$$

In Fig. 3.5 the absolute value of the $x$-dependent factor of a certain eigenfunction is represented for three cases: for rigid walls ($\zeta_x = \infty$), for mass-loaded walls with no energy loss ($\zeta_x = i$), and for walls with real impedance. In the second case, the standing wave is simply shifted together, but its shape remains unaltered. On the contrary, in the third case of lossy walls, there are no longer exact nodes and the pressure amplitude is different from zero at all points. This can easily be understood by keeping in mind that the walls dissipate energy, which must be supplied by waves travelling towards the

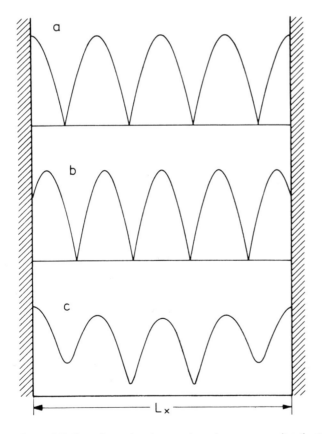

*Figure 3.5* One-dimensional normal mode, pressure distribution for $n_x = 4$: (a) $\zeta = \infty$ ; (b) $\zeta = i$; (c) $\zeta = 2$.

walls; thus, a pure standing wave is not possible. This situation is comparable to a standing wave in front of a single plane with a reflection factor less than unity as shown in Fig. 2.3.

## 3.4 Steady-state sound field

In Section 3.1 we saw that the steady-state acoustical behaviour of a room, when it is excited by a sinusoidal signal with angular frequency $\omega$, is described by a series of the form

$$p_\omega = \sum_n \frac{A_n}{\omega^2 - \omega_n^2 - 2i\delta\omega_n} \tag{3.30}$$

where we are assuming $\delta_n \ll \omega_n$ as in Section 3.1. By comparing this with our earlier eqn (3.10) we learn that the coefficients $A_n$ are functions of the source position, of the receiving position, and of the angular frequency $\omega$. If both points are considered as fixed, eqn (3.30) is the transfer function of the room for both positions.

Since we have supposed that the damping constants $\delta_n$ are small compared with the eigenfrequencies $\omega_n$, the predominant frequency dependence is that of the denominator. Whenever the frequency is close to one of the eigenfrequencies $\omega_n$ the corresponding series term will become very large, or, in other words, the system will resonate. Since it is the term $\omega^2 - \omega_n^2$ which is responsible for the strong frequency dependence, $\omega_n$ can be replaced by $\omega$ in the last term of the denominator without any serious error. The absolute value of the $n$th series term then becomes

$$\frac{|A_n|}{\left[(\omega^2 - \omega_n^2)^2 + 4\omega^2\delta_n^2\right]^{1/2}}$$

and thus agrees with the amplitude–frequency characteristics of a resonance system, according to eqn (2.29). Therefore the stationary sound pressure in a room and at one single exciting frequency proves to be the combined effect of numerous resonances. The half-widths of the resonance curves according to eqn (2.31) are:

$$(\Delta f)_n = \frac{1}{2\pi}(\Delta\omega)_n = \frac{\delta_n}{\pi} \tag{3.31}$$

In most rooms the damping constants lie between 1 and 20 $s^{-1}$. Therefore our earlier assumption concerning the relative magnitude of the damping constants seems to be justified. Furthermore, we see from this that the half-widths according to eqn (3.31), within which the various resonance terms assume their highest values, are of the order of magnitude of 1 Hz.

This figure is to be compared with the average spacing of eigenfrequencies on the frequency axis which is the reciprocal of $dN_f/df$ after eqn (3.21). If the mean spacing of resonance frequencies is substantially larger than the average half-width $\langle\delta_n\rangle/\pi$ we expect that most of the room resonances are well separated, and each of them can be individually excited and detected. In a tiled bathroom, for example, the resonances are usually weakly damped, and thus one can often find one or several of them by singing or humming. If, on the contrary, the average half-width of the resonances is much larger than the average spacing of the eigenfrequencies, there will be strong overlap of resonances and the latter cannot be separated. Instead, at any frequency several or many terms of the sum in eqn (3.30) will have significant values; hence several or many normal modes will contribute to the total sound pressure. According to Schroeder[4,5] a limiting frequency separating both cases can be defined by the requirement that on average three eigenfrequencies fall into one resonance half-width, or, with eqn (3.21):

$$\langle\Delta f_n\rangle = 3 \cdot \frac{c^3}{4\pi V f^2}$$

Introducing the average damping constant by eqn (3.31), the sound speed $c = 343$ m/s and solving for the frequency yields the limiting frequency, the so-called 'Schroeder frequency':

$$f_s \approx \frac{5500}{\sqrt{V\langle\delta_n\rangle}} \quad \text{Hz} \approx 2000\sqrt{\frac{T}{V}} \quad \text{Hz} \tag{3.32}$$

In this expression, the so-called 'reverberation time' $T = 6.91/\langle\delta\rangle$ has been introduced (see eqn (3.50)). The room volume $V$ has to be expressed in cubic metres.

In large halls the Schroeder frequency is typically below 50 Hz; hence there is strong modal overlap in the whole frequency range of interest, and there is no point in evaluating any eigenfrequencies. It is only in small rooms that a part of the important frequency range lies below $f_s$, and in this range the acoustic properties are determined largely by the values of individual eigenfrequencies. To calculate the expected number $N_{fs}$ of eigenfrequencies in the range from zero to $f_s$, eqn (3.32) is inserted into eqn (3.20) with the result:

$$N_{fs} \approx 800\sqrt{T^3/V} \tag{3.33}$$

Thus, in a classroom with a volume of 200 m³ and a reverberation time of 1 s, about 60–70 eigenfrequencies dominate the acoustical behaviour below the Schroeder frequency which is about 140 Hz. This example illustrates the somewhat surprising fact that the acoustics of small rooms are in a way more complicated than those of large ones.

It should be noted that eqn (3.32) can also be read as a criterion for the acoustical size of a room, again on the grounds of its modal structure. A given room can be considered as 'acoustically large' if

$$V > \left(\frac{2000}{f}\right)^2 T \qquad\qquad (3.32a)$$

($f$ in Hz, $T$ in seconds, $V$ in m$^3$). After the preceding discussion it is not surprising that this limit depends on the frequency.

For the rest of this section and for the next one we restrict our discussion to the frequency range above the Schroeder limit, $f > f_s$. Moreover, the observation point is supposed to be far enough from the sound source to make the direct sound component negligibly small. Hence, if the room under consideration is excited with a pure tone its steady-state response is made up by contributions of several or even many normal modes with randomly distributed phases. The situation may be elucidated by the vector diagram in Fig. 3.6. Each vector or 'phasor' represents the contribution of one term in eqn (3.30) (nine significant terms in this example). The resulting sound pressure is obtained as the vector sum of all components. For a different frequency or at a different point in the room, this diagram has the same general character, but it looks quite different in detail, provided that the change in frequency or location is sufficiently great.

Since the different components can be considered as mutually independent, the central limit theorem of probability theory can be applied to the real part as well as to the imaginary part of the resulting sound pressure $p_\omega$. According to this theorem both quantities are random variables obeying nearly a Gaussian distribution. This statement implies that the squared absolute value of the sound pressure $p$, divided by its frequency (or space)

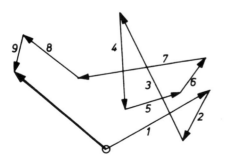

*Figure 3.6* Phasor diagram showing the components of the steady-state sound pressure in a room and their resultant for sinusoidal excitation.

average, $y = |p|^2/\langle|p|^2\rangle$, which is proportional to the energy density, is distributed according to an exponential law or, more precisely, the probability of finding this quantity between $y$ and $y + dy$ is given by

$$P(y)\,dy = \exp(-y)\,dy \qquad (3.34)$$

Its mean value and also its variance $\langle y^2\rangle - \langle y\rangle^2$ is 1, as is easily checked. The probability that a particular value of $y$ exceeds a given limit $y_0$ is

$$P_i(y > y_0) = \int_{y_0}^{\infty} P(y)\,dy = \exp(-y_0) \qquad (3.34a)$$

It is very remarkable that the distribution of the energy density is completely independent of the type of the room, i.e. on its volume, its shape or the treatment of its walls.

Figure 3.7a presents a typical 'space curve', i.e. the sound pressure level recorded with a microphone along a straight line in a room while the driving frequency is kept constant. Such curves express the space dependence of eqn (3.30). Their counterpart are 'frequency curves', i.e. representations of the sound pressure level observed at a fixed microphone position when the excitation frequency is slowly varied. They are based upon the frequency dependence of eqn (3.30). A section of such a frequency curve is shown in Fig. 3.7b. Recorded at another microphone position or in another room it would look quite different in detail; its general character, however,

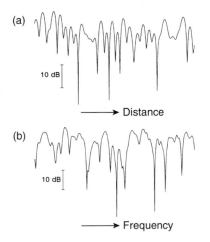

*Figure 3.7* Steady-state sound pressure (logarithmic representation): (a) along a straight line at constant frequency ('space curve'); (b) at a fixed position with slowly varying driving frequency ('frequency curve').

would be similar to the one shown. The same statement holds for space curves.

Both curves in Fig. 3.7 look quite similar: they are highly irregular and show flat peaks and deep valleys. A maximum of the pressure level occurs if in Fig. 3.6 many or all arrows happen to point in about the same direction, indicating similar phases of most contributions. Similarly, a minimum appears if these contributions more or less cancel each other. Therefore, the maxima of frequency curves are not related to particular room resonances or eigenfrequencies but are the result of accidental phase coincidences. The general similarity of space and frequency curves is not too surprising: both sample the same distribution of squared sound pressure amplitudes, namely that given by eqn (3.34). But since they do it in a different way, there are indeed differences which can be demonstrated by regarding their autocorrelation functions. As before, we consider $y = |p|^2/\langle|p|^2\rangle$ as the significant variable. However, in contrast to Section 1.4, the 'signals' to which the autocorrelation functions refer are not time functions but $y(x)$ and $y(f)$, respectively, with $x$ denoting the coordinate along the straight line where the pressure level is recorded. For space curves in rooms with an isotropic sound field the autocorrelation function reads:[6]

$$\phi_{yy}(\Delta x) = \langle y(x) \cdot y(x + \Delta x)\rangle = 1 + \left(\frac{\sin(k\Delta x)}{k\Delta x}\right)^2 \tag{3.35a}$$

while the autocorrelation function of frequency curves is given by[7]

$$\phi_{yy}(\Delta f) = \langle y(f) \cdot y(f + \Delta f)\rangle = 1 + \frac{1}{1 + \left(\pi \Delta f/\langle\delta_n\rangle\right)^2} \tag{3.35b}$$

The 1 in eqns (3.35) is due to the constant component of $y$, i.e. to mean value $\langle y\rangle$. Both autocorrelation functions are plotted in Fig. 3.8.

As long as the autocorrelation function $\phi_{yy}(\Delta x)$ is noticeably different from unity, it indicates that there is still some causal relationship between any two samples of $y$ taken at two points $\Delta x$ apart. We suppose that the samples can be considered as statistically independent if the variable part of $\phi_{yy}(\Delta x)$ is smaller than 0.2, corresponding to $k\Delta x > 2$; hence the reach of causality is about

$$\Delta x_{\text{corr}} \approx 2/k = \lambda/\pi \tag{3.36a}$$

A similar consideration applied to $\phi_{yy}(\Delta f)$ yields the critical frequency shift:

$$\Delta f_{\text{corr}} \approx 0.64\langle\delta_n\rangle \tag{3.36b}$$

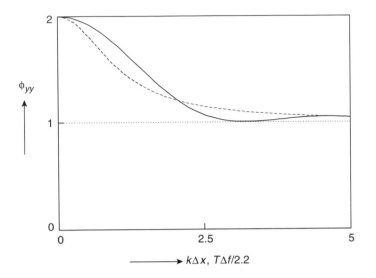

*Figure 3.8* Autocorrelation functions $\phi_{yy}$ of room responses: ————— $\phi_{yy}(\Delta x)$ after eqn (3.35*a*) (the abscissa is $k\Delta x$); --------- $\phi_{yy}(\Delta f)$ after eqn (3.35*b*) (the abscissa is $\pi \Delta f / \langle \delta \rangle = T\Delta f/2.2$).

Furthermore, it can be shown that the average distance of adjacent maxima of space curves in a diffuse sound field is

$$\langle \Delta x_{max} \rangle \approx 0.79\lambda \tag{3.37a}$$

Similarly, the mean spacing of frequency curve maxima is[5]

$$\langle \Delta f_{max} \rangle \approx \frac{\langle \delta_n \rangle}{\sqrt{3}} \approx \frac{4}{T} \tag{3.37b}$$

where again $T = 6.91/\langle \delta_n \rangle$ denotes the reverberation time, as in eqn (3.32).

A quantity which is especially important to the performance of sound reinforcement systems in rooms is the absolute maximum of a frequency curve within a given frequency bandwidth $B$. In order to estimate it we represent the frequency curve by $N$ equidistant and statistically independent samples. Then we define the absolute maximum $y_{max}$ by requiring that this value be exceeded by just one sample, i.e.

$$P_i(y > y_{max}) = \frac{1}{N}$$

Employing eqn (3.34a) we obtain the most likely value of the absolute maximum:

$$y_{max} = \ln N$$

The corresponding level difference between this value and the average of the energetic frequency response is

$$\Delta L_{max} = 10 \cdot \log_{10}(\ln N) = 4.34 \ln (\ln N) \quad dB$$

On the one hand, the samples should be close enough along the frequency axis to represent the frequency response of the room; on the other hand, their distance should be wide enough to assure their independence. A fair compromise between these conflicting requirements is achieved if we take four samples per frequency interval $\langle \Delta f_{max} \rangle$. This leads to

$$N = \frac{B}{\langle \Delta f_{max} \rangle / 4} \approx BT$$

Hence, the final expression for the expected maximum level in a frequency curve reads

$$\Delta L_{max} = 4.34 \ln [\ln (BT)] \quad dB = 10 \cdot \log_{10}(\log_{10} BT) + 3.62 \quad dB \tag{3.38}$$

This derivation is certainly not free of arbitrariness. On the other hand, the number $N$ is not too critical since the double logarithm varies only slowly with $N$. Schroeder[8] was the first to derive this formula, although in a somewhat different manner. A more rigorous derivation[9] led him to the expression

$$\Delta L_{max} = 10 \cdot \log_{10}\left[\log_{10}\left(\frac{BT}{22}\right)\right] + 6.3 \quad dB \tag{3.38a}$$

As an example, let us consider a room with a reverberation time of 1 s. According to eqn (3.38), the absolute maximum of its frequency curve in the range from 0 to 10 kHz is about 9.6 dB above its energetic average. The more exact formula (3.38a) yields a level difference of 10.5 dB.

If the driving frequency is slowly varied, both the amplitude of the sound pressure and its phase fluctuate in an irregular manner. However, a monotonic increase of the phase angle is superimposed on these quasi-statistical variations. The average phase shift per hertz is given by[10]

$$\left\langle \frac{d\psi}{df} \right\rangle = \frac{\pi}{\langle \delta_n \rangle} \approx 0.455 T \tag{3.39}$$

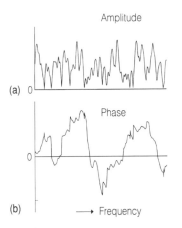

*Figure 3.9* (a) Magnitude and (b) phase of a room transfer function above the Schroeder limit.

Figure 3.9 shows in its upper part an amplitude–frequency curve. It is similar to that shown in Fig. 3.7b with the difference that the plotted quantity is not the sound pressure level but the absolute value of the sound pressure. The lower part plots the corresponding phase variation obtained after subtracting the monotonic change according to eqn (3.39). Since this figure represents the transfer function between two points within a room, the phase spectrum of any signal will be randomised when it is transmitted in the room. This can be demonstrated in the following way. A loudspeaker placed in a reverberant room is alternatively fed with two signals which have equal amplitude spectra, but different phase spectra. The first signal, for instance, may be a periodic sequence of rectangular impulses (see Fig. 3.10a), whereas the second one is a maximum length sequence (see Section 8.2) made up of rectangular impulses with quasi-randomly changing signs (see Fig 3.10b). If the listener is close to the loudspeaker he can clearly hear that both signals sound quite different provided the repetition rate $1/\Delta t$ is not too high. However, when he slowly steps away from the loudspeaker the room field will prevail over the direct sound signal; the perceived difference becomes smaller and smaller and finally disappears.

We close this section by emphasising again that the general properties of the room transfer function, especially the distribution of its absolute values and thus the depth of its irregularities, the sequence of maxima and the phase change associated with it, do not depend in a characteristic way on the room or on the point of observation. In particular, it is impossible to base a criterion of the acoustic quality on these quantities. This is in contrast to what has been expected in the past. From experience with transmission lines,

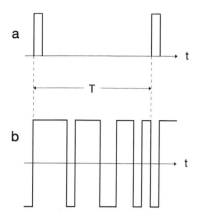

*Figure 3.10* Two periodic signals with equal amplitude spectrum but different phase spectra.

amplifiers and loudspeakers, for example, acousticians had the idea that a room would be the better in acoustical respect the smoother its frequency curve. This is not the case for several reasons. First, speech and music show such rapid variations in signal character that a large room does not reach steady-state conditions when excited by them except perhaps during very slow music passages. Second, it has been shown that our hearing organ is unable to perceive fluctuations of the frequency spectrum of a signal if these irregularities are spaced closely enough on the frequency axis (see Section 7.3).

## 3.5 Frequency and spatial averaging

The previous discussion concerned the steady-state response of a room to a sinusoidal excitation signal. In contrast, the present section deals with a room's response to signals of finite bandwidth. Obviously, the resulting response is obtained by some frequency averaging. Another important process is spatial averaging of room responses to sinusoidal excitation by which spatial fluctuations can be levelled out and which is of particular interest in some measuring techniques (see Section 5.5).

We start with a simple problem, namely the fluctuations of the room response when the room is simultaneously excited by $M$ sine signals with equal amplitudes but with different frequencies. As in the preceding section we denote the energetic room response at the frequency $f_n$ with $y_n = |p_n|^2/\langle|p|^2\rangle$. We are looking for the average and the variance of the sum $z = y_1 + y_2 + \ldots + y_M$, supposing that the excitation frequencies differ by more than the correlation interval $\Delta f_{corr}$ after eqn (3.36$b$). Then the samples $y_n$ are uncorrelated and, according to probability theory $\langle z \rangle = M \langle y \rangle = M$. Furthermore, the variance of the sum, Var $(z) = \langle z^2 \rangle - \langle z \rangle^2$ is $M$ times the variance of $y$, which is 1 (see preceding section). Accordingly, the relative variance of the sum is

$$\frac{Var\,(z)}{\langle z \rangle^2} = \frac{1}{M} \tag{3.40}$$

The same result is arrived at if one considers $M$ samples of the squared absolute value of the sound pressure taken at different points of the sound field, excited with a single sine signal and again under the condition that the mutual distances of these points exceed the correlation distance $\Delta x_{corr}$ after eqn (3.36$a$).

Matters are more complicated if we consider sound excitation with a signal having the finite bandwidth $B = f_2 - f_1$ because then we do not average uncorrelated samples.[11] The quantity we are interested in is

$$z = \frac{1}{B} \int_{f_1}^{f_2} y(f)\,df \qquad \text{with } y = \frac{|p(f)|^2}{\langle |p(f)|^2 \rangle} \tag{3.41}$$

Again, we have $\langle y \rangle = 1$ and $\langle z \rangle = 1$. The quadratic average of $z$ is

$$\langle z^2 \rangle = \frac{1}{B^2} \left\langle \left[ \int_{f_1}^{f_2} y(f)\,df \right]^2 \right\rangle = \frac{1}{B^2} \int_{f_1}^{f_2} df \int_{f_1}^{f_2} df'\, \langle y(f)y(f') \rangle$$

By comparing the integrand in the latter expression with eqn (3.35$b$) we see that it is equivalent to the autocorrelation function $\phi_{yy}$ with the argument $f - f'$. Hence

$$\langle z^2 \rangle = \frac{1}{B^2} \int_{f_1}^{f_2} df \int_{f_1}^{f_2} \phi_{yy}(f - f')\,df' = 1 + \frac{1}{B^2} \int_{f_1}^{f_2} df \int_{f_1}^{f_2} \frac{df'}{1 + \pi^2(f - f')^2/\langle \delta \rangle^2}$$

Evaluation of this double integral leads to the following expression for the 'relative variance' of $z$:

$$\frac{Var(z)}{\langle z \rangle^2} = \frac{2\langle \delta \rangle}{\pi B} \arctan\left(\frac{\pi B}{\langle \delta \rangle}\right) - \frac{\langle \delta \rangle^2}{\pi^2 B^2} \ln\left(1 + \frac{\pi^2 B^2}{\langle \delta \rangle^2}\right) \tag{3.42a}$$

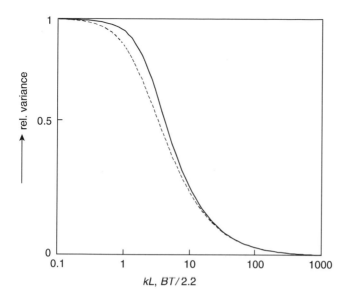

*Figure 3.11* Relative variance of space and frequency averaged room responses (squared magnitude of sound pressure), ———— space averaging over a distance $L$ after eqn (3.42b) (the abscissa is $kL$); --------- frequency averaging over a frequency interval $B$ after eqn (3.42a) (the abscissa is $\pi B/\langle \delta \rangle = BT/2.2$).

This quantity is plotted in Fig. 3.11 (dashed line), using the variable

$$\frac{\pi B}{\langle \delta \rangle} = \frac{BT}{2.2}$$

Again $T$ is the reverberation time (see next section). A useful approximation to eqn (3.42a) for large values of $BT$ is[12]

$$\frac{\mathrm{Var}(z)}{\langle z \rangle^2} \approx \frac{1}{1 + BT/6.9}$$

If the power spectrum $w(f)$ of the exciting signal is not constant within the exciting frequency band, the bandwidth $B$ is replaced by an 'equivalent bandwidth':

$$B_{eq} = \left[ \int_0^\infty w(f)\,df \right]^2 \bigg/ \int_0^\infty [w(f)]^2\,df$$

Basically the same procedure can be applied to calculate the space average over a straight path of length $L$. In this case we consider the average

$$z = \frac{1}{L} \int_0^L y(f)\,\mathrm{d}f$$

When calculating the quadratic average of $z$ we encounter the autocorrelation function $\phi_{yy}(x - x')$, and after inserting eqn (3.35a) we arrive at the relative variance of $z$:

$$\frac{\mathrm{Var}(z)}{\langle z \rangle^2} = \frac{1}{L^2} \int_0^L \mathrm{d}x \int_0^L \left[ \frac{\sin k(x - x')}{k(x - x')} \right]^2 \mathrm{d}x' \tag{3.42b}$$

The result obtained by executing the twofold integration is a somewhat lengthy formula, the content of which is plotted in Fig. 3.11 as a function of $kL$. Both curves of this diagram are quite similar, in contrast to the autocorrelation functions shown in Fig. 3.8.

## 3.6 Decaying modes, reverberation

If a room is excited not by a stationary signal as in the preceding sections but instead by a very short sound impulse emitted at time $t = 0$, we obtain, in the limit of vanishing pulse duration, the impulse response $g(t)$ at some receiving point of the room. According to the discussion in Section 1.4, this is the Fourier transform of the transfer function. Hence

$$g(t) = \int_{-\infty}^{\infty} p_\omega \exp(i\omega t)\,\mathrm{d}\omega$$

Generally the evaluation of this Fourier integral with $p_\omega$ after eqn (3.30), is rather complicated since both $\omega_n$ and $\delta_n$ depend on the driving frequency $\omega$, strictly speaking. At any rate the solution has the form

$$g(t) = \begin{cases} 0 & \text{for } t < 0 \\ \sum_n A'_n \exp(-\delta'_n t) \cos(\omega'_n t + \varphi'_n) & \text{for } t \geq 0 \end{cases} \tag{3.43}$$

It is composed of sinusoidal oscillations with different frequencies, each dying out with its own particular damping constant. This is plausible since each term of eqn (3.30) corresponds to a resonator whose reaction to an excitation impulse is a damped oscillation. If the wall losses in the room are not too large, the frequencies $\omega'_n$ and damping constants $\delta'_n$ differ only slightly from $\omega_n$ and $\delta_n$ as occur in eqn (3.30). As is seen from the more explicit representation (3.10), the coefficients $A'_n$ depend on the location of both the source and the receiving point.

If the room is excited not by an impulse but by a stationary signal $s(t)$, which is switched off at $t = 0$, the resulting room response $h(t)$ is, according to eqn (1.46),

$$h(t) = \int_{-\infty}^{0} s(\tau)g(t-\tau)\,d\tau$$

$$= \sum_{n} A'_n \exp(-\delta'_n t)\left[a_n \cos(\omega'_n t + \varphi'_n) + b_n \sin(\omega'_n t + \varphi'_n)\right] \quad \text{for } t \geq 0$$

where

$$a_n = \int_{-\infty}^{0} s(\tau)\exp\left(\delta'_n \tau\right)\cos\left(\omega'_n \tau\right)d\tau \quad \text{and}$$

$$b_n = \int_{-\infty}^{0} s(\tau)\exp\left(\delta'_n \tau\right)\sin\left(\omega'_n \tau\right)d\tau$$

The above expression for $h(t)$ can be written more simply as

$$h(t) = \sum_{n} c_n \exp(-\delta'_n t)\cos(\omega'_n t - \phi_n) \qquad \text{for } t \geq 0 \tag{3.44}$$

with

$$c_n = A'_n \sqrt{(a_n^2 + b_n^2)}$$

It is evident that only such modes can contribute to the general decay process whose eigenfrequencies are close to the frequencies contained in the spectrum of the driving signal. If the latter is a sinusoidal tone switched off at some time $t = 0$, then only such components contribute noticeably to $h(t)$, the frequencies $\omega'_n$ of which are separated from the driving frequency $\omega$ by not more than a half-width, i.e. by about $\delta'_n$.

The decay process described by eqn (3.44) is called the 'reverberation' of the room. It is one of the most important and striking acoustical phenomena of a room, familiar also to every layman.

An expression proportional to the energy density is obtained by squaring $h(t)$:

$$w(t) \propto \left[h(t)\right]^2 = \sum_n \sum_m c_n c_m \exp\left[-(\delta'_n + \delta'_m)t\right]$$
$$\cdot \cos\left(\omega'_n t - \phi_n\right) \cos\left(\omega'_m t - \phi_m\right) \tag{3.45}$$

This expression can be considerably simplified by short-time averaging, i.e. averaging is applied to the product of cosines only but not to the slowly-varying exponential. Thus, the products with $n \neq m$ will cancel, whereas each term $n = m$ yields a value $1/2$ . Thus, we obtain

$$w(t) = \sum_n c_n^2 \exp\left(-2\delta'_n t\right) \tag{3.46}$$

where all constants of no importance have been omitted.

Now we imagine that the sum is rearranged according to increasing damping constants $\delta'_n$. Additionally, the number of significant terms in eqn (3.46) is supposed to be very large. Then we can replace the summation by an integration by introducing a 'damping density' $H(\delta)$. This is done by including all terms with damping constants between $\delta$ and $\delta + d\delta$ by $H(\delta) d\delta$, and by normalising $H(\delta)$ so as to require

$$\int_0^\infty H(\delta) d\delta = 1$$

Then the integral envisaged becomes simply

$$w(t) = \int_0^\infty H(\delta) \exp\left(-2\delta t\right) d\delta \tag{3.47}$$

Just as with the coefficients $c_n$, the distribution of damping constants $H(\delta)$ depends on the sound signal, on the location of the sound source, and on the point of observation.

From this representation we can derive some interesting general properties of reverberation.

Usually reverberation measurements are based on the sound pressure level of the decaying sound field:

$$L_r = 10 \log_{10}\left(\frac{w}{w_0}\right) = 4.34 \ln\left(\frac{w}{w_0}\right)$$

Its decay rate is

$$\dot{L}_r = 4.34 \frac{\dot{w}}{w} \quad \text{dB/s} \tag{3.48a}$$

while the second derivative of the decay level is

$$\ddot{L}_r = 4.34 \frac{w\ddot{w} - \dot{w}^2}{w^2} \quad \text{dB/s}^2 \tag{3.48b}$$

(Each overdot in these formulae means differentiation with respect to time.) By invoking eqn (3.47) it is easy to show that the second derivative of the decay level $L_r$ is nowhere negative, which means that the decay curves are curved upwards. As a limiting case, they can be straight lines. The latter occurs if all damping constants involved in the decay process are equal, i.e. if the distribution $H(\delta)$ is a Dirac function.

For $t = 0$ the curve has its steepest part, its initial slope, as obtained from eqn (3.48a)

$$(\dot{L}_r)_{t=0} = -8.69 \int_0^\infty H(\delta)\delta \, d\delta = -8.69 \langle\delta\rangle \tag{3.49a}$$

which is determined by the mean value of the distribution $H(\delta)$. Furthermore, eqn (3.48b) leads to

$$(\ddot{L}_r)_{t=0} = -17.37 \left(\langle\delta^2\rangle - \langle\delta\rangle^2\right) \tag{3.49b}$$

This means, the second derivative of the level at $t = 0$, which is roughly the initial curvature of the decay curve, is proportional to the variance of the damping distribution $H(\delta)$.

In Fig. 3.12 are shown some examples of distributions of damping constants together with the corresponding logarithmic reverberation curves. The distributions are normalised so that their mean values (and hence the initial slopes of the corresponding reverberation curves) agree with each other. Only when all the damping constants are equal (Case d) are straight decay curves obtained.

Measured reverberation curves are often nearly straight, apart from some random or quasirandom fluctuations, as shown in Fig. 3.13. (These irregularities are due to beats between the decaying modes.) Then all decay constants can be replaced without much error by their average $\langle\delta\rangle$.

It is usual in room acoustics to characterise the duration of sound decay not by damping constants but by the 'reverberation time' or 'decay time' $T$,

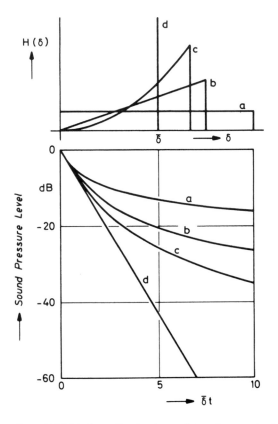

*Figure 3.12* Various distributions of damping constants and corresponding decay curves.

introduced by Sabine. It is defined as the time interval in which the decay level drops down by 60 dB. From

$$-60 = 10\log_{10}[\exp(-2\langle\delta\rangle T]$$

it follows that the reverberation time

$$T = \frac{3\ln(10)}{\langle\delta\rangle} \approx \frac{6.91}{\langle\delta\rangle} \qquad (3.50)$$

a relation which was already used in Section 3.4.

Typical values of reverberation times run from about 0.3 s (living rooms) up to 10 s (large churches, empty reverberation chambers). Most large halls have reverberation times between 0.7 and 2 s. Thus the average damping constants encountered in practice are in the range 1 to 20 s$^{-1}$.

The previous statements on the general shape of logarithmic decay curves, in particular on their curvature, are not valid for coupled rooms, i.e. for

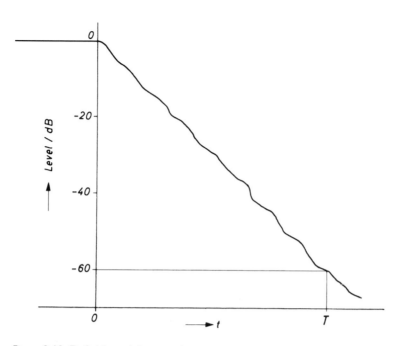

*Figure 3.13* Definition of the reverberation time.

rooms which are connected by relatively small coupling apertures or by partially transparent walls. The properties of such spaces will be discussed in Chapter 5.

## References

1 Morse PM, Feshbach H. Methods of Theoretical Physics. New York: McGraw-Hill, 1953. Chapter 6.
2 Morse PM, Ingard KU, Theoretical Acoustics. New York: McGraw-Hill, 1968. Chapter 9.
3 Zienkiewicz OC, Taylor RL. The Finite Element Method, London: McGraw-Hill, 1991.
4 Schröder M. Die statistischen Parameter der Frequenzkurven in großen Räumen. Acustica 1954; 4:594.
5 Kuttruff KH, Schroeder MR. On frequency response curves in rooms. Comparison of experimental, theoretical and Monte Carlos results for the average frequency spacing of maxima. J Acoust Soc Am 1962; 34:76.
6 Pierce AD. Acoustics—An Introduction to its Physical Principles and Applications. New York: McGraw-Hill, 1981.
7 Schroeder MR. Frequency-correlation functions of frequency responses in rooms. J Acoust Soc Am 1962; 34:1819.

8  Schroeder MR. Improvement of acoustic feedback stability in public address systems. Proc Third Intern Congr on Acoustics, Stuttgart, 1959. Amsterdam: Elsevier, 1961, p 771.

9  Schroeder MR. Improvement of acoustic feedback stability by frequency shifting. J Acoust Soc Am 1964; 36:1718.

10 Schroeder MR. Measurement of reverberation time by counting phase coincidences. Proc Third Intern Congr on Acoustics, Stuttgart, 1959. Amsterdam: Elsevier, 1961, p 897.

11 Schroeder MR. Effect of frequency and space averaging on the transmission responses of multimode media. J Acoust Soc Am 1969; 46:277.

12 Lubman D. Fluctuations of sound with position in a reverberant room. J Acoust Soc Am 1968; 44:1491.

# Chapter 4

# Geometrical room acoustics

The wave theoretical description of sound fields in closed spaces, as outlined in the preceding chapter, is certainly correct from a physical point of view. Nevertheless, it is not very profitable when it comes to solving practical problems in room acoustics such as designing a concert hall or analysing a given situation. This is because of the tremendous number of normal modes that need to be calculated in order to cover a wider frequency range and also the difficulty in calculating even one of them when the room has non-rigid walls or a more complicated shape (see Section 3.3). Furthermore, as we have seen, the knowledge of individual normal modes with the associated eigenfrequencies is pointless for the frequency range above the Schroeder frequency, i.e. in the frequency range which is usually of main interest in room acoustics.

We arrive at a much simpler and less abstract description—just as in geometrical optics—by employing the limiting case of vanishingly small wavelengths, i.e. we consider the limiting case of very high frequencies. This approach is the basis of what is called geometrical room acoustics; it is justified if the dimensions of the room and its walls are large compared with the wavelength of sound. This condition is not unrealistic in room acoustics; at a medium frequency of 1000 Hz, corresponding to a wavelength of 34 cm, the linear dimensions of the walls and the ceiling, and also the distances covered by the sound waves, are usually larger than the wavelength by orders of magnitude. Even if the reflection of sound from a balcony face is to be discussed, for instance, a geometrical description is applicable, at least qualitatively.

In geometrical room acoustics, the concept of a wave is replaced by the concept of a sound ray. The latter is an idealisation just as much as the plane wave. As in geometrical optics, we mean by a sound ray a small portion of a spherical wave with vanishing aperture which originates from a certain point. It travels in a well-defined direction and is subject to the same laws of propagation as a light ray, apart from the different propagation velocity. Thus, according to the above definition, the total energy conveyed by a ray remains constant provided the medium itself does not cause any energy losses. However, the intensity within a diverging bundle of rays falls as $1/r^2$, as in every spherical wave, where $r$ denotes the distance from its origin. Another fundamental fact is the manner in which a sound ray is reflected from a wall. In contrast, the transition to another medium, and the refraction accompanying it, does not occur in room acoustics, neither does the curvature of rays in an inhomogeneous medium. However, the finite velocity of propagation must be considered in many cases, since it is responsible for many important effects such as reverberation, echoes and so on.

Any diffraction phenomena are neglected in geometrical room acoustics, since propagation in straight lines is its main postulate. Likewise, interference is usually not considered, i.e. if several sound field components are superimposed their mutual phase relations are not taken into account; instead, simply their energy densities or their intensities are added. As explained in Section 2.5, this is permissible if the different components of the sound field are incoherent.

It is self-evident that geometrical room acoustics can reflect only a partial aspect of the acoustical phenomena occurring in a room. This aspect is, however, of great importance because of its conceptual simplicity and the ease of practical sound field computations.

## 4.1 Enclosures with plane walls, image sources

If a sound ray strikes a solid surface it is usually reflected from it. This process takes place according to the reflection law well known in optics. It states that the ray during reflection remains in the plane defined by the incident ray and the normal to the surface, and that the angle between the incident ray and reflected ray is halved by the normal to the wall. In vector notation, this law, which is illustrated in Fig. 4.1, reads:

$$\mathbf{u'} = \mathbf{u} - 2(\mathbf{un}) \cdot \mathbf{n} \tag{4.1}$$

Here $\mathbf{u}$ and $\mathbf{u'}$ are unit vectors pointing into the direction of the incident and the reflected sound ray, respectively, and $\mathbf{n}$ is the normal unit vector at the point where the arriving ray intersects the surface.

One simple consequence of this law is that any sound ray which undergoes a double reflection in an edge (corner) formed by two (three) perpendicular surfaces will travel back in the same direction, as shown in Fig. 4.2a, no matter from which direction the incident ray arrives. If the angle of the edge deviates from a right angle by $\delta$, the direction of the reflected ray will differ by $2\delta$ from that of the incident ray (Fig. 4.2b).

In this and the next two sections the law of specular reflection will be applied to enclosures, the boundaries of which are composed of plane and smooth walls. In this case one can benefit from the notion of image sources, which greatly facilitates the construction of sound paths within the enclosure. This is explained in Fig. 4.3. Suppose there is a point source A in front of

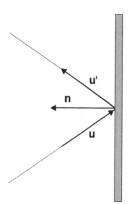

Figure 4.1  Illustration of vectors in eqn (4.1).

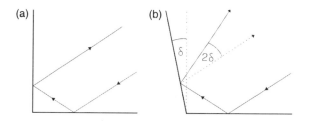

Figure 4.2  Reflection of a sound ray from a corner.

a plane wall or wall section. Then each ray reflected from this wall can be thought of as originating from a virtual sound source A′ which is located behind the wall, on the line perpendicular to the wall, and at the same distance from it as the original source A. Without the image source, the reflection path connecting the sound source A with a given point B could only be found by trial and error.

Once we have constructed the image source A′ associated with a given original source A, we can disregard the wall altogether, the effect of which is now replaced by that of the image source. Of course, we must assume that the image emits exactly the same sound signal as the original source and that its directional characteristics are symmetrical to those of A. If the extension of the reflecting wall is finite, then we must restrict the directions of emission of A′ accordingly or, put in a different way, for certain positions of the observation point B the image source may become 'invisible'. This is the case if the line connecting B with the image source does not intersect the actual wall.

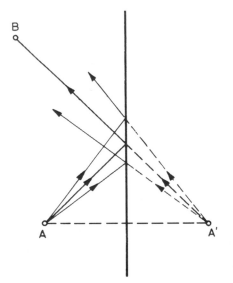

*Figure 4.3* Construction of an image source.

Usually not all the energy striking a wall is reflected from it; part of the energy is absorbed by the wall (or it is transmitted to the other side, which amounts to the same thing as far as the reflected fraction is concerned). The fraction of sound energy (or intensity) which is not reflected is characterised by the absorption coefficient $\alpha$ of the wall, which has been defined in Section 2.2 as the ratio of the non-reflected to the incident intensity. It depends generally, as we have seen, on the angle of incidence and, of course, on the frequencies of the spectral components of the incident sound. Thus the reflected ray generally has a different power spectrum and a lower intensity than the incident one. Using the picture of image sources, these circumstances can be taken into account by modifying the spectrum and the directional distribution of the sound emitted by A′. With such refinements, however, the usefulness of the concept of image sources is degraded considerably. It is more convenient to employ some mean value $\alpha$ of the absorption coefficient and accordingly to reduce the intensity of the reflected ray by a fraction $1 - \alpha$. As an alternative, the image source can be thought to emit a sound power reduced by this factor.

If a reflected sound ray strikes a second wall the continuation of the sound path can be found by repeating the mirroring process, as shown in Fig. 4.4. Accordingly, a second-order image A″ is constructed which is the mirror image of A′ with respect to that wall. Continuing in this way more and more image sources of increasing order are created.

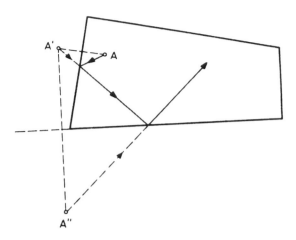

*Figure 4.4* Image sources of first and second order.

Strictly speaking, the concept of image sources is only exact if the reflecting boundary has the specific impedance +1 or −1. In all other cases the results obtained with it are not quite correct since quantities such as the reflection factor or the absorption coefficient are defined for plane waves only while the waves originating from the real sound source and its images are spherical. We expect that the errors are tolerable if the 'plane-wave condition' $kr \gg 1$ is fulfilled (see eqn (1.22)). This condition tells us that the curvature of the wavefronts of a spherical wave can be neglected if the distance of the receiving point from all sources—real as well as image sources—is large compared with the sound wavelength.

For a given enclosure and sound source position, the image sources can be constructed without referring to a particular sound path. Suppose the enclosure is made up of $N$ plane walls. Each of them is associated with one image of the original sound source. Now each of these image sources is mirrored by all walls, which leads to $N(N-1)$ new images of second order. By repeating this procedure again and again a rapidly growing number of images is generated with increasing distances from the original source. The number of images of order $i$ is $N(N-1)^{i-1}$ for $i \geq 1$; the total number of images of order up to $i_0$ is obtained by adding all these expressions:

$$N(i_0) = N\frac{(N-1)^{i_0} - 1}{N-2} \tag{4.2}$$

For enclosures with high symmetry (see Fig. 4.6, for instance) many of the higher-order images coincide. It should be noted, however, that each image source has its own directivity since it 'illuminates' only a limited solid angle, determined by the limited extension of the walls. Hence it may well happen that a particular image source is 'invisible' or rather 'inaudible' from a given receiving location. Moreover, only those image sources are valid which have been generated by mirroring at the inside of a wall (i.e. at the side facing the enclosure). These problems have been carefully discussed by Borish.[1] More will be said about this subject in Section 9.6.

These complications are not encountered with enclosures of high regularity, which in turn produce regular patterns of image sources. As a simple example, which may also illustrate the usefulness of the image model, let us consider a very flat room, the height of which is small compared with its lateral dimensions. For locations far from the side walls the effect of the latter may be totally neglected. Then we arrive at a space which is bounded by two parallel, infinite planes. We assume that both the sound source A and the observation point B are located in the middle between both planes and that the source radiates the power $P$ uniformly in all directions. The corresponding image sources (and image spaces) are depicted in Fig. 4.5. The source images form a simple pattern of equidistant points situated on a straight line, and each of them is a valid one, i.e. it is 'visible' from any observation point B. The path length from an image of $n$th order to $B$ is $(r^2 + n^2h^2)^{1/2}$, if $r$ is the horizontal distance of B from the original source A and $h$ is the 'height' of the room. Furthermore, if we assume, for the sake of simplicity, that both planes have the same absorption coefficient $\alpha$ independent of the angle of sound incidence, the energy density in B is given by the following expression:

$$w = \frac{P}{4\pi c} \sum_{-\infty}^{\infty} \frac{(1-\alpha)^{|n|}}{r^2 + n^2h^2} \tag{4.3}$$

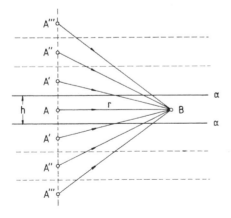

Figure 4.5 System of image sources of an infinite flat room: A is the original sound source; A', A'', etc., are image sources; B is the receiving point.

which can easily be evaluated with a programmable pocket calculator. A graphical representation of this formula will be found in Section 9.4.

Another example is the rectangular room, as depicted in Fig. 3.1. For this room shape certain image sources of the same order coincide but are complementary with respect to their directivity. The result is the regular pattern of image rooms, as shown in Fig. 4.6, each of them containing exactly one image source. So the four image rooms adjacent to the sides of the original rectangle contain one first-order image each, whereas those adjacent to its corners contain second-order images and so on. The lattice depicted in Fig. 4.6 has to be completed in the third dimension, i.e. we must imagine an infinite number of such patterns one upon the other at equal distances, one of them containing the original room.

In both examples, all image sources are visible. This is because the totality of image rooms, each of them containing one source image, fills the whole space without leaving uncovered regions and without any overlap. Enclosures of less regular shape would produce much more irregular patterns of image sources, and their image spaces would overlap each other in a complicated way. In these cases the validity or invalidity of each image source with respect to a given receiving point must be carefully examined. A more realistic example is shown in Fig. 4.7, along with a few image sources.

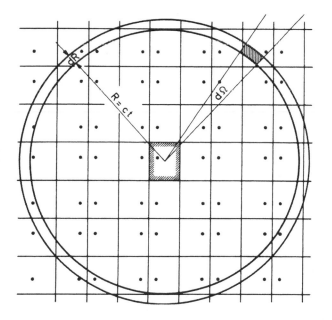

*Figure 4.6* Image sources in a rectangular room. The pattern continues in an analogous manner in the direction perpendicular to the drawing plane.

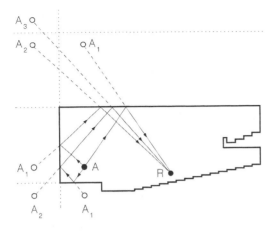

*Figure 4.7* Longitudinal section of an auditorium showing some image sources: A = sound source; $A_1$ = first-order image sources; $A_2$ = second-order image sources, etc.

When all image sources of a given enclosure have been found, the original room itself is no longer needed. The sound signal received at a given point R is then obtained as the superposition of the contributions of all significant image sources under the assumption that all sources including the original one emit the same sound signal simultaneously. Because of the different travelling distances, the waves (or rays) originating from these sources arrive at the receiving point with different delays and strengths. To obtain the signal at the receiving point, one has just to add the sound intensities of all contributions. As mentioned earlier, the strength of a particular contribution must also include the absorptivity of all walls which are crossed by the straight line connecting the image source with the receiving point.

If the absorption coefficients of all walls are frequency independent, the received signal $s'(t)$ is the superposition of infinitely many replicas of the original signal, each of them with its particular strength $A_n$ and delayed by its particular travelling time $t_n$:

$$s'(t) = \sum_n A_n s(t - t_n) \tag{4.4a}$$

Accordingly, the impulse response of the room reads:

$$g(t) = \sum_n A_n \delta(t - t_n) \tag{4.4b}$$

In reality, a Dirac impulse is deformed when it is reflected from a wall, i.e. the reflected signal is not the exact replica of the original impulse but is transformed into a somewhat different signal $r(t)$. This signal could be named the 'reflection response' of the surface, its Fourier transform is the frequency-dependent reflection factor $R$, as introduced in Section 2.1.

From the physical standpoint it would be correct to interpret $s$ and $s'$ in eqn (4.4$a$) as sound pressures. However, if the signals to be superimposed in $R$ have a wide frequency spectrum the 'signals' can be considered as incoherent and we can apply eqn (2.37), which tells us that it is sufficient to add just their energies. This is what has been done in deriving eqn (4.3). However, there are other situations where phase relations must not be neglected. We will take up this discussion once more in Section 9.6.

## 4.2 The temporal distribution of reflections

Each sound reflection received at a given point within an enclosure has three characteristic quantities: the time it arrives, the sound energy it carries and the direction from which it is received. This section deals with the first two of them. For this discussion it is sufficient to assume that the reflection factors of all walls are frequency independent.

If we mark the arrival times of the various reflections by perpendicular dashes over a horizontal time axis and choose the heights of the dashes proportional to the relative energy contained in them we obtain what is frequently called a 'reflection diagram' or 'echogram'. It contains all significant information on the temporal structure of the sound field at the considered room point. In Fig. 4.8 such a schematic reflection diagram is plotted. After

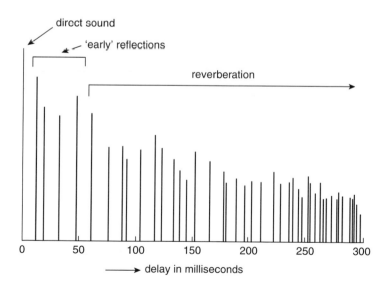

Figure 4.8 Schematic reflection diagram.

the direct sound, arriving at $t = 0$, the first strong reflections occur at first sporadically, later their temporal density increases rapidly, however; at the same time the reflections carry less and less energy. As we shall see later in more detail, the role of the first isolated reflections with respect to our subjective hearing impression is quite different from that of the very numerous weak reflections arriving at later times, which merge into what we perceive subjectively as reverberation. Thus we can consider the reverberation of a room not only as the common effect of decaying vibrational modes, as we did in Chapter 3, but also as the sum total of all reflections—except the very first ones.

A survey on the temporal structure of reflections and hence the law of reverberation in a rectangular room can easily be obtained by using the system of image rooms and image sound sources, as shown in Fig. 4.6. Suppose that at some time $t = 0$ all mirror sources generate impulses of equal energy $E_0$. In the time interval from $t$ to $t + dt$, all those reflections will arrive in the centre of the original room which originate from image sources whose distances to the centre are between $ct$ and $c(t + dt)$. These sources are located in a spherical shell with radius $ct$. The thickness of this shell (which is supposed to be very small as compared with $ct$) is $c \, dt$ and its volume is $4\pi c^3 t^2 \, dt$. In this shell volume, the volume $V$ of an image room is contained $4\pi c^3 t^2 \, dt / V$ times; this figure is also the number of mirror sources contained in the shell volume. Therefore the average temporal density of the reflections arriving at time $t$ is

$$\frac{dN_r}{dt} = 4\pi \frac{c^3 t^2}{V} \tag{4.5}$$

It is interesting to note, by the way, that the above approach is the same as applied to estimate the mean density of eigenfrequencies in a rectangular room (eqn (3.21)) with formally the same result. In fact, the pattern of mirror sources and the eigenfrequency lattice are in the same relation to each other as are the point lattice and the reciprocal lattice in crystallography: they are Fourier transforms of each other. Moreover, it can be shown that eqn (4.5) does not only apply to rectangular rooms but also to rooms with arbitrary shape as well.

Each reflection—considered physically—corresponds to a narrow bundle of rays originating from the respective image source in which the sound intensity decreases proportionally as $(ct)^{-2}$, i.e. as the square of the reciprocal distance covered by the rays. Furthermore, the rays are attenuated by absorption in the medium. According to eqn (1.16a) this effect can be taken into account by the factor $\exp(-mx/2)$, which describes the decrement of the pressure amplitude when a plane wave travels a distance $x$ in a lossy medium; hence, the intensity of a ray is reduced by another factor $\exp(-mx) = \exp(-mct)$. And finally, the intensity of a ray bundle is reduced by a factor $1 - \alpha$ whenever it crosses a wall of an image room; if this happens $\bar{n}$-times per second on the average, the intensity reduction due to wall absorption after $t$ seconds will be $(1 - \alpha)^{\bar{n}t} = \exp[\bar{n}t(1 - \alpha)]$. Therefore, a reflection received at time $t$ has the average intensity

$$\frac{E_0}{4\pi \, (ct)^2} \exp\left\{\left[-mc + \bar{n}\ln(1 - \alpha)\right]t\right\}$$

Dividing this expression by the sound velocity (see eqn (1.29)) and combining it with eqn (4.5) yields the time-dependent energy density for $t \geq 0$:

$$w(t) = w_0 \exp\left\{\left[-mc + \bar{n}\ln(1 - \alpha)\right]t\right\} \qquad \text{with } w_0 = E_0/V \qquad (4.6)$$

To complete this formula the average number of wall reflections or wall crossings per second must be calculated. For this purpose we refer to Fig. 4.9 and consider a sound ray whose angle with respect to the $x$-axis (i.e. to the horizontal axis) is $\beta_x$. It will cross a vertical mirror wall every $L_x/c \cos \beta_x$ seconds; $L_x$, $L_y$, $L_z$ are the room dimensions.

Therefore the number of such wall crossings is

$$n_x(\beta_x) = \left|\frac{c}{L_x} \cos \beta_x\right| \qquad (4.7)$$

Similar expressions hold for the crossings of walls perpendicular to the $y$-axis and the $z$-axis. Accordingly, the total number of wall crossings, i.e. of reflections which a ray with given direction undergoes per second, is

$$n\left(\beta_x, \beta_y, \beta_z\right) = n_x + n_y + n_z$$

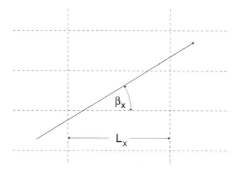

*Figure 4.9* Wall crossings of a sound ray in a rectangular room and its images.

with $\cos^2 \beta_x + \cos^2 \beta_y + \cos^2 \beta_z = 1$. This means that each sound ray decays at its own rate, leading to bent decay curves according to the discussion in Section 3.6.

One might be tempted to average $n(\beta_x, \beta_y, \beta_z)$ over all directions in order to arrive at a figure which is representative for the whole energy content of the room. This is only permissible, however, if the sound rays and the energy they transport change their direction once in a while.[2] Such changes will never happen in a rectangular room with smooth walls. But enclosures of more irregular shape or with diffusely reflecting walls or scattering bodies (see Section 2.6) do have the tendency to mix the directions of sound propagation. In the ideal case, such randomising effects could eventually result in what is called a 'diffuse sound field' in which the propagation of sound is completely isotropic.

Under this condition eqn (4.7) may be averaged over all directions:

$$\frac{c}{L_x} \langle |\cos \beta_x| \rangle = \frac{c}{L_x} \frac{1}{4\pi} 2\pi \cdot 2 \int\limits_0^{\pi/2} \cos \beta_x \sin \beta_x \, d\beta_x = \frac{c}{2L_x}$$

The same average is found for $n_y$ and $n_z$. Hence the total average of reflections per second is

$$\bar{n} = \frac{c}{2} \left( \frac{1}{L_x} + \frac{1}{L_y} + \frac{1}{L_z} \right) = \frac{cS}{4V} \qquad (4.8)$$

Here $S$ is the total boundary area of the room. By inserting this result into eqn (4.6) we arrive at a fairly general law of sound decay:

$$w(t) = w_0 \exp\left[ -ct \frac{4mV - S \ln(1-\alpha)}{4V} \right] \qquad \text{for } t \geq 0 \qquad (4.9)$$

The reverberation time, i.e. the time in which the total energy falls to one millionth of its initial value, is thus

$$T = \frac{1}{c} \cdot \frac{24 V \ln 10}{4mV - S \ln(1-\alpha)} \qquad (4.10)$$

or, if we insert 343 m/s for the sound velocity $c$ and express the volume $V$ in m$^3$, the area $S$ in m$^2$,

$$T = 0.161 \frac{V}{4mV - S \ln(1-\alpha)} \qquad (4.11)$$

In the preceding, we have derived by rather simple geometrical consider-ations the most important formula of room acoustics, which relates the reverberation time to some of its geometrical data and to the absorption coefficient of its walls. We have assumed tacitly that the latter is the same for all wall portions and that it does not depend on the angle at which a wall is struck by the sound rays. In the next chapter we shall look more closely into the laws of reverberation using a different approach, but the result will be essentially the same.

The exponential law of eqn (4.9) provides an approximate description of the decaying sound energy in a rectangular room. One must not forget, how-ever, that this formula is the result of two averaging processes. In reality, the reflections arrive at quite irregular times $t_n$, at the observation point—for instance, the ear of a listener—and also their energies vary in quite an irreg-ular way, even if the sound field is diffuse. Accordingly, the details of the decay process show considerable deviations from eqn (4.9) which may well be relevant for the acoustics of the room. So it may happen, for instance, that one particular reflection with long delay time is much stronger than the majority of its neighbours and stands out of the general reverberation. This can occur when several sound rays are reflected from a remote con-cave wall portion which concentrates the sound energy in a point near the receiving point. Such an isolated component is perceived as a distinct echo and is particularly disturbing if the portion of wall which is responsible is irradiated by loudspeakers. Another unfavourable condition is that of many reflections clustered together in a narrow time interval. Since our hearing has a limited time resolution and therefore performs some sort of short-time integration, this lack of uniformity may be audible and may give rise to unde-sirable effects which are similar to that of a single reflection of exceptional strength.

Particularly disturbing are reflections which form a periodic or a nearly periodic succession. This is true even if this periodicity is hidden in a great number of reflections distributed irregularly over the time axis, since our hearing is very sensitive to periodic repetitions of certain sound signals. For short periods, i.e. for repetition times of a few milliseconds, such periodic components are perceived as a 'colouration' of the reverberation; then the decay has a characteristic pitch and timbre. Hence speech or music in such a room will have its spectra changed. If the periods are longer, if they amount to 30, 50 or even 100 ms, the regular temporal structure itself becomes audible (see Section 7.3). This case, which is frequently referred to as 'flutter echo', is observed if sound is reflected repeatedly to and fro between parallel walls. Flutter echoes can be observed quite distinctly in corridors or other longish rooms where the end walls are rigid but the ceiling, floor and side walls are absorbent. They can also occur in rooms the shapes of which are less extreme, but then their audibility is mostly restricted to particular locations of the source and the observer.

## 4.3 The directional distribution of reflections, diffuse sound field

We shall now take into consideration the third property which characterises a reflection, namely the direction from which it reaches an observer. As before, we shall not attribute to each single reflection its proper direction, but we shall prefer a statistical description. This method commends itself not only because of the great number of reflections to be considered but also because we are unable to distinguish subjectively the directions of individual reflections. Nevertheless, whether the reflected components arrive uniformly from all directions or whether they all come from one single direction may have considerable bearing on what and how we hear in a room, especially in a concert hall. The directional distribution of sound is also important for certain measuring procedures.

First we assume that the sound source produces a stationary sound signal, and that the same holds for all image sources. Each of them, provided it is visible from the receiving point R, contributes to the received sound energy. Let us consider a narrow cone with the small solid angle $d\Omega$ around a direction characterized by the polar angle $\vartheta$ and the azimuth angle $\varphi$ (see Fig. 2.10 where the angles are named $\theta$ and $\phi$). Its tip is located at the receiving point R. The contribution of the sources within the 'directional cone' is $I(\varphi, \vartheta)$. This quantity, conceived as a function of the angles $\varphi$ and $\vartheta$, is called the directional distribution of sound. Of course it depends on the receiver's location. For a rectangular room the situation is illustrated in Fig. 4.6. Obviously, $I(\varphi, \vartheta)$ will only remain finite if there are some losses either caused by air attenuation or by non-vanishing wall absorption. Experimentally, $I(\varphi, \vartheta)$ can be determined by use of a directional microphone with high resolution, at least approximately (see Section 8.5).

If the sound source and its images emit a transient signal, the directional distribution will become time dependent. Again, the limiting case is that of an impulsive excitation signal with very short duration, idealised as a Dirac impulse. Then the intensity at time $t$, denoted by $I_t(\varphi, \vartheta)$, is due to those image sources which are lying within the directional cone and, at the same time, within a spherical shell with the thickness $c\,dt$. For the rectangular room in Fig. 4.6 this region corresponds to the cross-hatched area. Evidently, the relation between the time-dependent and the steady-state directional distribution is

$$I(\varphi, \vartheta) = \int_0^\infty I_t(\varphi, \vartheta)\,dt \qquad (4.12)$$

If the directional distribution $I(\varphi, \vartheta)$ does not depend in any way on the angles $\varphi$ and $\vartheta$, the stationary sound field is called 'diffuse' or isotropic. We have already encountered this important condition in Section 2.5.

If, in addition to this condition, $I_t(\varphi, \vartheta)$ is constant for all $t$, even the decaying sound field is diffuse.

In a certain sense the diffuse sound field is the counterpart of a plane wave. Just as certain properties can be attributed to plane waves, so relationships describing the properties of diffuse sound can be established. Some of these have already been encountered in Chapter 2. They are of particular interest to the whole of room acoustics, since, although the sound field in a concert hall or theatre is not completely diffuse, its directional structure resembles much more that of a diffuse field than that of a plane wave. Or, put another way, the sound field in an actual room, which always contains some irregularities in shape, can be approximated fairly well by a sound field with uniform directional distribution. By contrast, a single plane wave is hardly ever encountered in a real situation.

Finally we discuss a somewhat extreme example of a room with non-diffuse conditions: namely a strictly rectangular room, whose walls are perfectly rigid except for one which absorbs the incident sound energy completely. Its behaviour with respect to the formation and distribution of reflections is elucidated by the image room system depicted in Fig. 4.10, consisting of only two 'stores' since the absorbing wall generates no images of the room and the sound source. In the lateral directions, however (and perpendicular to the plane of the figure), the system is extended infinitely.

We denote the distance between the absorbing wall and its opposite wall by $L$ and the elevation angle by $\varepsilon$, measured from the point of observation (which may be located in the centre of the reflecting ceiling for

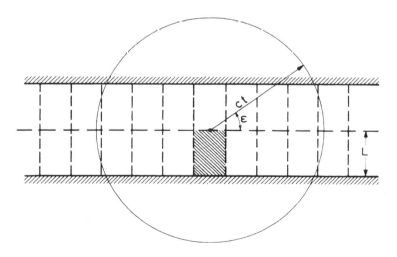

*Figure 4.10* Image rooms of a rectangular room, one side of which is perfectly sound absorbing. The original room is cross-hatched.

the sake of simplicity). The time-dependent directional distribution is then given by

$$I_t(\varphi, \varepsilon) = \begin{cases} \text{const} & \text{for } |\varepsilon| < \varepsilon_0 \\ 0 & \text{for } |\varepsilon| > \varepsilon_0 \end{cases} \tag{4.13}$$

where

$$\varepsilon_0(t) = \arcsin\left(\frac{L}{ct}\right) \qquad \text{for } t \geq \frac{L}{c} \tag{4.14}$$

The range of elevation angle subtended by the image sources contracts more and more with increasing time. With the presently used meaning of the angle $\varepsilon$, the element of solid angle becomes $\cos \varepsilon \, d\varepsilon \, d\varphi$; hence the integration over all directions yields the following expression for the decaying energy density:

$$w(t) = \frac{2}{c}\int_0^{2\pi} d\varphi \int_0^{\varepsilon_0} I_t(\varphi, \varepsilon)\cos \varepsilon \, d\varepsilon = \text{const} \cdot \frac{4\pi L}{ct} \qquad \text{for } t \geq \frac{L}{c} \tag{4.15}$$

As a consequence of the non-uniform distribution of the wall absorption the decay of the reverberant energy does not follow an exponential law but is inversely proportional to the time. The steady-state directional distribution is given by

$$I(\varepsilon) = \frac{\text{const}}{|\sin \varepsilon|} \tag{4.16}$$

## 4.4 Enclosures with curved walls

In this section we consider enclosures, the boundaries of which contain curved walls or wall sections. Practical examples are domed ceilings, as are encountered in many theatres or other performance halls, or the curved rear walls of many lecture theatres. Concavely curved surfaces in rooms are generally considered as critical or even dangerous in that they concentrate the sound energy in certain areas and thus impede its uniform distribution throughout the room.

Formally, the law of specular reflection as expressed by eqn (4.1) is valid for curved surfaces as well as for plane ones, since each curved surface can be approximated by many small plane sections. Keeping in mind the wave nature of sound, however, one should not apply this law to a surface the radius of curvature of which is not very large compared to the acoustical wavelength. Whenever the radius of curvature is comparable or even smaller than the wavelength the surface will scatter an impinging sound wave rather than reflect it specularly, as described in Section 2.7.

Very often, curved walls in rooms or halls are spherical or cylindrical segments, or they can be approximated by such surfaces. Then we can apply the laws of concave or convex mirrors, known from optics. In the following it should be noted that the direction of the ray paths can be inverted.

In Fig. 4.11a the section of a concave, spherical or cylindrical mirror is depicted. Its radius of curvature is $R$. A bundle of rays originating from a point S is reflected at the mirror and is focused into the point P from which it diverges. Focusing of this kind occurs when the distance of the source from the mirror is larger than $R/2$; if the incident bundle is parallel the focus is at distance $R/2$. The source distance $a$, the distance of the focus $b$ and the radius of the mirror are related by

$$\frac{1}{a} + \frac{1}{b} = \frac{2}{R} \tag{4.17}$$

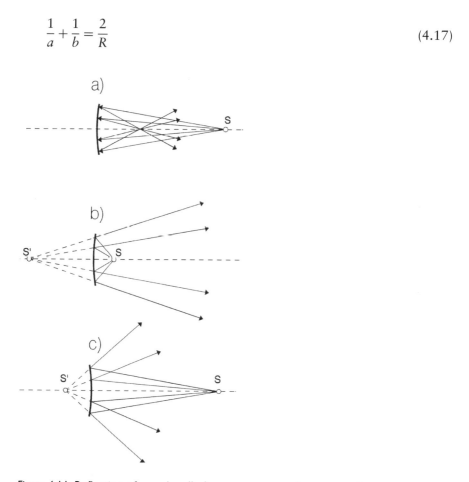

Figure 4.11 Reflection of a ray bundle from a concave and a convex mirror.

If the source is closer to the mirror than $R/2$ (see Fig. 4.11b), the reflected ray bundle is divergent (although less divergent than the incident one) and seems to originate from a point beyond the mirror. Equation (4.17) is still valid and leads to a negative value of $b$.

Finally, we consider the reflection at a convex mirror as depicted in Fig. 4.11c. In this case the divergence of any incident ray bundle is increased by the mirror. Again, eqn (4.17) can be applied to find the position of the 'virtual' focus after replacing $R$ with $-R$. As before, the distance $b$ is negative.

The effect of curved surfaces can be studied more quantitatively by comparing the intensity of the reflected ray bundle with that of a bundle reflected at a plane mirror. The latter is given by

$$I_0 = \frac{A}{|a+x|^n} \qquad (4.18a)$$

while the intensity of the bundle reflected from the mirror is

$$I_r = \frac{B}{|b-x|^n} \qquad (4.18b)$$

In both formulae $A$ and $B$ are constants; $x$ is the distance from the centre of the mirror, and the exponent $n$ is 1 for a cylindrical mirror and 2 for a spherical one. (It should be kept in mind that the distance $b$ has a negative sign for concave mirrors with $a < R/2$ and for convex mirrors.) At $x = 0$ both intensities must be equal, which yields $A/B = |a/b|^n$. Thus, the ratio of both intensities is

$$\frac{I_r}{I_0} = \left| \frac{1+x/a}{1-x/b} \right|^n \qquad (4.19)$$

Figure 4.12 plots the level $L_r = 10 \cdot \log_{10}(I_r/I_0)$ derived from this ratio for the cases depicted in Fig. 4.11 with $n = 2$ (spherical mirror). The focus occurring for $a > R/2$ (curve a) is clearly seen as a pole. Apart from this, there is a range of increased intensity in which $L_r > 0$. From eqns (4.17) and (4.19) it can be concluded that this range is given by

$$x < 2\left(\frac{1}{b} - \frac{1}{a}\right)^{-1} = \left(\frac{1}{R} - \frac{1}{a}\right)^{-1} \qquad (4.20)$$

Outside that range, the level $L_r$ is negative, indicating that the reflected bundle is more divergent than it would be when reflected from a plane mirror. If $a < R/2$ (curve b), the intensity is increased at all distances $x$. Finally, the convex mirror (curve c) reduces the intensity of the bundle everywhere.

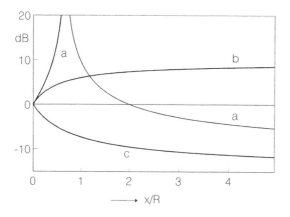

*Figure 4.12* Level difference in ray bundles reflected from a spherical and a plane reflector at distance *x*: (a) concave mirror, $a = 2R$; (b) concave mirror, $a = R/3$; (c) convex mirror, $a = 2R$.

According to eqn (4.19) the limit of $L_r$ for very large distances is

$$L_r \to 10n \cdot \log_{10} \left| \frac{b}{a} \right| \qquad \text{for } x \to \infty \tag{4.21}$$

From these findings a few practical conclusions may be drawn. A concave mirror may concentrate the impinging sound energy in certain regions, but it may also be an effective scatterer which distributes the energy over a wide angular range. Whether the one or the other effect dominates depends on the positions of the source and the observer. Generally, the following rule[3] can be derived from eqn (4.20). Suppose the mirror in Fig. 4.11a is completed to a full circle with radius *R*. Then, if both the sound source and the receiver are outside this volume, the undesirable effects mentioned at the beginning of this section are not to be expected.

The laws outlined above are valid only for narrow ray bundles, i.e. as long as the inclination of the rays against the axis is sufficiently small, or, what amounts to the same thing, as long as the aperture of the mirror is not too large. Whenever this condition is not met, the construction of reflected rays becomes more difficult. Either the surface has to be approximated piecewise by circular or spherical sections, or the reflected bundle must be constructed ray by ray. As an example, the reflection of a parallel bundle of rays from a concave mirror of large aperture is shown in Fig. 4.13. Obviously, the reflected rays are not collected just in one point; instead, they form an envelope which is known as a caustic. Next to the central ray, the caustic reaches the focal point in the distance $b = R/2$ in accordance with eqn (4.17) with $a \to \infty$. Another interesting shape is the ellipse or the ellipsoid, which has

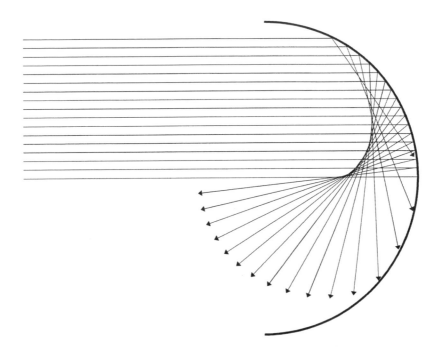

*Figure 4.13* Reflection of a parallel ray bundle from a spherical mirror of large aperture.

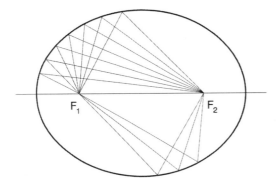

*Figure 4.14* Collection of sound rays in an elliptical enclosure.

two foci $F_1$ and $F_2$, as shown in Fig. 4.14. If a sound source S is placed in one of them, all the rays emitted by it are collected in the other one. For this reason, enclosures with elliptical floor plan are plagued by quite unequal sound distribution even if neither the sound source nor the listener are in a geometrical focus. The same holds, of course, for halls with a circular floor plan since the circle is a limiting case of the ellipse.

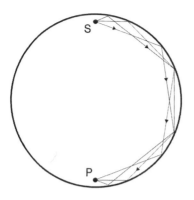

*Figure 4.15* Whispering gallery.

A striking experience can be made in such halls if a speaker is close to its wall. A listener who is also next to the wall, although distant from the sound source (see Fig. 4.15), can hear the speaker quite clearly even if the latter speaks in a very low voice or whispers. An enclosure of this kind is said to form a 'whispering gallery'. The explanation for this phenomenon is simple. If the speaker's head is more or less parallel to the wall, most of the sound rays hit the wall at grazing incidence and are repeatedly reflected from it. If the wall is smooth and uninterrupted by pillars, niches, etc., the rays remain confined within a narrow band: in other words, the wall conducts the sound along its perimeter. Probably the best known example is in St. Paul's Cathedral in London, which has a gallery at the circular base of the dome. Generally, a whispering gallery is an interesting curiosity, but if the hall is used for performances, the acoustical effects caused by it are rather disturbing.

## 4.5 Enclosures with diffusely reflecting walls, radiosity integral

In Section 2.7, surfaces with diffuse reflection brought about by structural surface irregularities, by abrupt changes of the wall impedance, etc., have been discussed. It is clear that this property of the boundary has a considerable influence on the acoustical behaviour of the room. Generally, diffuse wall reflections result in a more uniform distribution of the sound energy throughout the room.

The reflection at a surface is said to take place in a totally diffuse manner if the directional distribution of the reflected or the scattered energy does not depend in any way on the direction of the incident sound. In optics, this case can be realised quite well. In contrast, in acoustics and particularly in room acoustics, only partially diffuse reflections can be achieved. But, nevertheless, the assumption of totally diffuse reflections comes often closer to the properties of real walls than that of specular reflection, particularly if we are concerned not only with one but instead with many successive reflections of a ray from different walls or portions of walls. This is the case in reverberant enclosures.

Totally diffuse reflections from a wall take place according to Lambert's cosine law. Suppose a bundle of parallel or nearly parallel rays with the intensity $I_0$ hits an area element $dS$ under an angle $\vartheta_0$ (see Fig. 4.16). Then the intensity of the sound which is scattered in a direction characterised by an angle $\vartheta$, measured at some distance $r$ from $dS$, is given by

$$I(r, \vartheta) = I_0\, dS\, \frac{\cos \vartheta \cos \vartheta_0}{\pi r^2} = B_0\, dS\, \frac{\cos \vartheta}{\pi r^2} \tag{4.22}$$

$B_0$ is the energy falling per second on 1 m$^2$ of the wall. We call this quantity the 'irradiation strength'. Equation (4.22) neglects any absorption on $dS$, i.e. the incident energy is completely re-emitted. If this is not the case, the right-hand side has to be multiplied by an appropriate factor $\rho = 1 - \alpha$. As before, $\alpha$ is the absorption coefficient, whereas the quantity $\rho$ can be called the 'reflection coefficient' of the boundary. Figure 4.16 shows the wall element $dS$ and a circle representing the directional distribution of the scattered sound; the length of the arrow pointing to its periphery is proportional to $\cos \vartheta$.

According to eqn (4.22), each surface element can be considered as a secondary sound source, which is expressed by the fact that the distance $r$, which determines the intensity reduction due to propagation, must be measured from the reflecting area element $dS$. This is not so with specular reflection: here the distance in the $1/r^2$ law is the total length of the path

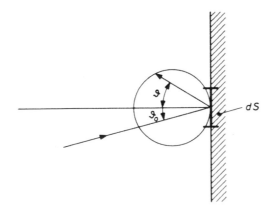

Figure 4.16 Ideally diffuse sound reflection from an acoustically rough surface.

between the sound source and the point of observation with no regard as to whether this path is bent or straight.

In the following we assume that the whole boundary of the considered enclosure reflects the impinging sound in a completely diffuse manner. This assumption enables us to describe the sound field within the room in a closed form, namely by an integral equation. To derive it, we start by considering two wall elements, dS and dS', of a room of arbitrary shape (see Fig. 4.17). Their locations are characterised by the vectors **r** and **r'**, respectively, each of them standing for three suitable coordinates. The straight line connecting them has the length $R$, and the angles between this line and the wall normals in dS and dS' are denoted by $\vartheta$ and $\vartheta'$.

Suppose the element dS' is irradiated with the energy $B(\mathbf{r'}) dS'$ per second, where $B$ again is the irradiation strength. The fraction it re-radiates into the space is given by the 'reflection coefficient' $\rho$.

To avoid unnecessary complication we assume that this quantity is independent of the angles $\vartheta$ and $\vartheta'$. According to eqn (4.22), the intensity of the sound received at dS from dS' is

$$dI = B(\mathbf{r'})\rho(\mathbf{r'})\frac{\cos\vartheta'}{\pi R^2}\,dS' \tag{4.23}$$

The total energy per second and unit area arriving at the element dS from the whole boundary is obtained by multiplying this equation with $\cos\vartheta$ and integrating it over all wall elements dS'. If the direct contribution

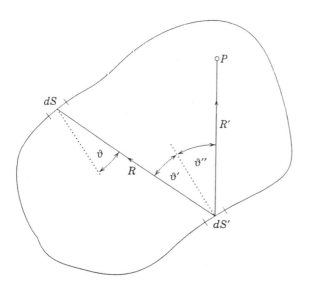

*Figure 4.17* Illustration of eqn (4.24).

$B_d$ produced by some sound source is added, the following relation is obtained:[4,5]

$$B(\mathbf{r}, t) = \iint_S \rho(\mathbf{r}')B(\mathbf{r}', t - R/c)\frac{\cos\vartheta\cos\vartheta'}{\pi R^2}\,dS' + B_d(\mathbf{r}, t) \qquad (4.24)$$

The argument $t - R/c$ takes regard of the delay due to the finite sound velocity.

Equation (4.24) is an inhomogeneous integral equation for the irradiation strength $B$ of the wall. It is fairly general in that it contains both the steady-state case (for $B_d$ and $B$ independent of time $t$) and that of a decaying sound field (for $B_d = 0$). Once it is solved, the energy density at a point P inside the room can be obtained from

$$w(\mathbf{r}, t) = \iint_S \rho(\mathbf{r}')B(\mathbf{r}', t - R'/c)\frac{\cos\vartheta''}{\pi c R'^2}\,dS' + w_d(\mathbf{r}, t) \qquad (4.25)$$

$R'$ is the distance of the inner point from the element $dS'$ while $\vartheta''$ denotes the angle between the wall normal in $dS'$ and the line connecting $dS'$ with the receiving point.

Generally, the integral equation (4.24) must be numerically solved. An example of its application to sound decay will be presented in Section 5.6. Closed solutions are available for a few simple room shapes only. One of them is the spherical enclosure, for which

$$\frac{\cos\vartheta\cos\vartheta'}{\pi R^2} = \frac{1}{S}$$

with $S$ denoting the surface of the sphere. With this relation the integral equation (4.24), with time-independent $B$, reads simply

$$B = \langle\rho B\rangle + B_d$$

where the brackets indicate averaging over the whole surface. By repeatedly replacing the $B$ in the brackets with $\langle\rho B\rangle + B_d$ one obtains the solution

$$B = \frac{\langle\rho B_d\rangle}{1 - \langle\rho\rangle} + B_d \qquad (4.26)$$

Of more practical interest is the flat room, as already described in Section 4.1. It consists of two parallel walls (ceiling and floor); in lateral directions it is unbounded. Many shallow factory halls or open plan bureaus may be treated as such a flat room as long as neither the sound source nor the observation point are close to one of its side walls. As in Section 4.1 (see Fig. 4.5) we assume that both walls have the same, constant absorption coefficient $\alpha$ or 'reflection coefficient' $\rho = 1 - \alpha$ and that the sound source is in the middle between both planes. But, in contrast to the earlier treatment, the walls are not assumed to be smooth but to produce diffuse reflections after eqn (4.22). Then the steady-state solution of eqn (4.26) for this situation reads[6]

$$w(r) = \frac{P}{4\pi c}\left(\frac{1}{r^2} + \frac{4\rho}{h^2}\int\limits_{0}^{\infty}\frac{\exp(-z)J_0\left(rz/h\right)}{1 - \rho z K_1(z)}z\,dz\right) \qquad (4.27)$$

In this expression, $J_0$ is the Bessel function of order zero, and $K_1$ is a modified Bessel function of first order (see Ref. 7). Here $r$ is the horizontal distance from a point source with the power output $P$.

This equation is certainly too complicated for practical applications. However, it can be approximated by a simpler formula, which will be presented in Section 9.4 along with a diagram explaining its content. Another problem having a closed solution is steady-state sound propagation in a cylindrical tube of infinite length, the inside of which is diffusely reflecting. Of course, to apply the aforementioned methods the diameter of the tube must be assumed to be very large compared with the acoustical wavelength.

The condition of a diffusely reflecting boundary of an enclosure is not as unrealistic as it may seem at first glance. As has been shown by Hodgson,[8] in virtually all halls a certain part of the sound energy is scattered in non-specular directions, even if the walls and the ceiling seem to be smooth.

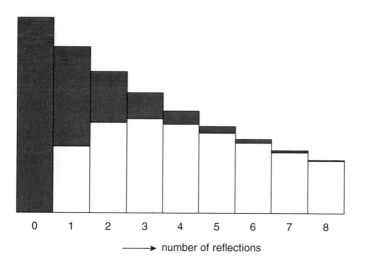

0    1    2    3    4    5    6    7    8

→ number of reflections

*Figure 4.18* Conversion of specularly reflected sound into diffuse sound by subsequent reflections ($\alpha = 0.15$, $s = 0.4$). The total height of bars corresponds to the totally reflected sound energy; the height of the white bars shows the diffusely reflected energy.

Another important fact is that the conversion of 'specular sound energy' into 'non-specular energy' by partially diffuse reflection is an irreversible process. Therefore repeated sound reflections lead to a relative increase of the diffuse component. To demonstrate this we split the sound energy reflected from a surface into two parts: the portion $s \cdot (1 - \alpha)$, which is scattered into non-specular directions; the portion $(1 - s) \cdot (1 - \alpha)$, which is reflected specularly. The parameter $s$ is called the 'scattering coefficient'. Then, after $n$ reflections, the specular component is reduced by a factor $[(1 - s) \cdot (1 - \alpha)]^n$, while the fraction of non-specular energy is $[1 - (1 - s)^n] \cdot (1 - \alpha)^n$. Both components are shown in Fig. 4.18. It is seen that the relative amount of diffuse energy increases monotonically with the order of the reflections; after a few reflections virtually all reflected energy has been converted into diffuse energy.

## References

1 Borish S. Extension of the image model to arbitrary polyhedra. J Acoust Soc Am 1984; 75:1827.
2 Cremer H, Cremer L. Über die theoretische Ableitung der Nachhallgesetze. Akust Zeitschr 1937; 2:225.
3 Cremer L, Müller HA. Principles and Applications of Room Acoustics, Vol. 1. London: Applied Science, 1982.
4 Kuttruff H. Simulierte Nachhallkurven in Rechteckräumen mit diffusem Schallfeld. Acustica 1971; 25:333; Kuttruff H. Nachhall und effektive Absorption in Räumen mit diffuser Wandreflexion. Acustica 1976; 35:141.
5 Joyce WB. Exact effect of surface roughness on the reverberation time of a uniformly absorbing spherical enclosure. J Acoust Soc Am 1978; 64:1429.
6 (1) Kuttruff H. Stationäre Schallausbreitung in Flachräumen. Acustica 1985; 57:62. (2) Stationäre Schallausbreitung in Langräumen. Acustica 1989; 69:53.
7 Abramowitz M, Stegun IA. Handbook of Mathematical Functions. New York: Dover, 1965.
8 Hodgson MJ. Evidence of diffuse surface reflections in rooms. J Acoust Soc Am 1991; 89:765.

# Chapter 5

# Reverberation and steady-state energy density

Probably the most characteristic acoustical phenomenon in a closed room is reverberation: i.e. the fact that sound produced in the room will not disappear immediately after the sound source is shut off but remains audible for a certain period of time afterwards, although with steadily decreasing loudness. For this reason, reverberation—or sound decay as it is also called—as yet yields the least controversial criterion for the judgement of the acoustical qualities of a room. It is this fact which justifies devoting the major part of a chapter to reverberation and to the laws which govern it.

The physical process of sound decay in a room depends critically on the structure of the sound field. Simple laws describing this process can be formulated only when all directions of sound propagation are equally likely and hence contribute equal sound energies to the sound field: in other words, when the sound field is diffuse (see Section 4.3). Likewise, simple relationships for the steady-state energy density in a room, as will be derived in Section 5.5, are also based on the assumption of a diffuse field. This is the reason why sound field diffusion plays a key role in this chapter although this condition is never perfectly fulfilled in practical situations. Even in certain measuring rooms such as reverberation chambers, where designers take pain to make the sound field as diffuse as possible (see Section 8.7), there is always some lack of directional diffusion. Nevertheless, the assumption of perfect sound field diffusion is a useful and important approximation of the actual sound field structure. In most instances in this chapter we shall therefore assume complete uniformity of sound field with respect to directional distribution.

In Chapter 3, we regarded reverberation as the common decaying of free vibrational modes. In Chapter 4, however, reverberation was understood to be the sum total of all sound reflections arriving at a certain point in the room after the room was excited by an impulsive sound signal. This chapter deals with reverberation from a more statistical point of view in that we consider the energy balance between the energy supplied by the sound source and that absorbed by the boundary. Furthermore, some extensions and generalisations are described, including sound decay in enclosures with imperfect sound field diffusion and in systems consisting of several coupled rooms. As in the preceding chapter, we shall consider the case of relatively high frequencies, i.e. we shall neglect interference and diffraction effects which are typical wave phenomena and which only appear in the immediate vicinity of reflecting walls or when obstacle dimensions are comparable with the wavelength.

We therefore suppose that the applied sound signals are of such a kind that the direct sound and all reflections from the walls are mutually incoherent, i.e. that they cannot interfere with each other (see Section 2.5). Consequently, their energies or intensities can simply be added together regardless of mutual phase relations. Under these assumptions sound behaves in much the same way as white light. We shall, however, not so much consider sound rays but instead we shall stress the notion of 'sound particles': i.e. of small energy packets which travel with a constant velocity $c$ along straight lines— except for wall reflections—and are supposed to be present in very large numbers. They cannot interact, in particular they will never collide with each other. If they strike a wall with absorption coefficient $\alpha$, only the fraction $1 - \alpha$ of their energy is reflected from the wall. One way to account for absorption is to assign reduced energy to the reflected particle. As an alternative, the absorption coefficient can be interpreted as an 'absorption probability'.

Of course, the sound particles considered in room acoustics are purely hypothetical and have nothing to do with the sound quanta or phonons known from solid state physics. To bestow some physical reality upon them we can consider the sound particles to be short sound pulses with a broad spectrum propagating along sound ray paths. Their exact shape is not important, but they must all have the same power spectrum.

## 5.1 Basic properties and realisation of diffuse sound fields, energy balance

As mentioned before, the uniform distribution of sound energy in a room is the crucial condition for the validity of some common and simple formulae which govern the sound decay or the steady-state energy in rooms. Therefore it is appropriate to deal first with some properties of diffuse sound fields and, furthermore, to discuss the circumstances under which we can expect them in enclosures.

Suppose we select from all sound rays crossing an arbitrary point P in a room a bundle within a very small solid angle $d\Omega$. Since the rays of the bundle are nearly parallel, an intensity $I(\varphi, \vartheta)\,d\Omega$ can be attributed to it with $\varphi$ and $\vartheta$ characterising their direction (see Fig. 5.1). The function $I(\varphi, \vartheta)$ is what has been called the directional distribution in Section 4.3.

*Figure 5.1* Bundle of nearly parallel sound rays.

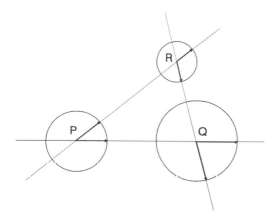

*Figure 5.2* Constancy of energy density in a diffuse sound field.

Furthermore, we can apply eqn (1.29) to these rays, according to which the energy density

$$dw = \frac{I(\varphi, \vartheta)}{c} \, d\Omega \qquad (5.1)$$

is associated with them.

Now the condition of a diffuse sound field means that the quantity $I$ is constant. Hence integrating over all directions is achieved just by multiplication by $4\pi$, and the total energy density is

$$w = \frac{4\pi I}{c} \qquad (5.2)$$

To prove the spatial constancy of the energy density, Fig. 5.2 shows three circles which represent the directional distributions of sound at the points P, Q and R. Each pair of points has exactly one sound ray in common. Since the energy propagated along a sound ray does not change with distance (see Chapter 4) it follows that they contribute the same amount of energy at both points. Therefore the circles must have equal diameters. Of course, this argument applies to all points of the space. Thus we can conclude that in a diffuse sound field the energy density is everywhere the same, at least under stationary conditions.

Another important property of a diffuse sound field has already been derived in Section 2.5. According to eqn (2.39), the energy incident on a wall element dS per second is $\pi I\,dS$ if $I$ does not depend on the angle of incidence. Hence the 'irradiation strength' $B$, already introduced in Section 4.5 as the energy incident per unit time and area, is

$$B = \pi I \tag{5.3}$$

From the constancy of the energy density $w$ it follows immediately that the irradiation strength $B$ of a wall element is independent of its location. Combining eqns (5.2) and (5.3) leads to the important relation

$$B = \frac{c}{4}w \tag{5.4}$$

which is to be compared with the corresponding eqn (1.29) for a plane wave.

We are now ready to set up an energy balance from which a simple law for the sound decay in a room can be derived. Suppose a sound source feeds the acoustical power $P(t)$ into a room. It is balanced by an increase of the energy content $Vw$ of the room and by the losses due to the absorptivity of its boundary which has the absorption coefficient $\alpha$:

$$P(t) = V\frac{dw}{dt} + B\alpha S$$

or, by using eqn (5.4) and replacing $\alpha S$ with $A$, the so-called 'equivalent absorption area' of the room:

$$P(t) = V\frac{dw}{dt} + \frac{cA}{4}w \tag{5.5}$$

When $P$ is constant the differential quotient is zero and we obtain the steady-state energy density:

$$w = \frac{4P}{cA} \tag{5.6}$$

If, on the other hand, the sound source is switched off at $t = 0$, if $P(t) = 0$ for $t = 0$, the differential equation (5.5) becomes homogeneous and has the solution

$$w(t) = w_0 \exp(-2\delta t) \qquad \text{for } t \geq 0 \tag{5.7}$$

with the damping constant or decay constant

$$\delta = \frac{cA}{8V} \tag{5.8}$$

The damping constant is related to the reverberation time $T$ by eqn (3.50). After inserting the numerical value of the sound speed in air, one obtains:

$$T = 0.161\frac{V}{A} \tag{5.9}$$

(all lengths expressed in metres). This is probably the best-known formula in room acoustics. It is due to W.C. Sabine,[1] who derived it first from the results of numerous ingenious experiments, and later on from considerations similar to the present ones. Nowadays, it is still the standard formula for predicting the reverberation time of a room, although it is obvious that it fails for high absorption. In fact, for $\alpha = 1$ it predicts a finite reverberation time although an enclosure without of any wall reflections cannot reverberate. The reason for the limited validity of eqn (5.8) is that the room is not—as assumed—in steady-state conditions during sound decay, and is less so the faster the sound energy decays. In the following sections more exact decay formulae will be derived which can also be applied to relatively 'dead' enclosures. Furthermore, eqn (5.8) and its more precise versions will be extended to the case of non-uniform absorptivity of its boundary.

In the rest of this section the circumstances will be discussed on which sound field diffusion depends. Strictly speaking, a real sound field cannot be completely diffuse, otherwise there would be no net energy flow within the enclosure. In a real room, however, the inevitable wall losses attract a continuous energy flow originating from the sound source. So, what we are discussing here is how the ideal condition can be approximated.

It is obvious that a diffuse sound field will not be established in enclosures whose walls have the tendency to concentrate the reflected sound energy in certain regions. In contrast, highly irregular room shapes help to establish a diffuse sound field by continuously redistributing the energy in all possible directions. Particularly efficient in this respect are rooms with acoustically rough walls, the irregularities of which scatter the incident sound energy in a wide range of directions, as has been already described in Section 2.7. Such walls are referred to as 'diffusely reflecting', either partially or completely. The latter case is characterised by Lambert's law, as expressed in eqn (4.22). Although this ideal behaviour is frequently assumed as a model of diffuse reflection, it will hardly ever be encountered in reality. Any wall or ceiling will, although it may be structured by numerous columns, niches, cofferings and other 'irregular' decorations, diffuse only a certain fraction of the incident sound whereas the remaining part of it is reflected into specular directions. The reader will be reminded of Fig. 2.16, which presented the scattering characteristics of an irregularly shaped ceiling.

But even if the boundary of an enclosure produces only partially diffuse reflections its contribution to sound field diffusion is considerable since each reflection converts some 'specular sound energy' into non-specular energy, whereas the reverse process, the conversion of diffuse energy into specular energy, never occurs. This has already been discussed at the end of Chapter 4 (see Fig. 4.18).

Another factor which influences sound field diffusion is the absorptivity of the boundary. To illustrate this, it is assumed that the boundary scatters the impinging sound according to Lambert's law eqn (4.22), which is now written as

$$I(\varphi, \vartheta) = \frac{1}{\pi} \rho B_0 \, d\Omega \tag{5.10}$$

where the wall absorption is accounted for by the reflection coefficient $\rho = 1 - \alpha$ and $d\Omega = dS \cos \vartheta / r^2$ is the solid angle subtended by the wall element $dS$. $I(\varphi, \vartheta)$ is the directional distribution of sound energy at a given point R, as introduced in Section 4.3. It is seen that a non-uniform distribution of wall absorption will extinguish potential ray paths and hence impede the formation of a diffuse sound. Suppose the reflectivity $\rho$ of a particular wall element is very small. Then it is not very likely that $B_0$ will make up for this deficiency in reflected energy because $B_0$ shows only slow variations along the boundary, due to the smoothing effect of wall scattering. The only exception is the spherical room with a diffusely reflecting wall: according to eqn (4.26) the irradiation strength $B$ is constant provided the direct irradiation strength $B_d$ is also kept constant. On the other hand, even a small diffusely reflecting wall portion is sufficient to establish a diffuse field in a room which is free of absorption no matter how it is shaped.

Quite a different method of increasing sound field diffusion is not to provide for rough or corrugated walls and thus to destroy specular reflections, but instead to disturb the free propagation of sound rays. This is achieved by suitable objects—rigid bodies or shells—which are suspended freely in the room at random positions and orientations, and which scatter the arriving sound waves or sound particles in all directions. This method is quite efficient even when applied to parts of the room only, or in enclosures with partially absorbing walls. Of course, no architect would agree to have the free space of a concert hall or a theatre filled with such 'volume diffusers'; therefore a uniform distribution of them can only be installed in certain measuring rooms, so-called reverberation chambers (see Section 8.7), for which achieving a diffuse sound field is of particular importance.

To estimate the efficiency of volume scatterers we assume $N$ of them to be randomly distributed in a room with volume $V$, but with constant mean density $\langle n_s \rangle = N/V$. The scattering efficiency of a single obstacle or diffuser is characterised by its 'scattering cross-section' $Q_s$, defined by eqn (2.46). If we again stress the notion of sound particles, the probability that a particle will travel a distance $r$ or more without being scattered by a diffuser is $\exp\left(-\langle n_s \rangle Q_s r\right)$ or $\exp\left(-r/\bar{r}_s\right)$ where

$$\bar{r}_s = \frac{1}{\langle n_s \rangle Q_s} \tag{5.11}$$

is the mean free path length of sound particles between successive collisions with diffusers. In Section 5.2 we shall introduce the mean free path of sound particles between successive wall reflections $\bar{\ell} = 4V/S$, with S denoting the wall surface of a room. Obviously the efficiency of volume diffusers depends on the ratio $\bar{r}_s/\ell_s$. If a diffuser is not small compared with the acoustical wavelength, its scattering cross-section $Q_s$ is roughly twice its visual cross-section. However, only half of this value represents scattering of energy in different directions; the other half corresponds to the energy needed to form the 'shadow' behind the obstacle by interfering with the sound wave. For non-spherical diffusers the cross-section has to be averaged over all directions of incidence.

The preceding discussion may be summarised in the following way: diffusely reflecting room walls help to establish isotropy of sound propagation in a room but they alone do not guarantee that a sound field is diffuse. At least of equal importance is the amount and distribution of wall absorption. Perfect or nearly perfect sound field diffusion is only possible if the absorption coefficient of the boundary is very small everywhere—or if the enclosure is a sphere.

## 5.2 Mean free path and average rate of reflections

In the following material we shall make use of the concept of 'sound particles', already introduced at the beginning of this chapter. We imagine that the sound field is composed of a very great number of such particles. Our goal is the evaluation of more general or exact laws according to which the sound energy decreases with time in a decaying sound field. For this purpose we have firstly to follow the 'fate' of one sound particle and subsequently to average over many of these fates.

In this connection the notion of the 'mean free path' of a sound particle is frequently encountered in literature on room acoustics. The notion itself appears at first glance to be quite clear, but its use is sometimes misleading, partly because it is not always evident whether it refers to the time average or the particle (ensemble) average.

We shall start here from the simplest concept: a sound particle is observed during a very long time interval $t$; the total path length $ct$ covered by it during this time is divided by $N$, the number of wall collisions which have occurred in the time $t$:

$$\bar{\ell} = \frac{ct}{N} = \frac{c}{\bar{n}} \tag{5.12}$$

where $\bar{n} = N/t$ is the average reflection frequency, i.e. the average number of wall reflections per second.

Here $\bar{\ell}$ and $\bar{n}$ are clearly defined as time averages for a single sound particle; they may differ from one particle to another. In order to obtain averages which are representative of all sound particles, we should average $\bar{\ell}$ and $\bar{n}$ once more, namely over all possible particle fates. In general, the result of such a procedure would depend on the shape of the room as well as on the chosen directional distribution of sound paths.

Fortunately we can avoid these complications by assuming the sound field to be diffuse. Then no additional specification of the directional distribution is required. Furthermore, no other averaging is needed. This is so because— according to our earlier discussions—a diffuse sound field is established by nearly non-predictable changes in the particle directions, caused either by a highly complex room shape, by diffuse wall reflections or by particles being scattered by obstacles within the room. In any case the sound particles change their roles and their directions again and again, and during this process they completely lose their individuality. Thus the distinction between time averages and particle or ensemble averages is no longer meaningful; it does not matter whether the quantities we are looking for are evaluated by averaging over many free paths traversed by one particle or by averaging for one instant over a great number of different particles. Or in short:

time average = ensemble average

Now we shall calculate the mean free path of sound particles in rooms of arbitrary shape with walls which reflect the incident sound in an ideally diffuse fashion, i.e. according to Lambert's cosine law of eqn (4.22). For the present purpose we shall formulate that law in a slightly different way: the probability of a sound particle being reflected or re-emitted into a solid angle element $d\Omega'$, which includes an angle $\vartheta$ with the wall normal, is

$$P(\vartheta)\,d\Omega' = \frac{1}{\pi}\cos\vartheta\,d\Omega' \tag{5.13}$$

It is thus independent of the particle's previous history.

We consider all possible free paths, i.e. all possible chords in the enclosure (see also Fig. 4.17) and intend to carry out an averaging of their lengths over all wall elements $dS$ as well as over its directions.

These paths are, however, not traversed with equal probabilities. When averaging over all directions we have to concede a greater chance to the small angles $\vartheta$ according to eqn (5.13) by using $P(\vartheta)$ as a weighting function. Therefore the averaging process is

$$\bar{\ell} = \frac{1}{S} \iint_S dS \frac{1}{\pi} \iint_{\Omega = 2\pi} R_{dS}(\vartheta) \cos \vartheta \, d\Omega'$$

$R_{dS}(\vartheta)$ is the length of the chord originating from dS under an angle $\vartheta$ with respect to the wall normal, and $S$ is the total wall area of the room. Now we interchange the order of integrations, keeping in mind, however, that $\vartheta$ is not measured against a fixed direction but against the wall normal, which changes its direction from one wall element dS to another. Thus we obtain

$$\bar{\ell} = \frac{1}{\pi S} \iint_{2\pi} d\Omega' \iint_S R_{dS}(\vartheta) \cos \vartheta \, dS$$

The second integrant is the volume of an infinitesimal cylinder with axis $R_{dS}$ and base dS; thus the second integral is twice the volume of the room. The remaining integral simply yields $2\vartheta$ and therefore our final result is

$$\bar{\ell} = \frac{4V}{S} \tag{5.14}$$

The direct calculation of the mean reflection frequency $\bar{n}$ is even simpler. Let us suppose that a single sound particle carries the energy $e_0$. Its contribution to the energy density of the room is

$$w = \frac{e_0}{V}$$

On the other hand, if the sound particle strikes the wall $\bar{n}$ times per second, it transports on average the energy per second and unit area

$$B = \bar{n} \frac{e_0}{S}$$

to the wall. By inserting these expressions into eqn (5.4) one arrives at the important relation

$$\bar{n} = \frac{cS}{4V} \tag{5.15}$$

This expression, which is the time average as well as the particle average, has already been derived in Section 4.2 for the rectangular room. It is evident that the present derivation is more general and, in a way, more satisfying than the previous one. By inserting its result into eqn (5.12) it again leads to the above expression for the mean free path length, eqn (5.14).

The actual free path lengths are distributed around their mean $\bar{\ell}$ in some way, of course. This distribution depends on the shape of the room, and the same is true for other characteristic values such as its variance $\overline{(\ell - \bar{\ell})^2} = \overline{\ell^2} - \bar{\ell}^2$. As an illustration, Fig. 5.3 shows the distributions of free path lengths for three different shapes of rectangular rooms with diffusely reflecting walls; the abscissa is the path length divided by its mean value $\bar{\ell}$. These distributions have been calculated by application of a Monte-Carlo method, i.e. by computer simulation of the sound propagation.[2]

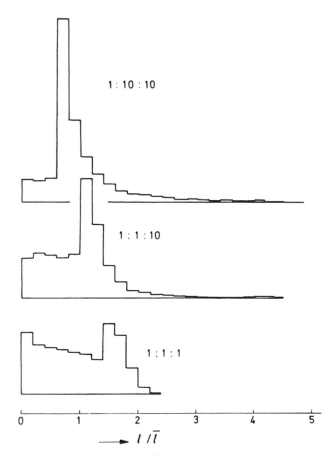

1 : 10 : 10

1 : 1 : 10

1 : 1 : 1

$\longrightarrow \ell / \bar{\ell}$

*Figure 5.3* Distribution of free path lengths for rectangular rooms with diffusely reflecting walls. The numbers indicate the relative dimensions of the rooms.

Table 5.1 Some Monte-Carlo results on the average and the relative variance of the path length distribution in rectangular rooms with diffusely reflecting walls

| Relative dimensions | Mean value ($\overline{\ell}_{MC}/\overline{\ell}$) | Relative variance $\gamma^2$ |
|---|---|---|
| 1:1:1 | 1.0090 | 0.342 |
| 1:1:2 | 1.0050 | 0.356 |
| 1:1:5 | 1.0066 | 0.412 |
| 1:1:10 | 1.0042 | 0.415 |
| 1:2:2 | 1.0020 | 0.363 |
| 1:2:5 | 1.0085 | 0.403 |
| 1:2:10 | 0.9928 | 0.465 |
| 1:5:5 | 1.0035 | 0.464 |
| 1:5:10 | 0.99930 | 0.510 |
| 1:10:10 | 1.0024 | 0.613 |

Typical parameters of path length distributions, evaluated in the same way for different rectangular rooms, are listed in Table 5.1. The first column contains the relative dimensions of the various rooms and the second column lists the results of the Monte-Carlo computation for the mean free path divided by the 'classical' value of eqn (5.14). These numbers are very close to unity and can be looked upon as an 'experimental' confirmation of eqn (5.14). The remaining deviations from 1 are insignificant and are due to the random errors inherent in the method. It may be added that a similar investigation of rooms with specularly reflecting walls, which are equipped with scattering elements in the interior, yields essentially the same result. Finally, the third column of Table 5.1 contains the 'relative variance' of the path length distributions:

$$\gamma^2 = \frac{\overline{\ell^2} - \overline{\ell}^2}{\overline{\ell}^2} \qquad (5.16)$$

The significance of this quantity will be discussed in Section 5.4.

## 5.3 Sound decay and reverberation time

As far back as Chapter 4, formulae have been derived for the time dependence of decaying sound energy and for the reverberation time of rectangular rooms (eqns (4.9) to (4.11)). In the preceding section it has been shown that the value $cS/4V$ of the mean reflection frequency, which we have used in Chapter 4, is valid not only for rectangular rooms but also for rooms of arbitrary shape provided that the sound field in their interior is diffuse. Thus the general validity of those reverberation formulae has been proven.

If the sound absorption coefficient of the walls depends on the direction of sound incidence, which will usually be the case, we must use the average value $\alpha_{uni}$ of eqn (2.41) instead of $\alpha$.

Further consideration is necessary if the absorption coefficient is not constant along the boundary but depends on the location of a certain wall element.

For the sake of simplicity we assume that there are only two different absorption coefficients in the room under consideration. The subsequent generalisation of the results for more than two different types of wall will be obvious.

We therefore attribute an absorption coefficient $\alpha_1$ to the wall portion with area $S_1$ and $\alpha_2$ to the portion with area $S_2$, where $S_1 + S_2 = S$. The situation is depicted in Fig. 5.4. We follow the path of a particular sound particle over $N$ wall reflections, among which there are $N_1$ reflections from $S_1$ and $N_2 = N - N_1$ reflections from $S_2$. These numbers can be assumed to be distributed in some way about their mean values

$$\overline{N}_1 = N\frac{S_1}{S} \quad \text{and} \quad \overline{N}_2 = N\frac{S_2}{S} \tag{5.17}$$

$S_1/S$ and $S_2/S$ are the a priori probabilities for the arrival of a sound particle at wall portion $S_1$ and $S_2$, respectively.

In a diffuse sound field subsequent wall reflections are stochastically independent from each other, i.e. the probability of hitting one or other portion of the wall does not depend on the past history of the particle. In this case the probability of $N_1$ collisions with wall portion $S_1$ among a total number of reflections $N$ is given by the binomial distribution

$$P_N(N_1) = \binom{N}{N_1} \cdot \left(\frac{S_1}{S}\right)^{N_1} \cdot \left(\frac{S_2}{S}\right)^{N-N_1} \tag{5.18}$$

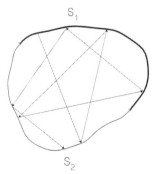

Figure 5.4 Enclosure with two differently absorbing types of boundary.

with

$$\binom{N}{N_1} = \frac{N!}{N_1!(N-N_1)!}$$

After $N_1$ collisions with $S_1$ and $N-N_1$ collisions with $S_2$, a sound particle has the remaining energy

$$E_N(N_1) = E_0 (1-\alpha_1)^{N_1} (1-\alpha_2)^{N-N_1} \qquad (5.19)$$

when $E_0$ is its initial energy. The expectation value of this expression with respect to the distribution (5.18) is

$$\langle E_N \rangle = \sum_{N_1=0}^{N} E_N(N_1) \cdot P_N(N_1) = E_0 \left[ \frac{S_1}{S}(1-\alpha_1) + \frac{S_2}{S}(1-\alpha_2) \right]^N$$

where we have applied the binomial theorem. Since $S_1 + S_2 = S$, this result can be written as

$$\langle E_N \rangle = E_0 (1-\overline{\alpha})^N = E_0 \exp\left[ N \ln(1-\overline{\alpha}) \right] \qquad (5.20)$$

with

$$\overline{\alpha} = \frac{1}{S}(S_1\alpha_1 + S_2\alpha_2) \qquad (5.20a)$$

This latter formula states that the significant quantity is the arithmetic mean of the absorption coefficients $\alpha_1$ and $\alpha_2$ with the respective areas as weighting factors. Finally, we replace the total number $N$ of wall reflections in time $t$ by its expectation value or mean value $\overline{n}t$ dt with $\overline{n}t = cS/4V$ and obtain for the energy of the 'average' sound particle and hence for the total energy in the room

$$E(t) = E_0 \exp\left[ \frac{cS}{4V}t \ln(1-\overline{\alpha}) \right] = E_0 \exp\left[ -\frac{cS}{4V}at \right] \qquad \text{for } t \geq 0$$

$$(5.21)$$

with $a$ denoting the 'absorption exponent':

$$a = -\ln(1-\overline{\alpha}) \qquad (5.22)$$

From this we can evaluate the reverberation time, i.e. the time interval $T$ in which the reverberating sound energy reaches one millionth of its initial value:

$$T = -\frac{24 V \cdot \ln 10}{cS \ln (1 - \overline{\alpha})}$$

or, after inserting $c = 343$ m/s:

$$T = 0.161\frac{V}{aS} \qquad (5.23)$$

As in eqns (4.6) to (4.11), the effect of air attenuation may be taken into account by an additional factor $\exp(-mct)$ in eqn (5.21). This leads, after inserting the numerical value for the sound velocity, to the final expression

$$T = 0.161\frac{V}{-S \ln (1 - \overline{\alpha}) + 4mV} \qquad (5.24)$$

with

$$\overline{\alpha} = \frac{1}{S}\sum_i S_i\alpha_i \qquad (5.24a)$$

which is a generalisation of eqn (5.20a) for any number of different absorption coefficients. In this formula all lengths have to be expressed in metres.

Equations (5.23) or (5.24) together with (5.24a) are known as Eyring's reverberation formula, although they have also been derived independently by Norris as well as by Schuster and Waetzmann.

In many practical situations the average absorption coefficient $\overline{\alpha}$ is small compared with unity. Then the logarithm in eqn (5.24) can be expanded into a series

$$-\ln(1 - \overline{\alpha}) = \overline{\alpha} + \frac{\overline{\alpha}^2}{2} + \frac{\overline{\alpha}^3}{3} + \cdots$$

and all terms except the first one may be neglected. This results in Sabine's famous reverberation formula:

$$T = 0.161\frac{V}{S\overline{\alpha} + 4mV} \qquad (5.25)$$

This formula agrees with that derived in the preceding section (see eqn (5.9)), however in a different way and without the term $4mV$, which can be neglected for small rooms.

In deriving eqn (5.20) together with eqn (5.20a) the quantities $N_1$ and $N_2 = N - N_1$ have been considered as statistical variables with the mean values or expectation values $NS_1/S$ and $NS_2/S$, both with respect to the probability distribution (5.18). If $N_1$ and $N_2$ are considered instead as the exact collision rates with the wall portions $S_1$ and $S_2$, i.e. if the mean values themselves are inserted for $N_1$ and $N_2$ in eqn (5.19), we get formally the same exponential law as in eqn (5.21). However, the absorption exponent $a$ in eqn (5.22) is to be replaced with

$$\bar{a}' = -\frac{1}{S}\sum_i S_i \ln(1-\alpha_i) \tag{5.26}$$

The resulting reverberation formula

$$T = 0.161 \frac{V}{S\bar{a}'} \tag{5.27}$$

could again be completed by adding a term $4mV$ to the denominator in order to account for the air attenuation. It is known as Millington–Sette's formula. It differs from eqn (5.23) or (5.24) in the manner in which the absorption coefficients of the various wall portions are averaged; here the logarithm of the average absorption coefficient is replaced by the average of the terms $-\ln(1-\alpha_i)$.

This has a strange consequence: suppose a room has a portion of wall, however small, with the absorption coefficient $\alpha_i = 1$. It would make the average (5.26) infinitely large and hence the reverberation time evaluated by eqn (5.27) would be zero. This is obviously an unreasonable result.

## 5.4 The influence of unequal path lengths

The averaging rule of eqn (5.26) was the result of replacing a probability distribution by its mean value. However, in the derivation of eqn (5.21) we have practised a similar simplification in that we have replaced the actual number of reflections in the time $t$ by its average $\bar{n}t$. For a more correct treatment one ought to introduce the probability $P_t(N)$ of exactly $N$ wall reflections occurring in a time $t$ and to calculate $E(t)$ as the expectation value of eqn (5.20) with respect to this probability distribution:

$$E(t) = E_0 \sum_{N=0}^{\infty} P_t(N)\cdot(1-\bar{\alpha})^N = E_0 \sum_{N=0}^{\infty} P_t(N)\cdot\exp(Na) \tag{5.28}$$

with the 'absorption exponent' $a$ as defined in eqn (5.22).

If, for the moment, $N$ is considered as a continuous variable, the function $\exp(-Na)$ in eqn (5.28) can be expanded in a power series for $(N-\bar{n}t)a$ by setting $N = \bar{n}t + (N-\bar{n}t)$. Truncating this series after its third term yields

$$\exp(-Na) \approx \exp(-\bar{n}ta)\left[1 - \frac{N-\bar{n}t}{1!}a + \frac{(N-\bar{n}t)^2}{2!}a^2\right]$$

Before inserting this expression into eqn (5.28) it should be kept in mind that

$$\sum_N P_t(N) = 1 \quad \text{and} \quad \sum_N (N - \bar{n}t) P_t(N) = 0$$

whereas

$$\sum_N (N - \bar{n}t)^2 P_t(N) = \sigma_N^2$$

is the variance of the distribution $P_t(N)$. Hence

$$E(t) = E_0 \exp(-\bar{n}ta)\left(1 + \frac{1}{2}\sigma_N^2 a^2\right) \approx E_0 \exp\left(-\bar{n}ta + \frac{1}{2}\sigma_N^2 a^2\right) \quad (5.29)$$

The latter approximation is permissible if $\sigma_N^2 a^2/2$ is small compared with unity. On the other hand, $\sigma_N^2$ is closely related to the relative variance $\gamma^2$ of the path length distribution, as defined in eqn (5.16). To illustrate this, Fig. 5.5 plots the individual 'fates' of three particles, i.e. the number $N$ of their wall collisions as a function of the distance $x$ they travelled. It is obvious that

$$\frac{\sigma_N}{\sigma_x} \approx \frac{\bar{n}t}{ct} = \frac{1}{\bar{\ell}}$$

Now suppose a particle travels $N$ successive and independent free paths $\ell_1, \ell_2, \cdots \ell_N$. According to basic laws of probability (see Ref. 3, for instance),

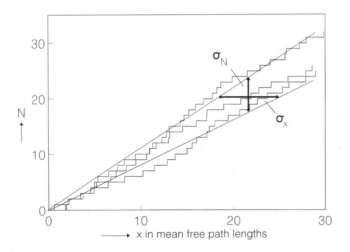

Figure 5.5 Time histories of three sound particles—relation between $\sigma_x$ and $\sigma_N$.

their sum $x$ has the variance

$$\sigma_x^2 = N\left(\overline{\ell^2} - \overline{\ell}^2\right) = N\overline{\ell}^2\gamma^2$$

This relation, combined with the preceding one, yields after replacing $N$ with $\overline{n}t$:

$$\sigma_N^2 \approx \overline{n}t\gamma^2 \qquad (5.30)$$

With this result we get from eqn (5.29):

$$E(t) = E_0 \exp\left[-\overline{n}ta\left(1 - \frac{\gamma^2}{2}a\right)\right] \qquad (5.31)$$

Accordingly, the reverberation time is

$$T = 0.161\frac{V}{Sa''} \qquad (5.32)$$

with the modified absorption exponent

$$a'' = -\ln(1-\overline{\alpha})\cdot\left[1 + \frac{\gamma^2}{2}\ln(1-\overline{\alpha})\right] \qquad (5.33)$$

In Fig. 5.6, the corrected absorption exponent $a''$ is compared with the absorption coefficient $\overline{\alpha}$ as is used in Sabine's formula (eqn (5.25) with $m = 0$). It plots the relative difference between both quantities for various parameters $\gamma^2$. The curve $\gamma^2 = 0$ corresponds to Eyring's formula (5.24) with $m = 0$. Accordingly, the latter is strictly valid for one-dimensional enclosures only where all paths have exactly the same length. For $\gamma^2 > 0$, $a''$ is smaller than $-\ln(1-\overline{\alpha})$. Hence the reverberation time is longer than that evaluated by the Eyring formula.

As long as the mean absorption coefficient is smaller than about 0.3, which is true for most rooms, the Eyring absorption exponent may be approximated by

$$-\ln(1-\overline{\alpha}) \approx \overline{\alpha} + \frac{\overline{\alpha}^2}{2}$$

which, after omitting all terms of higher than second order, yields

$$a'' \approx \overline{\alpha} + \frac{\overline{\alpha}^2}{2}\left(1 - \gamma^2\right) \qquad (5.33a)$$

With eqn (5.32) and (5.33) we have arrived for the first time at a reverberation formula in which the shape of the room is accounted for by the quantity

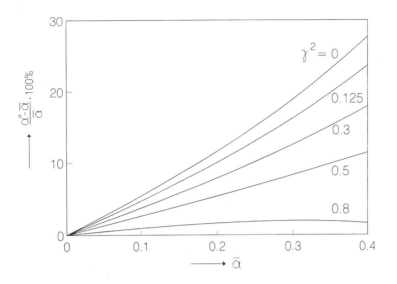

*Figure 5.6* Relative difference between $a''$ and $\overline{\alpha}$ in per cent after eqn (5.33).

$\gamma^2$. Unfortunately the latter can be calculated directly only for a limited number of room shapes. For a sphere, for instance, it turns out to be 1/8. For other shapes the relative variance $\gamma^2$ can be determined by computer simulation. Results obtained in this way for rectangular rooms have already been presented in Table 5.1. It is seen that for most shapes $\gamma^2$ is close to 0.4, and it is likely that this value can also be applied to other enclosures provided that their shapes do not deviate too much from that of a cube. It should be noted that for higher values of the relative variance $\gamma^2$ eqn (5.33) is not very accurate because of the approximations we have made in its derivation. Suppose, for instance, the free path lengths are exponentially distributed, i.e. according to

$$P(\ell)\,d\ell = \exp\left(-\ell/\overline{\ell}\right)d\ell \tag{5.34}$$

This distribution has the relative variance $\gamma^2 = 1$. On the other hand, the exponential distribution of path lengths is associated with the Poisson distribution of collision numbers,[3] i.e.

$$P_t(N) = \exp(-\overline{n}t)\frac{(\overline{n}t)^N}{N!}$$

If this expression is inserted into eqn (5.28) (first version) we get

$$E(t) = E_0 \exp(-\overline{n}t) \sum_{N=0}^{\infty} \frac{[\overline{n}t\,(1-\overline{\alpha})]^N}{N!} = E_0 \exp(-\overline{n}t\overline{\alpha})$$

since the sum represents the series expansion of the exponential function $\exp[\bar{n}t(1-\bar{\alpha})]$. The latter formula corresponds with Sabine's formula[4] (5.25) (with $m = 0$). It means that the correct curve for $\gamma^2 = 1$ in Fig. 5.6 should coincide with the horizontal axis.

For rooms with suspended 'volume diffusors' (see Section 5.1), the distribution of free path lengths is greatly modified by the scattering obstacles. The same applies to the relative variance $\gamma^2$ but not to the mean free path length.[2]

## 5.5 Enclosure driven by a sound source

In Section 5.1, a differential equation (5.5) was derived for the energy density $w$ in a room in which a sound source with time-dependent power output $P(t)$ is operated. This equation has the general solution:

$$w(t) = \frac{1}{V} \int_{-\infty}^{t} P(\tau) \exp\left[-2\delta(t-\tau)\right]d\tau = \frac{1}{V} \int_{0}^{\infty} P(t-\tau) \exp\left(-2\delta\tau\right)d\tau$$

(5.35)

with $\delta = cA/8V = 6.91/T$. Accordingly, the energy density is calculated by convolving the power output $P(t)$ with the 'energetic impulse response' of the enclosure:

$$w(t) = P(t) * \frac{1}{V} \exp\left(-2\delta t\right)$$

(5.35a)

As an application of this formula we consider, as shown in Fig. 5.7a, a sound source the power output of which varies sinusoidally with the angular frequency $\Omega$:

$$P(t) = P_0(1+\cos \Omega t)$$

Inserting this into eqn (5.35) and carrying out the integration leads after some obvious manipulations to

$$w(t) = \frac{P_0}{2\delta V}\left\{1 + m \cos\left[\Omega\left(t-t_0\right)\right]\right\}$$

(5.36)

with

$$m(\Omega) = \frac{2\delta}{\left(4\delta^2 + \Omega^2\right)^{1/2}}$$

(5.36a)

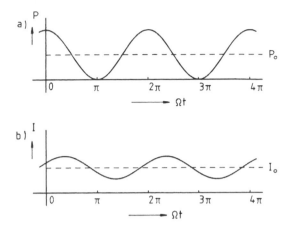

Figure 5.7 Flattening of the modulation of sound signals by transmission in a room: (a) power of the original sound signal; (b) received intensity.

and

$$t_0 = \frac{1}{\Omega} \arctan\left(\frac{\Omega}{2\delta}\right) = \frac{1}{\Omega} \arccos\left[\underline{m}(\Omega)\right] \qquad (5.36b)$$

Since $m(\Omega)$ is always smaller than unity, the reverberation of the room has a flattening and delaying effect on the modulation of the input power, as shown in Fig. 5.7b.

The function $m(\Omega)$ is known as the 'modulation transfer function' (MTF) since it expresses the way in which the modulation index $m$ is changed by the transient behaviour of the room. Its magnitude may be combined with the phase shift $\Omega t_0$ to the complex MTF, which in the present case reads:

$$m = \frac{1}{1 + i\Omega/2\delta} \qquad (5.36c)$$

It is easily checked that the magnitude of this function agrees with $m$ in eqn (5.36a).

Of course, the preceding formulae for the MTF are only true for enclosures the sound decay of which follow the idealized exponential law of eqn (5.7). In real rooms, the modulation transfer function may be quite different; in general it is a function not only of the modulation frequency $\Omega$ but also of the sound frequency $\omega$ or the frequency spectrum of the exciting sound.

An even more important—and simpler—case is that of a source with constant power output $P$. Here eqn (5.35a) yields immediately:

$$w_r = \frac{P}{2\delta V} = \frac{4P}{cA} \qquad (5.37)$$

with $\delta = cA/8V$ (see eqn (5.8)). This formula agrees with eqn (5.6), which was obtained directly from eqn (5.5) by setting the time derivative zero. The subscript r is to indicate that $w_r$ is the energy density of the 'reverberant field' excluding the contribution of the direct sound. For a point source with omnidirectional sound radiation, the direct sound intensity in distance $r$ is $I_d = cw_d = P/4\pi r^2$; hence the energy density of the direct component is

$$w_d = \frac{P}{4\pi c r^2} \tag{5.38}$$

In Fig. 5.8, $w_r$ and $w_d$ are presented schematically as functions of distance $r$ from the sound source. For a certain distance $r = r_c$ both energy densities are equal. This distance $r_c$ is called the 'critical distance' or 'diffuse-field distance' and is given by

$$r_c = \left(\frac{A}{16\pi}\right)^{1/2} \approx 0.1 \cdot \left(\frac{V}{\pi T}\right)^{1/2} \tag{5.39}$$

In the latter expression we have introduced the reverberation time $T$ from Sabine's formula (with $m = 0$); $V$ is to be measured in $m^3$.

Many sound sources have a certain directivity which can be characterised by their 'gain' or 'directivity' $g$. The latter is defined as the ratio of the

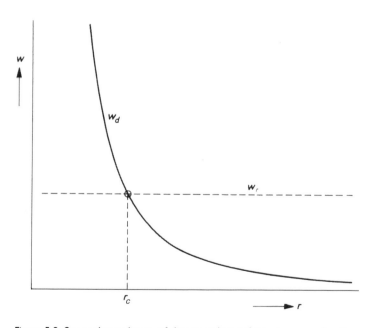

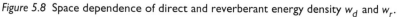

*Figure 5.8* Space dependence of direct and reverberant energy density $w_d$ and $w_r$.

maximum intensity and the intensity averaged over all directions, both at the same distance $r$:

$$g = \frac{I_{max}}{P/4\pi r^2} = 4\pi c r^2 \frac{w_{max}}{P} \tag{5.40}$$

(In the second expression $I_{max}$ has been replaced with $w_{max} = I_{max}/c$.) Then the critical distance depends on the direction of sound radiation; its maximum is

$$r_c = \left(\frac{gA}{16\pi}\right)^{1/2} \approx 0.1 \cdot \left(\frac{gV}{\pi T}\right)^{1/2} \tag{5.39a}$$

In any case, the total energy density can be expressed as

$$w = w_d + w_r = \frac{4P}{cA}\left(1 + \frac{r_c^2}{r^2}\right) \tag{5.41}$$

The practical application of eqn (5.37) and the related equations becomes questionable if the average absorption coefficient $\bar{\alpha} = A/S$ is not small compared with unity. This is because then the contribution of the very first reflections, which are not randomly distributed, is predominant in the total energy density. Therefore the range of validity of the above formulae with respect to $\bar{\alpha}$ is substantially smaller than that of the reverberation formulae developed in the preceding sections. This is one of the reasons why the absorption of a room and its reverberation time is usually determined by decay measurements and not by measuring the steady-state energy density and application of eqn (5.37) although this would be possible in principle.

On the other hand, however, measurements of the steady-state energy are a standard method for the evaluation of the total power $P$ of a sound source using eqn (5.41). The equivalent absorption area $A$ is determined by a reverberation measurement. This procedure is free from objection as long as the room utilised for this purpose is a 'reverberation chamber' with a long reverberation time and hence with low absorption. To increase the precision of the result it is advisable to apply some space averaging as described in Section 3.5. We should bear in mind that in the measurement of acoustic power an error of 10% (corresponding to 0.4 dB) can usually be tolerated, but not so in the determination of absorption or reverberation time.

## 5.6 Application of the radiosity integral

In a way, the theory of reverberation as outlined in the preceding sections is inconsistent in that it is based upon the condition of perfect sound field diffusion, although we know from Section 5.1 that such a sound field requires the absence of any boundary absorption and hence cannot decay, strictly speaking. It is amazing that despite the practical impossibility of creating a diffuse field the decay formulae derived on this basis have proved to be quite useful.

To avoid this contradiction we now replace the requirement for a perfect sound diffusion by a less stringent one: namely, that the boundary reflects the impinging sound energy in a perfectly diffuse, i.e. Lambertian, manner. This assumption is also an ideal one, but is not entirely unrealistic, as discussed at the end of Chapter 4.

In this case the propagation of sound energy within the enclosure can be described by the integral equation (4.24) for the 'irradiation strength' $B$ of the boundary, as discussed in Section 4.5. Since we are interested only in sound decay, we can omit the term due to direct irradiation by a sound source. Then the integral equation reads:

$$B(\mathbf{r}, t) = \iint_S \rho(\mathbf{r}) B(\mathbf{r}', t - R/c) \frac{\cos \vartheta \cos \vartheta'}{\pi R^2} \, dS' \qquad (5.42)$$

Figure 5.9 presents a few decay curves obtained by numerically solving this equation for a rectangular room with relative dimensions 1:2:3. For this purpose the boundary is subdivided into a number of equal area elements on which the irradiation density $B$ and the reflection coefficient $\rho = 1 - \alpha$ is assumed as constant. By this discretisation the integral equation is converted into a system of linear equations. The 'floor' (i.e. one of the walls with dimensions 2:3) was assumed to be totally absorbent ($\rho = 0$), whereas the remaining walls are free of absorption ($\rho = 1$). It is evident that this distribution of wall absorption causes severe deviations from diffuse field conditions. The curves show the decay of the irradiation strengths at three different locations. At the beginning there are some fluctuations which, however, gradually fade out leaving a straight line according to an exponential decay of the sound energy. It should be noted that the final slope is the same for all three curves. (Generally, the independence of the final slope on the receiver position has been proven by Miles[5]). However, it differs from that predicted by Eyrings's or Sabine's reverberation formulae, eqns (5.24) and (5.25). In the example of Fig. 5.9, the slope is about $-1.55$ dB per mean free path (MFP), whereas after Eyring's formula (5.24) it is only $10 \cdot \log_{10}(1 - 3/11) = -1.38$ dB/MFP.

Deviations from Eyring's predictions are typical for enclosures with imperfect sound field diffusion. To illustrate this, Fig. 5.10 presents the rate of overall sound decay in rectangular rooms of various shapes. The method used here was ray-tracing, a procedure which will be explained in Section 9.6. As before, one of the six walls is assumed to be totally absorbent ($\alpha = 1$), while the remaining walls are supposed to be free of absorption ($\alpha = 0$). Each room is characterised by its relative dimensions; the first two numbers refer to the

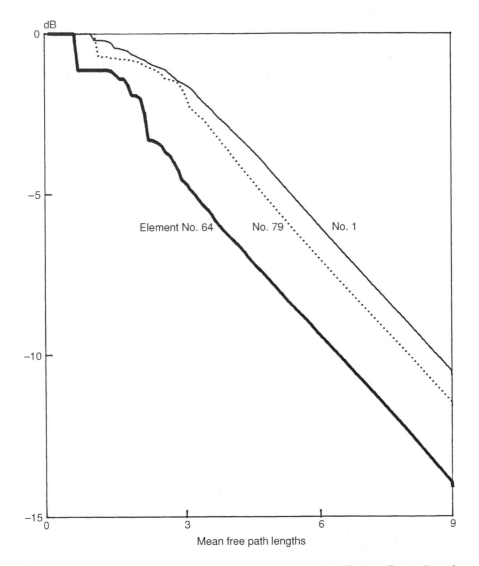

*Figure 5.9* Logarithmic decay curves in a rectangular room with diffusely reflecting boundary, calculated for three different receiver positions by solving eqn (5.42). The
relative room dimensions are 1:2:3; the 'floor' with dimensions 2:3 has the
absorption coefficient 1 while the remaining walls are free of absorption. The
abscissa is $ct/\bar{\ell}$, the number of mean free path lengths which a sound ray travels
within a given time $t$.

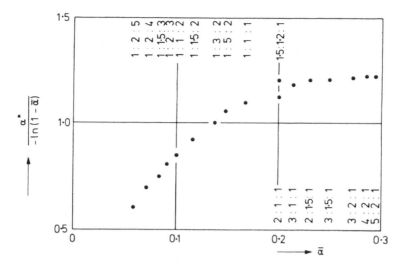

Figure 5.10 Effective absorption exponent $\alpha^*$, divided by the Eyring absorption exponent $-\ln(1 - \bar{\alpha})$, as obtained from Monte-Carlo computations for rectangular rooms with $\alpha = 1$ for one wall and $\alpha = 0$ for the remaining walls. Numbers in the figure indicate the relative room dimensions; the first two numbers refer to the absorbing wall.

absorbing wall. The plotted quantity is the effective absorption exponent $a^*$ defined by

$$E(t) = E_0 \exp\left(-\frac{cS}{4V}a^*t\right) = E_0 \exp\left(-\bar{n}a^*t\right) \qquad (5.43)$$

and divided by its Eyring value $a_{ey} = -\ln(1 - \bar{\alpha})$; the abscissa is the mean absorption coefficient $\bar{\alpha}$. It is seen that the results deviate from the ordinate 1 in both directions; in particular, for relatively low rooms with a highly absorbing floor (see right side of the figure) the absorption exponent is higher, and hence the reverberation time is shorter than predicted by the Eyring formulae. This is of practical interest since virtually all auditoria are of this general type because of the high absorption of occupied seats.

How can we understand such deviations? Let us have a look at Fig. 5.11 in which a section through a rectangular room is depicted. In Fig. 5.11a it is assumed that the 'floor' is highly absorbing while the remaining walls are rigid or nearly rigid. As indicated, a floor point P is 'irradiated' from all directions while a point Q at the ceiling does not receive energy from the floor because of its high absorption. Hence the floor receives and absorbs more energy than it would under diffuse field conditions. In Fig. 5.11b the absorbing surface is a side wall. In this case it is obvious that sound particles

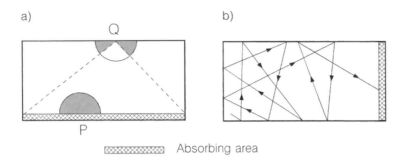

*Figure 5.11* Sound absorption in a longish room: (a) with absorbing 'floor'; (b) with one absorbing side wall.

in the left part of the enclosure are not much affected by the absorbing wall. Therefore we expect that in the right half of the enclosure there is less energy than in the left one, and hence the absorbing wall is hit by less sound particles than it would be in a diffuse field. Obviously this effect predominates the one responsible for the elevated absorption in the case of Fig. 5.11a.

To predict the slope of the final decay and hence the reverberation time, several numerical iteration schemes have been developed.[6,7] Although such methods may be too complicated for the practical design of a hall, one of them will be briefly described in the following.

At first, the irradiation strength is assumed to decay according to the exponential law in eqn (5.43) (second version) with unknown $a^*$. Next, the integral equation (5.42) is converted in a symmetrical form by introducing $\beta(\mathbf{r}) = B(\mathbf{r})\sqrt{\rho(\mathbf{r})}$:

$$\beta(\mathbf{r}) = \iint_S \kappa(\mathbf{r}, \mathbf{r}') \beta(\mathbf{r}') \exp\left(\overline{n}a^* R/c\right) dS' \tag{5.44}$$

with

$$\kappa(\mathbf{r}, \mathbf{r}') = \frac{\cos \vartheta \cos \vartheta'}{\pi R^2} \sqrt{\rho(\mathbf{r})\rho(\mathbf{r}')} \tag{5.45}$$

Now we require that the quantity $\beta$ is normalized according to $\int \beta_{n-1}^2 dS = C$, where $C$ denotes some constant. Suppose the iteration process has already supplied a $(n-1)$th approximation $\beta_{n-1}$ along with the eigenvalue $a^*_{n-1}$. Then both quantities are inserted into the right side of eqn (5.44) from which we obtain a new function $\beta'_n$ which, however, is not necessarily normalised but yields $\int \beta_n'^2 dS = \mu^2 C$. If $\mu < 1$, the exponential in eqn (5.44) and hence $a^*_{n-1}$ is obviously too small and vice versa. Accordingly, the $n$th approximation must be $\beta_n = \mu\beta'_n$. Furthermore, the factor $\mu$ yields a correction to the eigenvalue:

$$a_n^* \approx a_{n-1}^* + \ln \mu \tag{5.46}$$

As starting values, $a_0^* = 0$ and $\beta_0 = \sqrt{C/S}$ can be used. This process converges relatively fast, and a sufficiently accurate result is arrived at after a few iterations.

## 5.7 Sound propagation as a diffusion process

Quite a different description of sound fields in enclosures, including their decay, is arrived at by conceiving sound propagation as a diffusion process. (It should be noted that the term 'diffusion' in this context has nothing to do with the diffuse sound field but means the overall behaviour of an ensemble of particles.) In fact, the sound particles in an enclosure which represent its acoustical energy content can be imagined as the 'molecules' of a gas. If the boundary of the room scatters the impinging sound particles rather than reflecting them specularly these molecules will move in an almost unpredictable way. However, in contrast to the molecules of a regular gas, all sound particles travel with the same velocity, namely the sound velocity. Therefore it is only the directions of their motion which are randomly distributed. Furthermore, there are no collisions between sound particles. If we assign the same energy to all particles the energy density is proportional to the number or particles per unit volume.

From this idea it is only a small step to treat the motion of the 'sound gas' as a diffusion process, in which the particles tend to flow from a place of higher density to a place of lower density. Hence the flow intensity (in watts/m²) is $I = -D \cdot \mathrm{grad}\, w$, where $w$ is the energy density and $D$ the so-called diffusion constant:[8]

$$D = \frac{1}{3} c \bar{\ell} \tag{5.47}$$

As earlier, $\bar{\ell}$ is the mean free path length. Such processes are governed by the differential equation

$$D \Delta w - \frac{\partial w}{\partial t} = -P \delta(\mathbf{r} - \mathbf{r}_s) \tag{5.48}$$

In this equation $P$ is the power output of a point source located at $\mathbf{r}_s$. (The same equation also describes the conduction of heat, by the way.) The boundary condition is $I_n = \alpha B$ or, with $I_n = -D(\partial w/\partial n)$ and $B = (c/4) \cdot w$ (see eqn (5.4)):

$$D \frac{\partial w}{\partial n} + \frac{c}{4} \alpha w = 0$$

with the absorption coefficient $\alpha$. It should be noted that this condition employs a relation which holds for isotropic fields only.

It was Ollendorff[9] who proposed this approach as early as 1969 as a generalization of the Sabine theory. Since then, it has been applied by several authors to various problems in room acoustics.[10,11] The solution of eqn (5.48) is by no means easier than that of the wave equation (3.1); in both cases we have to determine eigenvalues and eigenfunctions. However, since the diffusion equation deals with the energy neglecting any phase relations the required number of eigenfunctions is often much smaller than that needed for calculating the sound pressure distribution in an enclosure. This holds in particular for the calculation of energy decay. In this case we can usually content ourselves with the lowest eigenvalue since the higher ones are associated with components which fade out very rapidly.

To calculate the reverberation time of the enclosure, we set $P = 0$ in eqn (5.48) and assume that the energy decays according to $\exp(-2\delta t)$. Then, eqn (5.48) reads

$$D\Delta w + 2\delta w = 0 \qquad\qquad (5.48a)$$

For very small absorption coefficients the damping constants calculated with this method, using the diffuse field expression $\bar{\ell} = 4V/S$ for the mean free path length agree with those from the Sabine theory, eqn (5.8). Generally, it can be shown that the damping constant obtained with the diffusion equation is equal to or smaller than the Sabine value, i.e. the reverberation times are longer than $0.161\,V/A$ (see eqn (5.9)).

It should be mentioned that the diffusion equation (5.48) is itself an approximation valid only if the flow density $D \cdot \mathrm{grad}\,w$ is much smaller than $c \cdot w$.[8] This condition corresponds to that of a nearly isotropic angular distribution of the sound particles, or, expressed in acoustical terms, of a nearly diffuse sound field. Nevertheless, the diffusion model is more general than the Sabine theory in that it allows for internal energy flows and hence for local variations of the energy density.

## 5.8  Coupled rooms

Sometimes a room is shaped in such a way that it can barely be considered as a single enclosure but rather as a collection of partial rooms separated by virtually non-transparent walls. The only communication is by relatively small apertures in these walls. The same situation is encountered if the partition walls are totally closed but are not completely rigid, and have some slight sound transparency. In this section the acoustical behaviour of such 'coupled rooms' will be dealt with from a geometrical and statistical point of view.

Let us suppose that the sound field in every partial room is diffuse and that reverberation would follow a simple exponential law if there were no interaction between them. Then for the $i$th partial room we have after eqn (5.5) (with $P_i = 0$):

$$\frac{dw_i}{dt} = -2\delta_{i0}w_i$$

$\delta_{i0} = cA_i/4V_i$ being the damping constant of that room without coupling.

The coupling elements permit an energy exchange between the partial rooms. If there are wall apertures with areas $S_{ij} = S_{ji}$ (with $S_{ii} = 0$) between rooms $i$ and $j$, the energy loss in room $i$ per second due to coupling is $\sum_j B_i S_{ij} = cS_i' w_i/4$ with $S_i' = \sum_j S_{ij}$. On the other hand, the energy increase contributed by room $j$ per second is $cS_{ij}w_j/4$. Hence the energy balance yields for the $i$th room

$$\frac{dw_i}{dt} + \left(2\delta_{i0} + \frac{cS_i'}{4V_i}\right)w_i = \sum_j \frac{cS_{ij}}{4V_i}w_j \tag{5.49}$$

The summation index is running from 1 to $N$, where $N$ is the total number of partial rooms. Evidently, the coupling areas are treated as 'open windows' having an absorption coefficient 1.

This system of linear differential equations can be further simplified to read

$$\frac{dw_i}{dt} = \sum_{j=1}^{N} k_{ij}w_j \qquad (i = 1, 2, \ldots, N) \tag{5.50}$$

where

$$k_{ii} = -\left(2\delta_{i0} + \frac{cS_i'}{4V_i}\right) \qquad \text{and for } i \neq j: \qquad k_{ij} = \frac{cS_{ij}}{4V_i}$$

First we look for the conditions under which the decay process obeys an exponential law with one common damping constant $\delta'$. Then we have $dw_i/dt = -2\delta'w_i$ for all partial rooms. Inserting this into eqns (5.50) yields $N$ homogeneous linear equations for the energy densities $w_i$:

$$k_{i1}w_1 + k_{i2}w_2 + \cdots + (k_{ii}w_i + 2\delta') + \cdots + k_{iN}w_N = 0 \qquad (i = 1, 2, \ldots, N) \tag{5.51}$$

which have non-vanishing solutions only if their coefficient determinant is zero. This condition is an equation of degree $N$ (characteristic equation) for $\delta'$ which has $N$ roots $\delta_1', \delta_2', \ldots, \delta_N'$. These are the eigenvalues of the system.

If they are inserted one after the other into eqns (5.51), the energy densities associated with a certain root, for instance with the $r$th root, can be determined except for a common factor

$$w_1^{(r)}, w_2^{(r)}, \ldots, w_N^{(r)} \qquad (r = 1, 2, \ldots, N)$$

For a certain root $\delta_r'$ the ratios of the energy densities $w_i^{(r)}$ remain constant during the whole decay process and are equal to the ratios of the initial values $w_{i0}^{(r)}$. This result can be summarised as follows: There are at most (since in principle several roots $\delta_i'$ can happen to coincide) $N$ possibilities for the reverberation to follow the same exponential law throughout the whole room system. A particular set of initial energy densities is required for each of these possibilities.

The general solution of the system is obtained as a linear superposition of the $N$ particular solutions evaluated above. Thus, the decay of sound energy in the $i$th partial room is

$$w_i(t) = \sum_{r=1}^{N} c_r w_{i0}^{(r)} \exp(-2\delta_r' t) \qquad (i = 1, 2, \ldots, N) \tag{5.52}$$

If the initial energy densities $w_i(0)$ of all partial rooms are given, the constants $c_r$ can be determined uniquely from eqn (5.52) by setting $t = 0$. The decay of sound energy in the $i$th partial room is then represented by an expression of the form

$$w_i(t) = \sum_{j=1}^{N} a_{ij} \exp(-2\delta_j' t) \tag{5.52a}$$

which is similar to eqn (3.46). In contrast to the coefficients $c_n^2$ of the latter, however, some of the $a_{ij}$ may become negative and so we cannot draw the conclusion (as we did in Section 3.6) that the logarithmic decay curves are either straight lines or curved upwards.

If the system of coupled rooms is driven by sound sources, a term $P_i/V_i$ must be added on the right-hand side of each of the equations (5.50) with $P_i$ denoting the power output of the source operating in the partial room $i$. For constant powers $P_i$ the steady-state energy densities are obtained by setting the time derivatives to zero. This yields the equations:

$$-\frac{P_i}{V_i} = \sum_{j=1}^{N} k_{ij} w_j \qquad (i = 1, 2, \ldots, N) \tag{5.53}$$

They can be used to determine the initial values $w_i(0)$ for the energy decay which starts after all sound sources have been stopped at $t = 0$, according to eqn (5.52).

If the subroom $i$ is not driven by a sound source ($P_i = 0$), the sum on the right-hand side of eqn (5.53) is zero, and the same is true for the time derivative in eqn (5.50) immediately after switching off all sources:

$$\left(\frac{\mathrm{d}w_i}{\mathrm{d}t}\right)_{t=0} = 0$$

i.e. the decay curve of that room starts with zero slope. In the same way it may be inferred that not only the initial slope of the decay curve but also its initial curvature vanishes if neither the considered room nor one of its adjacent neighbours are driven by a sound source.

These findings may be illustrated by an example depicted in Fig. 5.12. It refers to three coupled rooms of equal volumes in a line with $S_{12} = S_{23}$, $S_{13} = 0$. Only one of them is driven directly by a sound source. Accordingly the steady-state level from which the decay process starts is highest in room 1, and only in this room does the decay curve begin with a negative slope. The decay rate in this room becomes gradually less steep since the adjacent rooms feed some of the energy stored in them back to room 1. The decay curves in

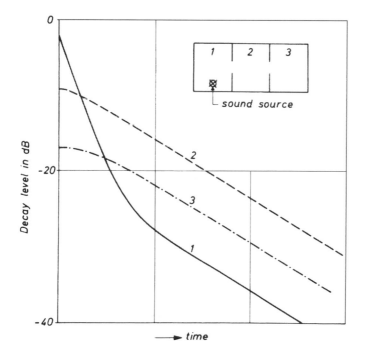

Figure 5.12 Sound decay in a system consisting of three coupled rooms of equal size: $\delta_2 = k_{12} = \delta_1/10$, $\delta_3 = \delta_1/2$, $k_{13} = 0$, $k_{23} = \delta_1/5$.

the rooms 2 and 3 have horizontal tangents at $t = 0$; moreover, the initial curvature of decay curve 3 is zero.

On the whole we can see that in coupled rooms the variety of possible decay curves is considerably greater than it is in an isolated room.

The occurrence of coupling effects as outlined above is not restricted to room systems of a special geometrical structure. They can also be observed in apparently 'normal' rooms in which one or several normal modes are excited whose eigenfrequencies are close to those of other modes but which do not exchange much energy. Convexly curved decay curves (viewed from above ) are a safe indication of the presence of several energy stores which are only weakly coupled to each other.

Typical coupling phenomena are to be expected if the probability of a sound particle escaping to a neighbouring partial room is small compared with the probability of it being absorbed by the boundary. Expressed by the characteristic parameters of the system, this condition is

$$\frac{cS_i'}{4V_i} \ll 2\delta_{i0} \tag{5.54}$$

In this case the characteristic solutions $\delta_i'$ of eq. (5.51) are very close to the decay constants of the $i$th room:

$$2\delta_i' \approx 2\delta_{i0} + \frac{cS_i'}{4V_i} = -k_{ii} \tag{5.55}$$

Increasing the ratio $|k_{ij}/k_{ii}|$ marks the transition from a system of coupled rooms to single rooms having just one damping constant.

Sometimes, when calculating the reverberation time of a room, it is advisable to take coupling effects into account. This aspect will be discussed in Chapter 9.

## References

1 Sabine WC. The American Architect 1900. See also Collected Papers on Acoustics No. 1. Cambridge: Harvard University Press, 1923.
2 Kuttruff H. Weglängenverteilung und Nachhallverlauf in Räumen mit diffus reflektierenden Wänden. Acustica 1970; 23:238; Kuttruff H. Weglängenverteilung in Räumen mit schallzerstreuenden Elementen. Acustica 1971; 24:356.
3 Papoulis A. Probability, Random Variables, and Stochastic Processes, 2nd edn, Int student edition. New York: McGraw-Hill, 1985.
4 Schroeder MR. Some new results in reverberation theory and measurement methods. Proc Fifth Intern Congr on Acoustics, Liege, 1965, paper G31.
5 Miles RN. Sound field in a rectangular enclosure with diffusely reflecting boundaries. J Sound Vibr 1984; 92(2):203.
6 Gilbert EN. An iterative calculation of reverberation time. J Acoust Soc Am 1981; 69:178.

7 Kuttruff H. A simple iteration scheme for the computation of decay constants in enclosures with diffusely reflecting boundaries. J Acoust Soc Am 1995; 98:288.
8 Morse PM, Feshbach H. Methods of Theoretical Physics, Section 2.4. New York: McGraw-Hill, 1953.
9 Ollendorff F. Statistische Raumakustik als Diffusionsproblem. Acustica 1969; 21:236.
10 Valeau V, Picaut J, Hodgson M. On the use of a diffusion equation for room-acoustic predictions. J Acoust Soc Am 2006; 119:1504.
11 Picaut J, Simon L, Polack JD. A mathematical model of diffuse sound field based on a diffusion equation. Acta Acustica 1997; 83:614.

# Chapter 6

# Sound absorption and sound absorbers

Of considerable importance to the acoustics of a room are the loss mechanisms which reduce the energy of sound waves when they are reflected from its boundary as well as during their free propagation in the air. They influence the strengths of the direct sound and of all reflected components and therefore all acoustical properties of the room.

The attenuation of sound waves in the free medium becomes significant only in large rooms and at relatively high frequencies; for scale model experiments, however, it causes serious limitations. We have to consider it inevitable and something which cannot be influenced by the efforts of the acoustician. Nevertheless, in reverberation calculations it has to be taken into account. Therefore it is sufficient in this context to give a brief description of the causes of air attenuation and to present the important numerical values.

The situation is different in the case of the absorption to which sound waves are subjected when they are reflected from the boundary of a room. The magnitude of wall absorption and its frequency dependence varies considerably from one material to another. There is also an unavoidable contribution to the wall absorption which depends on certain physical properties of the medium, but it is so small that in most cases it can be neglected.

Since the boundary absorption is of decisive influence on the sound field in a room, the acoustical designer should understand the various absorption mechanisms and know the various types of sound absorbers. In fact, the well-considered use of sound absorbers is one of his most important design tools. In particular, sound absorbers are usually employed to meet one of the following objectives:

- to adapt the reverberation of the room, for instance of a concert hall, to the type of performances which are to be staged in it
- to suppress undesired sound reflections from remote walls which might be heard as echoes
- to reduce the acoustical energy density and hence the sound pressure level in noisy rooms such as factories, large bureaus, etc.

This chapter will discuss the principles and mechanisms of the most important types of sound absorbers. For a more comprehensive account the reader is referred to Ref. 9 in Chapter 2.

## 6.1 The attenuation of sound in air

In the derivation of the wave equation (1.5) it was tacitly assumed that the changes in the state of the air, caused by the sound waves, occurred without loss. This is not quite true, however. We shall refrain here from a proper modification of the basic equations and from a quantitative treatment of attenuation. Instead the most prominent loss mechanisms are briefly described in the following. As earlier, the propagation losses will be formally characterised by the attenuation constant $m$ as defined in eqn (1.16a). The decrease of the sound intensity $I$, which is proportional to the square of the sound pressure in a plane sound wave is expressed by:

$$I(x) = I_0 \cdot \exp(-mx) \tag{6.1}$$

The attenuation in air is mainly caused by the following effects:

(a) Equation (1.4) is based on the assumption that the changes in the state of a volume of gas take place adiabatically, i.e. there is no heat exchange between neighbouring volume elements. The equation states that a compressed volume element has a slightly higher temperature than an element which is rarefied by the action of the sound wave. Although the temperature differences occurring at normal sound intensities amount to small fractions of a degree centigrade only, they cause a heat flow because of the finite thermal conductivity of the air. This flow is directed from the warmer to the cooler volume elements. The changes of state are therefore not taking place entirely adiabatically. According to basic principles of physics, the energy transported by these thermal currents cannot be reconverted completely into mechanical, i.e. into acoustical energy; some energy is lost to the sound wave. The corresponding portion of the attenuation constant $m$ increases with the square of the frequency.

(b) In a plane sound wave each volume element becomes periodically longer or shorter in the direction of sound propagation. This distortion of the original element can be considered as a superposition of an omnidirectional compression or rarefaction and of a shear deformation, i.e. of a pure change of shape. The medium offers an elastic reaction to the omnidirectional compression proportional to the amount of compression, whereas the shear is controlled by viscous forces which are proportional to the shear velocity. Hence—as with every frictional process—mechanical energy is irreversibly converted into heat.

This 'viscous portion' of the attenuation constant $m$ also increases proportionally with the square of the frequency.

(c) Under normal conditions the above-mentioned causes of attenuation in air are negligibly small compared with the attenuation caused by what is called 'thermal relaxation'. It can be described briefly as follows. Under equilibrium conditions the total thermal energy contained in a certain quantity of a uniform polyatomic gas is distributed among several energy stores (degrees of freedom) of the gas molecules, namely as translational, vibrational and rotational energy of the molecules. If the gas is suddenly compressed, i.e. if its energy is suddenly increased, the whole additional energy will be stored at first in the form of translational energy. Afterwards a gradual redistribution among the other stores will take place. Or, in other words, the establishment of a new equilibrium requires a finite time. If compressions and rarefactions change periodically as in a sound wave, a thermal equilibrium can be maintained—if at all—only at very low frequencies; with increasing frequency the actual energy content of certain molecular stores will lag behind the external changes and will accept or deliver energy at the wrong moments.

This is a sort of 'internal heat conduction', which weakens the sound wave just like the normal heat conduction with which we are more familiar. The relative amount of energy being dissipated along one wavelength has a maximum when the duration of one sound period is comparable with a specific time interval, the so-called 'relaxation time' being characteristic of the time lag in internal energy distribution.

In Fig. 6.1, the attenuation constant $m$ multiplied by the wavelength for a single relaxation process is plotted in arbitrary units as a function of the product of frequency and relaxation time. The very broad frequency range

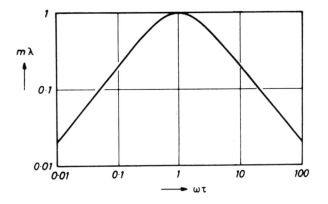

*Figure 6.1* Wavelength-related attenuation constant $m\lambda$ as typical for a relaxation process. The abscissa is the product of angular frequency $\omega$ and the relaxation time $\tau$.

Table 6.1 Attenuation constant of air at 20°C and normal atmospheric pressure, in $10^{-3}$ m$^{-1}$ (after Bass et al.[1])

| Relative humidity (%) | Frequency (kHz) | | | | | | |
|---|---|---|---|---|---|---|---|
| | 0.5 | 1 | 2 | 3 | 4 | 6 | 8 |
| 40 | 0.60 | 1.07 | 2.58 | 5.03 | 8.40 | 17.71 | 30.00 |
| 50 | 0.63 | 1.08 | 2.28 | 4.20 | 6.84 | 14.26 | 24.29 |
| 60 | 0.64 | 1.11 | 2.14 | 3.72 | 5.91 | 12.08 | 20.52 |
| 70 | 0.64 | 1.15 | 2.08 | 3.45 | 5.32 | 10.62 | 17.91 |

in which it appears is characteristic of a relaxation process. (For comparison we refer to the resonance curve in Fig. 2.9b.) Moreover, the relaxation of a medium causes not only a substantial increase in absorption but also a slight change in sound velocity, but this is not of importance in this connection.

For mixtures of polyatomic gases such as air, which consists mainly of nitrogen and oxygen, matters are much more complicated because there are many more possibilities of internal energy exchange which we shall not discuss here.

Because of their importance in room acoustics and, in particular for the calculation of reverberation time, a few numerical values of the intensity-related absorption constant $m$ are listed in Table 6.1.

## 6.2 Unavoidable wall absorption

Even if the walls, the ceiling and the floor of a room are completely rigid and free from pores, they cause a sound absorption which is small but different from zero. It only becomes noticeable, however, when there are no other absorbents or absorbent portions of wall in the room, no people, no porous or vibrating walls. This is the case for measuring rooms which have been specially built to obtain a high reverberation time (reverberation chambers; see Section 8.7). Physically this kind of absorption is again caused by the heat conductivity and viscosity of the air.

According to eqn (1.4), the periodic temperature changes caused by a sound wave are in phase with the corresponding pressure changes—apart from the slight deviations discussed in the preceding section. Therefore the maximum sound pressure amplitude which is observed immediately in front of a rigid wall should be associated with a maximum of 'temperature amplitude', which in turn is possible only if the wall temperature can completely follow the temperature fluctuations produced by the sound field. In reality the contrary is true: because of its high thermal capacity the wall surface remains virtually at a constant temperature. Therefore, in some boundary layer adjacent to the wall, strong temperature gradients will develop and hence a periodically alternating heat flow will be directed to and from the wall. This energy transport occurs at the expense of the sound energy, since the heat which was produced by the wave in a compression phase can only be partly reconverted into mechanical energy during the rarefaction phase.

A related effect becomes evident when we consider a sound wave imping-ing obliquely onto a rigid plane. In this case the normal component of the particle velocity vanishes at the boundary, but the parallel component does not, at least if calculated without accounting for the viscosity of air. This, however, cannot be true, since in a real medium the molecular layer immedi-ately on the wall is fixed to the latter, which means that the parallel velocity component vanishes as well. For this reason our assumption of a perfectly reflecting wall is not correct, in spite of its rigidity. In reality, another bound-ary layer is formed between the region of unhindered parallel motion in the air and the wall; at oblique incidence, high viscous forces and hence a substantial conversion of mechanical energy into heat takes place in this layer.

The effect of both loss processes can be accounted for by ascribing an absorption coefficient to the wall. It can be shown that the thickness of both boundary layers is inversely proportional to the square root of the frequency. Since, on the other hand, the gradients of the temperature and of the parallel velocity component increase proportionally with frequency, both contributions to the absorption coefficient are proportional to the square root of the frequency. Their dependence on the angle of incidence, however, is different. The viscous portion is zero for normal sound incidence, whereas the heat flow to and from the wall does not vanish at normal incidence.

Both effects are very small even at the highest frequencies relevant in room acoustics. For practical design purposes they can be safely neglected.

## 6.3 Sound absorption by membranes and perforated sheets

For the acoustics of a room it does not make any difference whether the apparent absorption of a wall is physically brought about by dissipative processes, i.e. by conversion of sound energy into heat, or by parts of the energy penetrating through the wall into the outer space. In this respect an open window is a very effective absorber, since it acts as a sink for all the arriving sound energy.

A less trivial case is that of a wall or some part of a wall forced by a sound field into vibration with a substantial amplitude. (Strictly speak-ing, this happens more or less with any wall, since completely rigid walls cannot be constructed.) Then a part of the wall's vibrational energy is re-radiated into the outer space. This part is withdrawn from the incident sound energy, viewed from the interior of the room. Thus the effect is the same as if it were really absorbed. It can therefore also be described by an absorption coefficient. In practice this sort of 'absorption' occurs with doors, windows, light partition walls, suspended ceilings, circus tents and similar 'walls'.

This process, which may be quite involved especially for oblique sound incidence, is very important in all problems of sound insulation. From the viewpoint of room acoustics, it is sufficient, however, to restrict discussions to the simplest case of a plane sound wave impinging perpendicularly onto the wall, whose dynamic properties are completely characterised by its mass inertia. Then we need not consider the propagation of bending waves on the wall.

Let us denote the sound pressures of the incident and of the reflected waves on the surface of a wall (see Fig. 6.2a) by $p_1$ and $p_2$, and the sound pressure of the transmitted wave by $p_3$. The total pressure acting on the wall is then $p_1 + p_2 - p_3$. It is balanced by the inertial force $i\omega M' v$, where $M'$ denotes the mass per unit area of the wall and $v$ the velocity of the wall vibrations. This velocity is equal to the particle velocity of the wave radiated from the rear side, for which $p_3 = \rho_0 c v$ holds. Therefore we have $p_1 + p_2 - \rho_0 c v = i\omega M' v$, from which we obtain

$$Z = i\omega M' + \rho_0 c \qquad (6.2)$$

for the wall impedance. Inserting this into eqn (2.8) yields

$$\alpha = \left[ 1 + \left( \frac{\omega M'}{2\rho_0 c} \right)^2 \right]^{-1} \approx \left( \frac{2\rho_0 c}{\omega M'} \right)^2 \qquad (6.3)$$

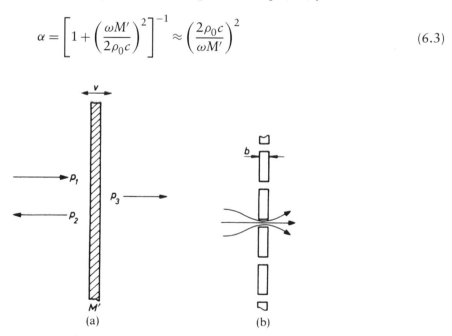

*Figure 6.2* (a) Pressures acting on a layer with mass $M'$ per unit area. (b) Perforated panel.

The latter approximation is permissible if the characteristic impedance of air is small compared with the mass reactance of the wall. In any case, the 'absorption' of a light wall or partition becomes noticeable only at low frequencies.

At a frequency of 100 Hz the absorption coefficient of a glass pane with 4 mm thickness is—according to eqn (6.3)—as low as 0.02 approximately. For oblique or random incidence this value is a bit higher due to the better matching between the air and the glass pane, but it is still very low. Nevertheless, the increase in absorption with decreasing frequency has the effect that rooms with many windows sometimes sound 'crisp' since the reverberation at low frequency is not as long as it would be in the same room without windows.

The absorption caused by vibrations of single-leaf walls and ceilings is thus very low. Matters are different for double-leaf or more complex walls, provided that the partition on the side of the room under consideration is mounted in such a way that vibrations are not hindered and provided that it is not too heavy. Because of the interaction between the leaves and the enclosed volume of air such a system behaves as a resonance system. This will be discussed in the next section.

It is a fact of great practical interest that a rigid perforated plate or panel has essentially the same properties as a membrane or foil. Let us look at Fig. 6.2b. Each hole in a plate may be considered as a short tube or channel with length $b$; the mass of air contained in it, divided by the cross-section, is $\rho_0 b$. Because of the contraction of the air stream passing through the hole, the air vibrates with a greater velocity than that in the sound wave remote from the wall, and hence the inertial forces of the air included in the hole are increased. The increase is given by the ratio $S_2/S_1$, where $S_1$ is the area of the hole and $S_2$ is the area per hole. Hence the equivalent mass of the perforated panel per unit area is

$$M' = \frac{\rho_0 b'}{\sigma} \tag{6.4}$$

with

$$\sigma = \frac{S_1}{S_2} \tag{6.5}$$

$\sigma$ is the perforation ratio of the plate; sometimes it is also called 'porosity' (generally in a different context).

In eqn (6.4) the geometrical length of the hole has been replaced with an "effective length"

$$b' = b + 2\Delta b \tag{6.6}$$

The correction $2\Delta b$, known as the "end correction", accounts for the fact that the streamlines (see Fig. 6.2b) cannot contract or diverge abruptly but only gradually when entering or leaving the hole.

For circular apertures with radius $a$ and with relatively large lateral distances the end correction is given by

$$\Delta b = \frac{\pi}{4}a \tag{6.7}$$

Finally, the absorption coefficient of a perforated panel is obtained from eqn (6.3) by substituting $M'$ from eqn (6.4).

Very often perforated plates are so light that they will vibrate as a whole when a sound waves strikes them. In this case $M'$ in eqn (6.5) must be replaced by the effective mass

$$M'' = \left( \frac{1}{M'} + \frac{1}{M'_0} \right) \tag{6.8}$$

$M'$ is the equivalent mass per square metre of the panel itself after eqn (6.4) and $M'_0$ is the specific mass of the panel material.

If frictional forces within the holes and other loss processes are disregarded, the absorption coefficient given by eqn (6.3) represents the fraction of incident sound energy which is transmitted by a wall or a perforated panel. It thus characterises the sound transparency of the wall. Let us illustrate this by an example: to yield a transparency of 90%, $\omega M'/2\rho_0 c$ in eqn (6.3) must have the value 1/3. At 1000 Hz this is the case with a mass layer with an (equivalent) mass per unit area of about 45 g/m². If realised by a perforated panel, this can be achieved, for instance, by a 1-mm thick sheet with 7.5% perforation and with holes having a diameter of 2 mm.

## 6.4 Resonance absorbers

In this section, we come back to the idealised resonator discussed in Section 2.4. It consists basically of a membrane with mass $M'$ per unit area, which is mounted in front of a rigid wall and parallel to it. Under the influence of an impinging sound wave, the membrane will perform vibrations which are controlled by its mass and by the air cushion behind. Furthermore, there are vibrational losses which can be represented by some loss resistance $r_s$ (see Section 2.4).

The absorption coefficient of this configuration can be calculated by using eqn (2.32). The result is plotted in Fig. 6.3 for various ratios $r_s/\rho_0 c$ and under the additional assumption $M'\omega_0 = 10\rho_0 c$; the abscissa is the angular frequency divided by the resonance frequency. A maximum absorption coefficient of 1 is only reached for exact matching, i.e. for $r_s = \rho_0 c$.

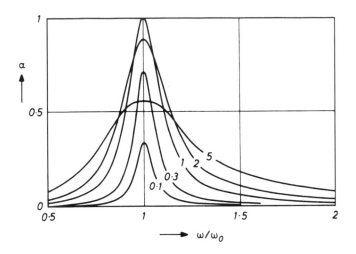

*Figure 6.3* Absorption coefficient (calculated) of resonance absorbers as a function of frequency at normal sound incidence, for $M'\omega = 10\rho_0 c$. Parameter is the ratio $r_s/\rho_0 c$.

For $r_s > \rho_0 c$ the maximum absorption is less than unity and the curves are broadening. This is qualitatively the same behaviour as that of a porous layer which is arranged at some distance and in front of a rigid wall (compare Fig. 2.7). The frequency-selective absorption characteristics make this device a useful tool for the control of reverberation, mainly in the low- and mid-frequency range. In practical applications the 'membrane' consists usually of a panel of wood, chipboard or gypsum (see Fig. 6.4a). The vibrational losses occurring in this system may have several physical reasons. One of them has to do with the fact that any kind of panel must be fixed at certain points or along certain lines to a supporting construction which forces the panel to be bent when it vibrates. Now all elastic deformations of a solid, including those by bending, are associated with internal losses that depend on the material and other circumstances. In metals, for instance, the intrinsic losses of the material are relatively small, but they may be substantial for plates made of wood or of plastic. If desired the losses may be increased by certain surface layers, or by porous materials placed in the space between the panel and the rigid rear wall.

According to the discussion of the preceding section, the mass layer can also be realised as a perforated or slotted sheet (see Fig. 6.4b). This device is often called a Helmholtz resonator, although the original Helmholtz resonator consists of a single, rigid-walled cavity with a narrow opening, as described in the next section. If the apertures in the sheet are very narrow, the frictional losses occurring in them may be sufficient to ensure the low

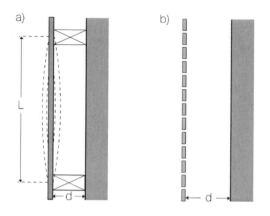

*Figure 6.4* Resonance absorbers: (a) with vibrating panel; (b) with perforated panel.

Q-factors that are needed for high efficiency of the absorber. Such devices are known as microperforated absorbers and have found considerable attention during the past two decades,[2] prompted by the technical progress in drilling numerous thin holes. When applied to transparent materials such as acrylic glass, they offer the possibility of manufacturing transparent resonance absorbers, although it may prove difficult in practice to keep apertures with diameters in the sub-millimeter range free of dust particles and other obstructions.

For wider holes it is usually necessary to provide for additional losses. This can be achieved, for instance, by covering the holes with a porous fabric. Another method of adapting the magnitude of $r_s$ to a desired value is to fill the air space behind the panel partially or completely with porous material.

For normal sound incidence the (angular) resonance frequency of this absorber is given by eqn (2.28), provided the space behind the membrane or perforated panel is empty. For its practical application, the following form may be more useful:

$$f_0 = \frac{600}{\sqrt{(M'd)}} \quad \text{Hz} \tag{6.9}$$

where $M'$ is in kg/m$^2$ and $d$ in cm. If the air space behind the panel is filled with porous material, the changes of state of the air will be rather isothermal than adiabatic. This means that eqn (1.4) is no longer valid and has to be replaced with $p/p_0 = \delta\rho/\rho_0$; the sound velocity becomes $c_{iso} = \sqrt{(p_0/\rho_0)}$. Accordingly, the numerical value in eqn (6.9) is reduced by a factor $\sqrt{\kappa}$, i.e. by about 20%, and is 500.

This formula is relatively reliable if the mass layer consists of a perforated panel, as in Fig. 6.4b, or a flexible membrane. Then the way in which the mass layer is fixed has no influence on its acoustical effect, at least as long as the sound waves arrive frontally. Matters are different for non-perforated wall linings made of panels with noticeable bending stiffness. Since such panels must be fixed in some way, for instance on battens which are mounted on the wall (Fig. 6.4a), their vibrations are controlled not only by the air cushion behind but also by their bending stiffness. Accordingly, the resonance frequency will be higher than that given by eqn (6.9), namely

$$f_0' = \sqrt{(f_0'^2 + f_1'^2)} \qquad (6.10)$$

with $f_1$ denoting the lowest bending resonance frequency of a panel supported (not clamped) at two opposite sides. The typical range of $f_1$ is 10–30 Hz, whereas $f_0$ is typically 50–100 Hz. This shows that the influence of the bending stiffness on the resonance frequency can be neglected in most practical cases, and eqn (6.9) may be applied to give at least a clue to the actual resonance frequency.

Another consequence of the strong lateral coupling between adjacent elements of an unperforated panel is that the angle dependence of sound absorption is more complicated than that in eqn (2.17) (with frequency-independent $\zeta$). In fact, a sound wave with oblique incidence excites a forced bending wave in the panel, even if it is of infinite extension. The propagation of this wave is strongly affected by the supporting construction and the air layer behind the panel. Since general statements on the way it influences the absorption coefficient and its dependence on the incidence angle are not possible, we shall not discuss this point in any further detail.

For resonators with perforated panels, lateral coupling of surface elements is affected by lateral sound propagation in the air space behind the panels. It can be hindered by lateral partitions made of rigid material, or by filling the air space with porous and hence sound-absorbent materials like glass or mineral wool. Neighbouring elements of the panel can then be regarded as independent; the wall impedance and similarly the resonance frequency is independent of the direction of sound incidence. In any case it is difficult to assess correctly the losses of a resonance absorber which determine its absorption coefficient. Therefore, the acoustical consultant must rely on his experience or on a good collection of typical absorption data. In cases of doubt it may be advisable to measure the absorption coefficient of a wall lining by putting a sufficiently large sample of it into a reverberation chamber (see Section 8.7).

Resonance absorbers of the described type are typically mid-frequency or low-frequency absorbers. Their practical importance stems from the possibility of choosing their significant data (dimensions, materials) from a wide range so as to give them the desired absorption characteristics. By a suitable combination of several types of resonance absorbers in a room the acoustical designer is able to achieve a prescribed frequency dependence of the reverberation time. The most common application of vibrating panels is to effect a low-frequency balance for the strong absorption of the audience at medium and high frequencies, and thus to equalise the reverberation time. This is the reason for the generally favourable acoustical conditions which are frequently met in halls whose walls are lined with wooden panels or are equipped with similar components, i.e. walls or suspended decoration ceilings made of thin plaster. Thus it is not, as is sometimes believed by laymen, a sort of 'amplification' caused by 'resonance' which is responsible for the good acoustics of many concert halls lined with wooden panels. Likewise, audible decay processes of the wall linings, which are sometimes also believed to be responsible for good acoustics, do not occur in practical situations although they might be possible in principle. If a resonance system with the relative half-width (reciprocal of the $Q$-factor) $\Delta\omega/\omega_0$ is excited by an impulsive signal, its amplitude will decay with a damping constant $\delta = \Delta\omega/2$ according to eqn (2.31); thus the 'reverberation time' of the resonator is

$$T' = \frac{13.8}{\Delta\omega} = \frac{2.2}{\Delta f} \tag{6.11}$$

To be comparable with the reverberation time of a room which is of the order of magnitude of 1 s, the frequency half-width $\Delta f$ must be about 2 Hz. However, the half-width of a resonating wall lining is larger by several orders of magnitude.

In Fig. 6.5, the absorption coefficients of a wooden wall lining and of a resonance absorber with perforated panels are plotted as functions of the frequency, measured at omnidirectional sound incidence.

## 6.5 Helmholtz resonators

Sometimes sound-absorbent elements are not distributed so as to cover uniformly the wall or the ceiling of a room but instead they are single or separate objects arranged either on a wall or in free space. Examples of this are chairs, small wall openings or lamps; musical instruments too can absorb sound. To sound absorbers of this sort we cannot attribute an absorption coefficient, since the latter always refers to a uniform surface. Instead their absorbing power is characterised by their 'absorption cross-section' or their equivalent absorption area, which is defined as the ratio of sound energy being absorbed per second by them and the intensity which the incident sound wave would have at the place of the absorbent object

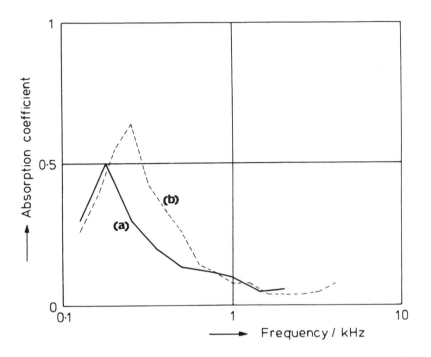

*Figure 6.5* Absorption coefficient of resonance absorbers at random sound incidence (measured in a reverberation chamber, see Section 8.7): (a) wooden panel, 8 mm thick, $M' = 5$ kg/m$^2$, 30 mm away from rigid wall, with 20 mm rock wool behind; (b) panels, 9.5 mm thick, perforated at 1.6% (diameter of holes 6 mm), 5 cm distant from rigid wall, air space filled with glass wool.

if it were not present:

$$A = \frac{P_{\text{abs}}}{I_0} \tag{6.12}$$

This definition is similar to that of the scattering cross-section in eqn (2.46).

When calculating the reverberation time of a room, absorption of these types of absorbers is taken into account by adding their absorption areas $A_i$ to the sum $\Sigma \alpha_i S_i$ (see eqn (5.24a)). The same holds for all other formulae and calculations in which the total absorption or the mean absorption coefficient $\bar{\alpha}$ of a room appears, as for instance for the calculation of the steady-state energy density as described in Section 5.5.

In this section we are discussing discrete sound absorbers with pronounced resonant behaviour. Their characteristic feature is an air volume which is enclosed in a rigidly walled cavity and which is coupled to the surrounding space by an aperture, as shown in Fig. 6.6a. The latter may equally well be a channel or a 'neck'. The whole structure is assumed to be small compared with the wavelength of sound and thus it has one single resonance in the interesting frequency range. It is brought about by interaction of two elements: the air contained in the neck or in the aperture which acts as a mass load, and the air within the cavity which can be regarded as a spring counteracting the motion of the air in the neck. Arrangements of this type are called 'Helmholtz resonators'; examples of these are all kinds of bottles, vases and similar vessels. In ancient times, as 'Vitruv's sound vessels', they played an unknown, possibly only a surmised, acoustical role in antique theatres and other spaces.

Figure 6.6 depicts a Helmholtz resonator along with its schematic presentation. The neck has a length $l$ and a cross-sectional area $S$; its opening is flush with the surface of a rigid wall of infinite extension. The basic parameters of the resonator are the mass $M = \rho_0 l s$ of the air enclosed in the neck, and the volume $V$ of the attached cavity. The shape of the latter is of no relevance. Furthermore, there are some internal losses represented by a resistance $R_0$,

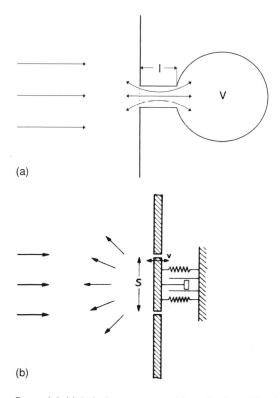

(a)

(b)

*Figure 6.6* Helmholtz resonator: (a) realisation; (b) schematic.

and the opening is loaded by its radiation impedance $R_r$ defined by

$$\overline{P_r} = R_r \overline{v^2}$$

where—as usual—the bars indicate time averaging. ($P_r$ = radiated power, $v$ = air velocity in the neck). Since the lateral dimensions of the opening are small compared with the wavelength, $P_r$ can be substituted from eqn (1.31) using $\hat{Q}^2 = (Sv)^2 = 2S^2\overline{v^2}$. However, we must keep in mind that the re-radiation of sound from the aperture is restricted to the half space only and hence is a factor of 2 higher than in eqn (1.31). Thus the radiation resistance is

$$R_r = \frac{\rho_0 \omega^2 S^2}{2\pi c} = 2\pi\rho_0 c \left(\frac{S}{\lambda}\right)^2 \tag{6.13}$$

Suppose the resonator is working at its resonance frequency $\omega_0$. In this case all the reactive parts of the mechanical load will mutually cancel each other. Hence, the ratio of the total force $F$ acting on the piston and its velocity $v$ is real:

$$\frac{F}{v} = R_0 + R_r \tag{6.14}$$

The energy converted to heat per second by the internal friction is

$$P_{abs} = R_0 \overline{v^2} = \frac{R_0}{(R_0 + R_r)^2} \overline{F^2} \tag{6.15}$$

For a given radiation resistance, $P_{abs}$ is maximum if $R_0$ is equal to $R_r$. The force exerted externally on the piston arises from the sound field. If $p$ denotes the sound pressure, the force is $F = 2pS$. The factor 2 takes into account the reflection from the rigid wall surrounding the piston. By application of eqn (1.28) we can express the sound pressure and hence the force in terms of the intensity $I$ of the incident sound wave:

$$\overline{F^2} = 4S^2\overline{p^2} = 4\rho_0 cS^2 I \tag{6.16}$$

Now we are ready to evaluate eqns (6.12) and (6.15) with $R_0 = R_r$ by substituting from eqns (6.13) and (6.16), and we obtain as a final result

$$P_{abs} = \frac{\lambda_0^2}{2\pi} I \quad \text{and} \quad A_{max} = \frac{\lambda_0^2}{2\pi} \tag{6.17}$$

where $\lambda_0$ is the wavelength corresponding to the resonance frequency.

For the frequency dependence of the absorption area, we can essentially adopt eqn (2.32):

$$A(\omega) = \frac{A_{\max}}{1 + Q_A^2 \left(\omega/\omega_0 - \omega_0/\omega\right)^2} \qquad (6.18)$$

with

$$Q_A = \frac{M\omega_0}{2R_r} \qquad (6.18a)$$

The (angular) resonance frequency of the system is given by the general formula

$$\omega_0^2 = \frac{s}{M}$$

Here $s$ is the elastic stiffness of the air enclosed in the resonator volume. To calculate it we suppose the air in the neck to be displaced by $\delta x$; the corresponding change in air pressure is $p_i$. According to the usual definition of the stiffness,

$$s = \frac{\text{force}}{\text{displacement}} = -\frac{p_i S}{\delta x}$$

On the other hand, the pressure change is associated with a change $\delta\rho$ in air density, which in turn is due to the volume change $S\delta x$. These quantities are related to each other by (see eqns (1.4) and (1.6))

$$\frac{S\delta x}{V} = -\frac{\delta\rho}{\rho_0} = -\frac{p_i}{\rho_0 c^2}$$

Thus we obtain for the stiffness of the air cushion:

$$s = \frac{\rho_0 c^2 S^2}{V}$$

Now we can insert $M = s/\omega_0^2 = \rho_0 c^2 S^2/\omega_0^2 V$ into eqn (6.18a) and express $R_r$ by eqn (6.13) with the result

$$Q_A = \frac{\pi}{V}\left(\frac{c}{\omega_0}\right)^3 = \frac{\pi}{V}\left(\frac{\lambda_0}{2\pi}\right)^3 \qquad (6.19)$$

We observe that the maximum absorption area of a resonator matched to the sound field is fairly large, according to eqn (6.17). On the other hand, the $Q$-factor given by eqn (6.19) is very large too, which means that the relative frequency half-width, which is the reciprocal of the $Q$-factor, is very small, i.e. large absorption will occur only in a very narrow frequency range. This is clearly illustrated by the following numerical example: suppose a resonator is tuned to a frequency of 100 Hz corresponding to a wavelength of 3.43 m. This can be achieved conveniently by a resonator volume of 1 litre. If it is matched, the resonator has a maximum absorption area, which is as large as 1.87 m². Its $Q$-factor is—according to eqn (6.19)—about 500; the relative half-width is thus 0.002. This means, it is only in the range from 99.9 to 100.1 Hz that the absorption area of the resonator exceeds half its maximum value. Therefore the very high absorption in the resonance is paid for by the exceedingly narrow frequency bandwidth. This is why the application of such weakly damped resonators does not seem too useful. It is more promising to increase the losses and hence the bandwidth at the expense of maximum absorption.

Finally, we investigate the problem of audible decay processes, which we have already touched on in the preceding section. The 'reverberation time' of the resonator can again be calculated by the relation (6.11)

$$T' = \frac{13.8}{\Delta\omega} = 13.8\frac{Q_A}{\omega_0}$$

In many cases this time cannot be ignored when considering the reverberation time of a room. What about the perceptibility of the decay process?

It is evident from the derivation of the absorption area, eqn (6.17), that the same amount of energy per second which is converted to heat in the interior of the resonator is being reemitted by it, since we have assumed $R_0 = R_r$. Its maximum radiation power is thus $I(\lambda_0^2/2\pi)$, $I$ being the intensity of the incident sound waves. Thus the intensity of the re-radiated sound at distance $r$ is

$$I_s = \left(\frac{\lambda_0}{2\pi r}\right)^2 I = \left(\frac{c}{r\omega_0}\right)^2 I \tag{6.20}$$

Both intensities are equal at a distance

$$r = \frac{c}{\omega_0} = \frac{55}{f_0} \quad \text{metres} \tag{6.21}$$

The decay process of the resonator is therefore only audible in its immediate vicinity. In the example mentioned above this critical distance would be 0.55 m; at substantially larger distances the decay cannot be heard.

## 6.6 Sound absorption by porous materials

Nearly all practically used sound absorbers contain some porous material which is exposed in one or another way to the arriving sound wave. One example was already mentioned in Section 6.4. This section presents a more detailed discussion of the dissipation process taking place in such materials.

In Section 6.2, inevitable reflection losses of sound waves impinging on smooth surfaces were discussed; these are caused by viscous and thermal processes and occur within a boundary layer next to the surface, produced by the sound field. The thickness of this layer is typically in the range of 0.01 to 0.2 millimetres, depending on the frequency.

The absorption due to these effects is negligibly small if the surface is smooth (see Fig. 6.7a). It is larger, however, at rough surfaces since the roughness increases the volume of the zone in which the losses occur (Fig. 6.7b). And it is even more pronounced if the material contains pores, channels and voids connected with the air outside. A material of this kind is drawn schematically in Fig. 6.7c. Then the pressure fluctuations associated with the external sound field give rise to alternating air flows in the pores and channels which will be filled more or less by the lossy boundary layer, so to speak. The consequence is that a significant amount of mechanical energy is withdrawn from the external sound field and is converted into heat.

This mechanism of sound absorption concerns all porous materials with pores accessible from the outside. Examples are the 'porous layers' which the reader encountered in Section 2.4 and which can be thought of as woven fabrics or thin carpets. The standard materials for reverberation control in rooms acoustics are mineral wool and glass wool. They are manufactured from anorganic fibres by pressing them together, often with the addition of suitable binding agents. They are commercially available in the form of

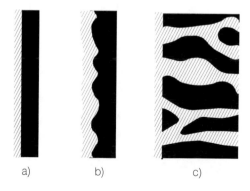

a)       b)       c)

*Figure 6.7* Lossy boundary layer: (a) in front of a smooth surface; (b) in front of a rough surface; (c) in front of and within a porous material.

plates or loose blankets and also find wide application in sound and heat insulation. Other common materials are porous plaster, or foams of certain polymers. The latter should have open cells, otherwise their absorption is rather low. It should be mentioned that a brick wall may show noticeable absorption as well.

The characteristic parameters of a porous material are mainly the porosity $\sigma$ and the specific flow resistance $\Xi$. The porosity is the fraction of volume which is not occupied by the solid structure. It can be determined by immersing a sample in a suitable liquid. A more reliable way is by comparing the isothermal compressibility of air contained in a solid chamber before and after inserting a sample of the material into it. For most materials of practical interest the porosity exceeds 0.5 and may come close to unity.

The definition of the specific flow resistance is similar to that of the flow resistance of a porous layer in eqn (2.18). Suppose a constant air flow with velocity $v_s$ is forced through a sample with the thickness or length $d$. Then the specific flow resistance is

$$\Xi = \frac{p - p'}{v_s d} \tag{6.22}$$

Where $p$ and $p'$ are the air pressures on both sides of the sample. The unit of specific flow resistance is $1 \text{ kg m}^{-3} \text{ s}^{-1} = 1 \text{ Pa s/m}^2 = 10^{-3} \text{ Rayl/cm}$. It may be more practical to express the specific flow resistance in relation to the characteristic impedance, i.e. $\Xi/\rho_0 c$. For most porous material this quantity lies between 10 and 200 $\text{m}^{-1}$.

This is certainly not the place to present a comprehensive and exact description of sound absorption in porous materials which is applicable to the full variety of such materials. In order to understand the basic processes it is sufficient to restrict the discussion to an idealised model of a porous material, the so-called Rayleigh model, which qualitatively exhibits the essential features. Furthermore, we shall see the viscous processes in the foreground, i.e. we neglect the less prominent effects which are due to heat conduction.

The Rayleigh model consists of a great number of similar, equally spaced and parallel channels which traverse a skeleton material considered to be completely rigid (see Fig. 6.8). It is supposed that the surface of that system, being located at $x = 0$, is perpendicular to the axes of the channels; in the positive $x$-direction the model is assumed unbounded. (A practical realisation of the Rayleigh model is the microperforated absorber mentioned in Section 6.4.)

First we consider the sound propagation in a single channel. We suppose that it is so narrow that the profile of the air stream is determined almost completely by the viscosity of the air and not by any inertial forces. This is always the case at sufficiently low frequencies. Then the same lateral distribution of

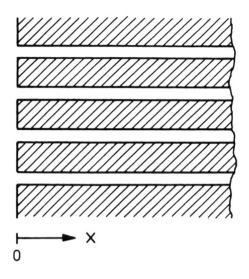

X

0

*Figure 6.8* Rayleigh model.

flow velocities prevails in the interior of the channel as for constant (i.e. for non-alternating) flows; and likewise the flow resistance of the channel per unit length has virtually the same value as for constant flow velocity. From eqn (6.22) it follows that this quantity is

$$\Xi' = -\frac{1}{\bar{v}} \frac{\partial p}{\partial x} \tag{6.23}$$

(In this and the following formulae $\bar{v}$ is the flow velocity averaged over the cross-section of the channel.) For a flow between parallel planes at distance $b$, $\Xi'$ is given by

$$\Xi' = 12 \frac{\eta}{b^2} \tag{6.24a}$$

whereas for narrow channels with circular cross-section (radius $a$)

$$\Xi' = 8 \frac{\eta}{a^2} \tag{6.24b}$$

In these formulae $\eta$ denotes the viscosity of the streaming medium. For air under normal conditions $\eta = 1.8 \times 10^{-5}$ kg m$^{-1}$s$^{-1}$.

In order to study the sound propagation within one channel, consider a length element d$x$ of it (see Fig. 6.9). The net force acting on the air in it in a positive $x$-direction is $-(\partial p/\partial x)\,\mathrm{d}x\,\mathrm{d}S$, with d$S$ denoting the cross-sectional area of the channel. It is kept at equilibrium by an inertial

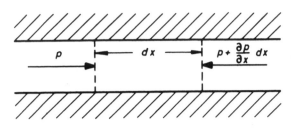

*Figure 6.9* Forces acting on a small length element in a rigid channel.

force $\rho_0 (\partial \bar{v}/\partial t)\,dx\,dS$ and a frictional force $\Xi'\bar{v}\,dx\,dS$, according to eqn (6.23). Thus the force equation which is analogous to the one-dimensional version of eqn (1.2) reads

$$-\frac{\partial p}{\partial x} = \rho_0 \frac{\partial \bar{v}}{\partial t} + \Xi' \bar{v} \tag{6.25}$$

This relation assumes that the flow velocity (or the particle velocity) is constant over the whole cross-section and equal to $\bar{v}$. This is of course not true. In reality a certain velocity distribution develops across and in the channel. However, a more thorough investigation shows that the representation (6.25) can be applied with sufficient accuracy as long as

$$\frac{\rho_0 \omega}{\Xi'} \leq 4 \tag{6.26}$$

The equation of continuity (1.3) is not affected by the viscosity; thus in one-dimensional formulation

$$\rho_0 \frac{\partial \bar{v}}{\partial x} = -\frac{\partial \rho}{\partial t} = -\frac{1}{c^2}\frac{\partial p}{\partial t} \tag{6.27}$$

Into eqns (6.25) and (6.27) we insert the following plane wave relations:

$$p = \hat{p}\exp\left[i\left(\omega t - k'x\right)\right], \qquad \bar{v} = \hat{\bar{v}}\exp\left[i\left(\omega t - k'x\right)\right]$$

and obtain two homogeneous equations for $p$ and $\bar{v}$:

$$k'p - \left(\omega\rho_0 - i\Xi'\right)\bar{v} = 0, \qquad \omega p - \rho_0 c^2 k'\bar{v} = 0 \tag{6.28}$$

Setting the determinant, formed of the coefficients of $p$ and $\bar{v}$, equal to zero yields the complex propagation constant

$$k' = \beta' - i\gamma' = \frac{\omega}{c}\left(1 - \frac{i\Xi'}{\rho_0 \omega}\right)^{1/2} \tag{6.29}$$

with the phase constant $\beta' = 2\pi/\lambda'$ and the attenuation constant $\gamma' = m'/2$ (compare eqn (1.17)).

If we insert this result into one of eqns (6.28), we obtain the ratio of sound pressure to velocity, i.e. the characteristic impedance in the channel:

$$Z_0' = \frac{p}{v} = \rho_0 c \left( 1 - \frac{i\Xi'}{\rho_0 \omega} \right)^{1/2} \tag{6.30}$$

For very high frequencies—or for very wide channels—these expressions approach the values $\omega/c$ and $\rho_0 c$, valid for free sound propagation, because then the viscous boundary layer occupies only a very small fraction of the cross-section. In contrast to this, at very low frequencies eqn (6.29) yields

$$\beta' = \gamma' \approx \left( \frac{\omega \Xi'}{2\rho_0 c^2} \right)^{1/2} \tag{6.31}$$

since $\sqrt{i} = (1+i)/\sqrt{2}$. The attenuation in this range is thus considerable: a wave is reduced in its amplitude by $20 \cdot \log_{10}[\exp(2\pi)] - 54.6$ dB per wavelength.

From the characteristic impedance $Z_0'$ inside the channels we pass to the 'average characteristic impedance' $Z_0$ of a porous material by use of the porosity $\sigma$, as already introduced in eqn (6.5), which is the ratio of the cross-sectional area of one channel and the surface area per channel:

$$Z_0 = \frac{Z_0'}{\sigma} \tag{6.32a}$$

Likewise, the 'outer flow resistance' $\Xi$, which can be measured directly by forcing air through a test sample of the material (see eqn (6.22)), is related to $\Xi'$ by

$$\Xi_0 = \frac{\Xi_0'}{\sigma} \tag{6.32b}$$

For highly porous materials such as rock wool or mineral wool the porosity is close to unity, and the specific flow resistance $\Xi$ is mostly in the range from 5000 to $10^5$ Pa s/m$^2$.

Now we apply the above relations to a typical arrangement consisting of a homogeneous layer of porous material mounted directly against a rigid wall. The thickness of the layer is $d$ (see Fig. 6.10). A sound wave arriving at the surface of the layer will be partially reflected from it; the remaining part of the sound energy will penetrate into the material and again reach the surface after its reflection from the rigid rear wall. Then it will again split up into one portion penetrating the surface and another one return-ing to the rear wall, and so on. This qualitative consideration shows that

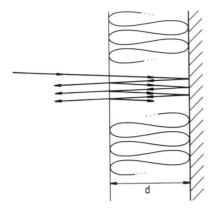

*Figure 6.10* Porous layer in front of a rigid wall.

the reflected sound wave can be thought of as being made up of an infinite number of successive contributions, each of them weaker than the preceding one because of the attenuation of the interior wave. Furthermore, it shows that the total reflection factor and hence the absorption coefficient may show maxima and minima, depending on whether the various components interfere constructively or destructively at a given frequency.

For a quantitative treatment which will be restricted, however, to normal sound incidence we refer to eqn (2.21), according to which the 'wall impedance' of an air layer of thickness $d$ in front of a rigid wall is $-i\rho_0 c \cdot \cot(kd)$. By replacing $\rho_0 c$ with $Z_0$ and $k$ with $k'$, we obtain for the wall impedance of the porous layer

$$Z_1 = -iZ_0 \cot(k'd) \tag{6.33}$$

From this expression the reflection factor and the absorption coefficient can be calculated using eqns (2.7) and (2.8) with $Z = Z_1$.

For the following discussion it is useful to separate the real and the imaginary part of the cotangent:

$$\cot(k'd) = \frac{\sin(2\beta'd) + i\sinh(2\gamma'd)}{\cosh(2\gamma'd) - \cos(2\beta'd)} \tag{6.34}$$

Then one can draw the following qualitative conclusions:

(a) For a layer which is thin compared with the sound wavelength, i.e. for $k'd \ll 1$, $\cot(k'd)$ can be replaced with $1/k'd$. Hence the surface of the layer has a very large impedance and accordingly low absorption.

In other words, substantial sound absorption cannot be achieved by just applying some kind of paint to a wall.

(b) If the sound waves inside the porous material undergo strong attenuation during one round trip, i.e. if $\gamma'd \gg 1$, the cotangent becomes i, according to eqn (6.34), and the wall impedance is the same as that of an infinitely thick layer:

$$Z_1 = \frac{Z'_0}{\sigma} = \frac{\rho_0 c}{\sigma}\left(1 - \frac{i\sigma\,\Xi}{\rho_0\omega}\right)^{1/2} \tag{6.35}$$

Figure 6.11a shows the absorption coefficient calculated from eqn (6.35). For high frequencies it approaches asymptotically the value

$$\alpha_\infty = \frac{4\sigma}{(1+\sigma)^2} \tag{6.36}$$

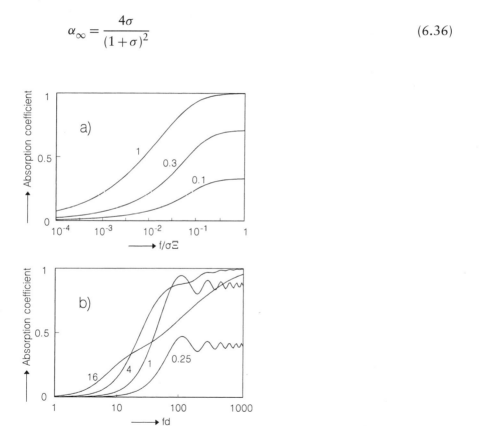

Figure 6.11 Absorption coefficient of a porous layer (Rayleigh model) in front of a rigid wall, normal sound incidence: (a) infinite thickness (frequency in Hz, $\Xi'$ in Pa s/m²); (b) finite thickness, $\sigma = 0.95$ (fd in Hz m). The parameter is $\Xi'd/\rho_0 c$. For $\Xi'd/\rho_0 c = 0.25$, the condition (6.26) is only partially fulfilled.

(c) If, on the contrary, $\gamma'd \ll 1$, i.e. if the attenuation inside the material is small, the periodicity of the trigonometric functions in eqn (6.34) will dominate in the frequency dependence of the cotangent, and the same holds for that of the absorption coefficient represented in Fig. 6.11b. The latter has a maximum whenever $\beta'd$ equals $\pi/2$ or an odd multiple of it, i.e. at all frequencies for which the thickness $d$ of the layer is an odd integer of $\lambda'/4$, with $\lambda' = 2\pi/\beta'$ denoting the wavelength inside the material. For higher values of $\Xi'd$ the fluctuations fade out. As a figure-of-merit we can consider $(fd)_{0.5}$, i.e. the value of the product $fd$ for which the absorption coefficient equals 0.5. Its minimum value is $(fd)_{0.5} = 23$, which is achieved for $\Xi'd/\rho_0 c = 6$.

If a second porous layer with different acoustical properties is arranged in front of the one we considered so far, the wall impedance $Z_2$ of the combination is calculated by means of eqn (2.11) in which $Z_r$ must be replaced with $Z_1$ after eqn (6.33):

$$Z_2 = Z_{01} \frac{Z_1 + iZ_{01} \tan(k''d')}{iZ_1 \tan(k''d') + Z_{01}} \tag{6.37}$$

where $d'$ is the thickness of the additional layer. As before, the complex propagation constant $k''$ and the characteristic impedance $Z_{01}$ of the second material is calculated from eqns (6.29), (6.30) and (6.32a) using its specific flow resistance $\Xi''$ and porosity $\sigma'$. The absorption coefficient of the combination is obtained from eqn (2.8) with $Z = Z_2$.

As is evident from Fig 6.11b, a homogeneous porous layer is basically a high-frequency absorber. One way to increase the absorption coefficient in the low-frequency range is to increase the thickness $d$ of the layer. Another way is to abandon the concept of a homogeneous absorber or, more specifically, to provide for some air space between the rear side of the absorber and the rigid backing. For frontal sound incidence, the absorption coefficient of this configuration can be calculated with eqn (6.37) by setting $Z_1 = -i\rho_0 c \cot(kd)$ with $k = \omega/c$. Qualitatively, the effect of the air space can be explained using Fig. 6.12. When the porous layer is fixed immediately on the wall (Fig. 6.12a) it is close to a zero of the particle velocity (dashed line) and almost no air is forced through the pores. The air backing (Fig. 6.12b) shifts the active layer to a position where the particle velocity may be larger, causing some air motion within the pores. In the limiting case of a very thin sheet we arrive at the stretched fabric, which we have already dealt with in Section 2.4.

We want to emphasise here that the Rayleigh model, even at normal sound incidence, is only useful for a qualitative understanding of the effects in a porous material but not for a quantitative calculation of the acoustical properties of real absorbent materials. The assumed rigidity of its skeleton is a

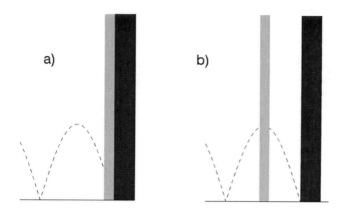

*Figure 6.12* Porous layer in front of a rigid wall: (a) without air backing; (b) with air backing. (Dashed line: particle velocity without layer.)

simplification which is not entirely justified in practice. Furthermore, these materials do not contain well-separated and distinguishable channels but rather irregularly shaped cavities which are mutually connected. For this reason the surface of a porous layer is not locally reacting to the sound field except at very low frequencies where the attenuation within the material is high enough to prevent lateral coupling. In addition, the pores or channels are often so narrow that there must surely be some heat exchange between the air contained in the channels and the walls. Therefore the changes of state of the air occur neither adiabatically nor according to an isothermal law but somehow in between, which causes an additional complication. In the past attempts have been made to take these effects, which are not covered by the simple Rayleigh model, into account, for instance by introducing a 'structure factor' which can, however, only be evaluated experimentally. From a practical point of view it therefore seems more advantageous to omit a rigorous treatment and to determine the absorption coefficient by measurement. This procedure is recommended all the more because the performance of porous absorbers depends only partially on the properties of the material and to a greater extent on its arrangement, on the covering and on other constructional details, which vary substantially from one situation to another.

When porous absorbent materials are used for reverberation control in a room, it will usually be necessary to cover these materials in some way on the side exposed to the room. Many of these materials will in the course of time shed small particles which must be prevented from polluting the air in the room. If the absorbent portions of wall are within the reach of people, a suitable covering is desirable too as a protection against unintentional thoughtless damage of the materials which are not usually very hard wearing. And, finally, the architect usually wishes to hide the rock wool layer, which is not aesthetically pleasing, behind a surface which can be treated according to his wishes. Very often, thin and highly perforated or slotted panels of wood, metal or gypsum board are employed for this purpose. Their surfaces can be painted and cleaned from time to time; however, care must be taken that the openings are not blocked.

To prevent purling (or to keep water away from the pores, as for instance in swimming baths) it is sufficient to bag the absorbent materials in very thin plastic foils. Furthermore, purling can be avoided by a somewhat denser porous front layer on the bulk of the material.

According to Section 6.3, foils, as well as perforated or slotted panels, have a certain sound transmissibility which may be close to unity at low frequencies and which is identical to the absorption coefficient in eqn (6.3). It would be wrong, however, to calculate the absorption of the combination of porous material plus covering just by multiplying the latter absorption coefficient with that of the uncovered layer. Instead one has to add the wall impedance $i\omega M'$ of the perforated panel or of the foil to the impedance of the porous layer. The resulting impedance can be better matched in certain frequency ranges to the characteristic impedance of the air, in which case the absorption is increased by the covering. This will occur if the wall impedance of the uncovered arrangement has a negative imaginary part, i.e. if the distance between the surface of the porous layer and the rigid wall behind it is less than a quarter wavelength. The added mass reduces this imaginary part and this results in a higher absorption coefficient according to Fig. 2.2: in other words, the absorbent layer has been changed into a resonance absorber, as discussed in Section 6.4.

We close this section by presenting a few typical results as measured in the reverberation chamber, i.e. at random sound incidence. This method will be described in more detail in Section 8.7. It is very useful, but has the peculiarity that the results it yields for highly absorptive test samples are occasionally in excess of unity, although absorption coefficients beyond 1 are physically meaningless. Reasons for this strange behaviour will be discussed in Section 8.7.

In Fig. 6.13 the absorption coefficient of two homogeneous rock wool layers, which are 50 mm thick, is shown as a function of the frequency. Both materials differ in their densities and hence in their flow resistances. Obviously the denser material exhibits an absorption coefficient close to 1 even at lower frequencies.

Figure 6.14 shows the absorption coefficient of a porous sheet with 30 mm thickness and a density of 46.5 kg/m³ which is mounted directly in front of a rigid wall in the first case (solid curve) in the other case there is an air space of 50 mm between the sheet and the wall (dotted curve). The latter is partitioned off by wooden lattices with a pattern of 50 cm × 50 cm.

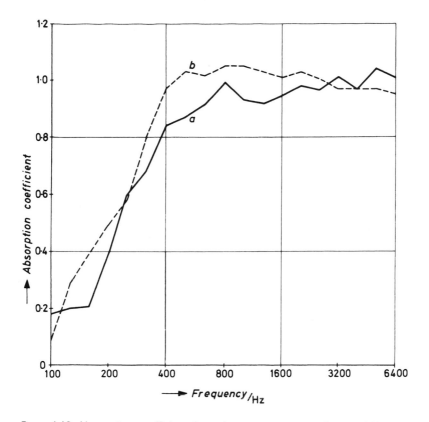

*Figure 6.13* Absorption coefficient (reverberation room, see Section 8.7) of rock wool layers, 50 mm thick, immediately on concrete: (a) density 40 kg/m$^3$, $\Xi =$ 12.7 × 10$^3$ Pa s/m$^2$; (b) density 100 kg/m$^3$, $\Xi = 22 \times 10^3$ Pa s/m$^2$.

Evidently, the air space achieves a significant shift of the steep absorption increase towards lower frequencies. Thus an air space behind the absorbent material considerably improves its effectiveness or helps to save material and costs. The partitioning is to prevent lateral sound propagation within the air space at oblique sound incidence.

The influence of a perforated covering is demonstrated in Fig. 6.15. In both cases the porous layer has a thickness of 50 mm and is mounted directly onto the wall. The fraction of hole areas, i.e. the perforation, is 14%. The mass load corresponding to it is responsible for an absorption maximum at 800 Hz which is not present for the bare material. This resonance absorption can sometimes be very desirable. At a higher degree of perforation this influence is much less pronounced, and with a perforation of 25% or more the effect of the covering plate can virtually be neglected.

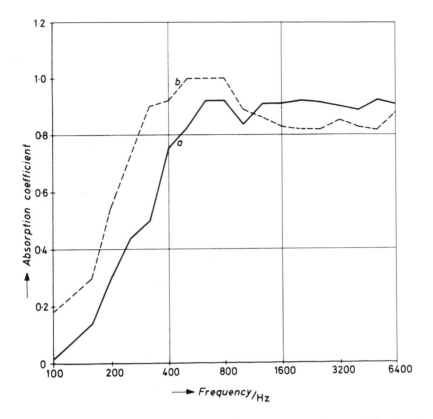

*Figure 6.14* Absorption coefficient (reverberation room, see Section 8.7) of a rock wool layer 30 mm thick, density 46.5 kg/m³, $\Xi = 12 \times 10^3$ Pa s/m²; (a) mounted immediately on concrete; (b) mounted with 50 mm air backing, laterally partitioned.

## 6.7 Audience and seat absorption

The purpose of most medium- to large-size halls is to accommodate a large number of spectators or listeners and thus enable them to watch events or functions of common interest. This is true for concert halls and lecture rooms, for theatres and opera houses, for churches and sports halls, cinemas, council chambers and entertainment halls of every kind. The important acoustical properties are therefore those which are present when the rooms are occupied or at least partially occupied. These properties are largely determined by the audience itself, especially by the sound absorption effected by the people. The only exceptions are broadcasting and television studios, which are not intended to be used with an audience present, and certain acoustical measuring rooms.

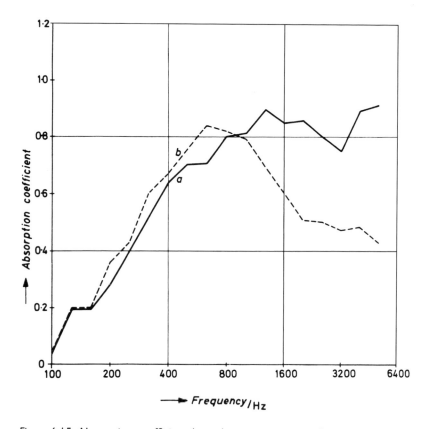

*Figure 6.15* Absorption coefficient (reverberation room, see Section 8.7) of 50 mm glass wool, mounted immediately on concrete: (a) uncovered; (b) covered with a protective panel, 5 mm thick, 14% perforation.

The absorption effected by an audience is due mainly to people's clothing and its porosity. Since clothing is not usually very thick, the absorption is considerable only at medium and high frequencies; in the range of low frequencies it is relatively small. Since people's clothing differs from individual to individual, only average values of the audience absorption are available and it is quite possible that these values are changing with the passage of time according to changing fashion or season. Furthermore, audience absorption depends on the kind of seats and their arrangement, on the occupancy density, on the way in which the audience is exposed to the incident sound, on the interruption of 'blocks' by aisles, stairs, etc., and not least on the structure of the sound field. It is quite evident that a person seated at the rear of a narrow box with a small opening, as was typical in 18th- to 19th-century theatres, absorbs much less sound energy than a person sitting among steeply raked rows of seats and who is thus well exposed to the sound. Therefore it is not surprising that there are considerable differences in the data on audience absorption which have been given by different authors.

There are two ways to determine experimentally the sound absorption of audience and seats. One is to place seats and/or persons into a reverberation chamber and evaluate their absorption from the change in reverberation time they cause. This has the advantage that sound field diffusion, which is a prerequisite for the applicability of the common reverberation formulae, can be established by adequate means. On the other hand, it may be doubtful whether a 'block' consisting 20 or 25 seats is representative for an extended area covered with occupied or unoccupied seats. In the second method, completed concert halls are used as reverberation chambers, so to speak: the absorption data are derived from reverberation times measured in them. The structure of the sound field is unknown in the second method, but—provided the shape of the hall is not too unusual—it can at least be considered as typical for such halls.

The absorption of persons standing or seated singly is characterised most appropriately by their absorption cross-section or equivalent absorption area $A$, already defined in eqn (6.12) in Section 6.5. The absorption area of each person is added to the sum of eqn (5.24a), which in this case reads

$$\bar{\alpha} = \frac{1}{S}\left(\sum_i S_i\alpha_i + N_pA\right) \tag{6.38}$$

$N_p$ being the number of persons.

Table 6.2 lists some absorption areas as a function of the frequency measured by Kath and Kuhl[3] in a reverberation chamber.

When people are seated close together (which is usual in occupied rooms) it seems to be more correct[4] to relate the absorption of audience and areas covered with equal chairs, not to the number of 'objects' but to their area, i.e. to characterise it by an apparent absorption coefficient, since this figure seems to be less dependent on the density of chairs or listeners. In Table 6.3 absorption coefficients of seated audience and of unoccupied seats, as measured in a reverberation chamber, are listed. Although the absorption of an audience

Table 6.2 Equivalent absorption area of persons, in $m^2$ (after Kath and Kuhl[3])

| Kind of person | Frequency (Hz) | | | | | |
|---|---|---|---|---|---|---|
| | 125 | 250 | 500 | 1000 | 2000 | 4000 |
| Male standing in heavy coat | 0.17 | 0.41 | 0.91 | 1.30 | 1.43 | 1.47 |
| Male standing without coat | 0.12 | 0.24 | 0.59 | 0.98 | 1.13 | 1.12 |
| Musician seated, with instrument | 0.60 | 0.95 | 1.06 | 1.08 | 1.08 | 1.08 |

*Table 6.3* Absorption coefficients of audience and chairs (reverberation chamber data)

| Type of seats | Frequency (Hz) | | | | | | |
|---|---|---|---|---|---|---|---|
| | 125 | 250 | 500 | 1000 | 2000 | 4000 | 6000 |
| Audience seated on wooden chairs, two persons per m² | 0.24 | 0.40 | 0.78 | 0.98 | 0.96 | 0.87 | 0.80 |
| Audience seated on wooden chairs, one person per m² | 0.16 | 0.24 | 0.56 | 0.69 | 0.81 | 0.78 | 0.75 |
| Audience seated on moderately upholstered chairs, 0.85 m × 0.63 m | 0.72 | 0.82 | 0.91 | 0.93 | 0.94 | 0.87 | 0.77 |
| Audience seated on moderately upholstered chairs, 0.90 m × 0.55 m | 0.55 | 0.86 | 0.83 | 0.87 | 0.90 | 0.87 | 0.80 |
| Moderately upholstered chairs, unoccupied, 0.90 m × 0.55 m | 0.44 | 0.56 | 0.67 | 0.74 | 0.83 | 0.87 | 0.80 |

may in a particular hall be different from the data shown, the latter at least demonstrate the general features of audience absorption: at increasing frequencies the absorption coefficients increase at first. For frequencies higher than 2000 Hz, however, they decrease. This decrease is presumably due to mutual shadowing of absorbent surface areas by the backrests or listeners' bodies. At high frequencies this effect becomes more prominent, whereas at lower frequencies the sound waves are diffracted around the listeners' heads and shoulders.

The effect of upholstered chairs essentially consists of an increase in absorption at low frequencies, whereas at frequencies of about 1000 Hz and above there is no significant difference between the absorption of audiences seated on upholstered or on unupholstered chairs.

Large collections of data on seat and audience absorption have been published by Beranek[5] and by Beranek and Hidaka,[6] the latter publication being obviously an update of the former one. These authors determined absorption coefficients from the reverberation times of completed concert halls using the Sabine formula, eqn (5.25) with (5.24a). It should be emphasised that their calculations are based upon the 'effective seating area' $S_a$, which includes not only the floor area covered by chairs but also a strip of 0.5 m around the actual area of a block of seating except for the edge of a block when it is adjacent to a wall or a balcony face. This correction is to account for the so-called 'edge effect', i.e. diffraction of sound which generally occurs at the edges of an absorbent area (see Section 8.7). Absorption coefficients for closed blocks of seats—unoccupied as well as occupied—are listed in Table 6.4. The data shown are averages over three groups of halls with different types of seat upholstery. They show the same general frequency dependency as the absorption coefficients listed in Table 6.3, apart from the slight decrease towards very high frequencies for occupied seats which is missing in Table 6.4.

*Table 6.4* Absorption coefficients of unoccupied and occupied seating areas in concert halls (after Beranek and Hidaka[6])

| Type of seats | | Frequency (Hz) | | | | | |
|---|---|---|---|---|---|---|---|
| | | 125 | 250 | 500 | 1000 | 2000 | 4000 |
| Heavily | Unoccupied (seven halls) | 0.70 | 0.76 | 0.81 | 0.84 | 0.84 | 0.81 |
| upholstered | Occupied (seven halls) | 0.72 | 0.80 | 0.86 | 0.89 | 0.90 | 0.90 |
| Medium | Unoccupied (eight halls) | 0.54 | 0.62 | 0.68 | 0.70 | 0.68 | 0.66 |
| upholstered | Occupied (eight halls) | 0.62 | 0.72 | 0.80 | 0.83 | 0.84 | 0.85 |
| Lightly | Unoccupied (four halls) | 0.36 | 0.47 | 0.57 | 0.62 | 0.62 | 0.60 |
| upholstered | Occupied (six halls) | 0.51 | 0.64 | 0.75 | 0.80 | 0.82 | 0.83 |

First group: 7.5 cm upholstery on front side of seat back, 10 cm on top of seat bottom, arm rest upholstered. Second group: 2.5 cm upholstery on front side of seat back, 2.5 cm on top of seat bottom, solid arm rests. Third group: 1.5 cm upholstery on front side of seat back, 2.5 cm on top of seat bottom, solid arm rests.

*Table 6.5* Residual absorption coefficients $a_r$ from concert halls (after Beranek and Hidaka[6])

| Type of hall | Frequency (Hz) | | | | | |
|---|---|---|---|---|---|---|
| | 125 | 250 | 500 | 1000 | 2000 | 4000 |
| Group A: halls lined with wood, less than 3 cm thick, or with other thin materials (six halls) | 0.16 | 0.13 | 0.10 | 0.09 | 0.08 | 0.08 |
| Group B: halls lined with heavy material, i.e. with concrete, plaster more than 2.5 cm thick, etc. (three halls) | 0.12 | 0.10 | 0.08 | 0.08 | 0.08 | 0.08 |

Evidently, the influence of seat upholstery is particularly pronounced in the low-frequency range. Beranek and Hidaka had the opportunity of assembling reverberation data from several halls before and after the chairs were installed. From these values they evaluated what they called the 'residual absorption coefficients', $\alpha_r$, i.e. the total absorption of all walls, the ceiling, balcony faces, etc., except the floor, divided by their area $S_r$. Since these data are interesting in their own right we present their averages in Table 6.5. The residual absorption coefficients include the absorption of chandeliers, ventilation openings and other typical installations, and they show remarkably small variances.

It has been known for a long time that an audience does not only absorb the impinging sound waves, thus reducing the reverberation time of the room, but that it also attenuates the sound waves propagating parallel or nearly parallel to the audience. About the same holds for unoccupied seats. Attenuation in excess of the $1/r$ law of eqn (1.19) or (1.21) is actually

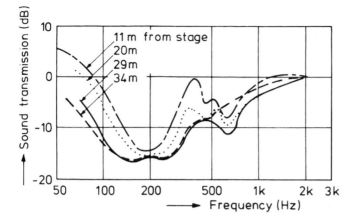

*Figure 6.16* Transmission characteristics of direct sound, measured at various seats of the main floor in Boston Symphony Hall.[7] The numbers in the figure indicate the distance from the stage.

observed whenever a wave propagates along an absorbent surface and may be attributed to some sort of refraction which partially directs the sound wave into the absorbent material.

Even sound waves travelling over the empty seating rows of a large hall are subject to a characteristic attenuation, which is known as the 'seat dip effect' and has been observed by many researchers[7,8] The additional attenuation has its maximum typically in the range of 80–200 Hz, depending on the angle of sound incidence and other parameters. An example is shown in Fig. 6.16, for which the attenuation due to spherical divergence has been subtracted from the data. The minimum in these curves is usually attributed to a vertical λ/4 resonance of the space between seating rows. Some authors have found a seat dip also with occupied seats, others have not. Figure 6.17 presents the level reduction (shaded areas) caused by the audience, as measured by Mommertz[9] in a large hall with a horizontal floor, by employing maximum length sequence techniques (see Section 8.2). The figures indicate the number of the row where the microphone was located. An evaluation of these data shows that there is a linear level decrease from front to rear seats, indicating an excess attenuation of the audience of roughly 1 dB/metre in the range 500–2000 Hz. It should be noted that not only the direct sound but also reflections from vertical side walls are subject to this kind of attenuation.

The selective attenuation of sound propagating over the audience cannot be considered as an acoustical fault since it takes place in good concert halls as well as in poorer ones. Obviously it is not the spectral composition of the direct sound and of the side wall reflections which is responsible for the tonal balance but rather that of all components.

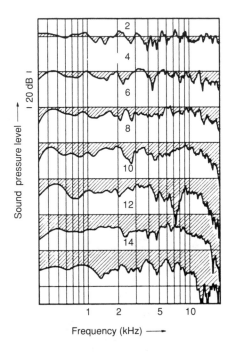

*Figure 6.17* Sound pressure level over audience, relative to the level at free propagation (horizontal lines). The figures indicate the number of the seating row. The source is 2.5 m from the first row and 1.4 m high (after Mommertz[9]).

Generally, listening conditions will be impaired if the sight lines from the listeners to the sound source are obscured by the heads of other listeners sitting in front of them. An old and well-proved way to avoid this is by sloping the audience area upwardly. This has the effect of exposing the listeners more freely to the direct sounds and avoids grazing sound incidence (see Sections 9.1 and 9.2).

## 6.8 Miscellaneous objects

The first 'sound absorber' to be discussed in this section is a plane drapery of a flexible and porous material hanging in a room, well away from all its walls. Its acoustical properties are characterised by its flow resistance $r_s$ and its mass $M'$ per unit area. These quantities are combined in an impedance $Z_r$, which we encountered already in eqn (2.26):

$$Z_r = \frac{p - p'}{v_s} = \frac{r_s}{1 - i\omega_s/\omega} \tag{6.39}$$

with $\omega_s = r_s/M'$. The difference of sound pressures on both sides of the layer is $p - p'$ and the normal component of the particle velocity at its surface is denoted by $v_s$.

Consider a plane sound wave with the pressure amplitude $\hat{p}_0$, arriving at the porous sheet at an angle $\theta$. According to Section 2.3, the total sound pressure at the side of sound incidence is $p = (1 + R)\hat{p}_0$ while the normal component of the particle velocity is

$$v_s = \hat{p}_0(1 - R) \cos \theta / \rho_0 c$$

with $R$ denoting the reflection factor of the layer. The sound pressure at the rear side is that of the incident wave multiplied by a factor $T$, the so-called 'transmission factor': $p' = T\hat{p}_0$. The same holds for the particle velocity,

$$v_s = \frac{\hat{p}_0}{\rho_0 c} T \cos \theta / \rho_0 c$$

Since both expressions for the particle velocity must be equal, we have

$$T = 1 - R$$

A second relation between $R$ and $T$ is obtained by inserting the expression for $v_s$ and $p - p' = \hat{p}_0(1 + R - T)$ into eqn (6.39):

$$Z_r = \rho_0 c \frac{1 + R - T}{T \cos \theta}$$

From both relations the reflection factor $R$ and the transmission factor $T$ can be evaluated:

$$R(\theta) = \frac{\zeta_r \cos \theta}{\zeta_r \cos \theta + 2} \tag{6.40}$$

and

$$T(\theta) = \frac{2}{\zeta_r \cos \theta + 2} \tag{6.41}$$

with $\zeta_r = Z_r/\rho_0 c$ .

In contrast to the situation considered in Section 6.3, the curtain is not a boundary of the room. Therefore, the sound energy transmitted through it is not lost but remains still within the room. Hence the energy dissipated within the drapery is obtained by subtracting both the reflected and the transmitted sound energy from the incident one. Therefore we have to apply

$$\alpha = 1 - |R|^2 - |T|^2$$

instead of eqn (2.1). This yields, after inserting eqns (6.40) and (6.41)

$$\alpha(\theta) = \frac{4\mathrm{Re}\left(\zeta_r\right)\cos\theta}{\left|\zeta_r\cos\theta + 2\right|^2} = \frac{\mathrm{Re}\left(\zeta_r\right)\cos\theta}{\left(\left|\zeta_r\right|/2\right)^2\cos^2\theta + \mathrm{Re}\left(\zeta_r\right)\cos\theta + 1^2} \tag{6.42}$$

Next we have to average eqn (6.42) over all incidence angles $\theta$ according to eqn (2.41). Since the last expression in eqn (6.42) has the same structure as eqn (2.17) we can immediately adopt eqn (2.42) as the result of this averaging after replacing the factor 8 with a factor 4 and $\zeta$ with $\zeta_r/2$. Therefore the absorption coefficient of a curtain in an isotropic sound field reads:

$$\alpha_{\mathrm{uni}} = \frac{16}{\left|\zeta_r\right|^2}\cos\mu\left[\frac{\left|\zeta_r\right|}{2} + \frac{\cos(2\mu)}{\sin\mu}\arctan\left(\frac{\left|\zeta_r\right|\sin\mu}{2 + \left|\zeta_r\right|\cos\mu}\right)\right.$$

$$\left. - \cos\mu\ln\left(1 + \left|\zeta_r\right|\cos\mu + \frac{\left|\zeta_r\right|^2}{4}\right)\right] \tag{6.43}$$

As in Section 2.5, $\mu = \arctan(\mathrm{Im}\,\zeta_r/\mathrm{Re}\,\zeta_r)$. In Fig. 6.18a, $\alpha_{\mathrm{uni}}$ of the curtain is plotted as a function of the frequency ratio $f/f_s = \omega/\omega_s$ for various values of the flow resistance $r_s$. Far below the characteristic frequency $f_s$ the absorption of the layer is very small since at these frequencies it follows nearly completely the vibrations imposed by the sound field. With increasing frequency, the inertia of the curtain becomes more and more relevant, leading to an increasing motion of the air relative to the fabric. For high frequencies the porous layer stays practically at rest, and the absorption coefficient

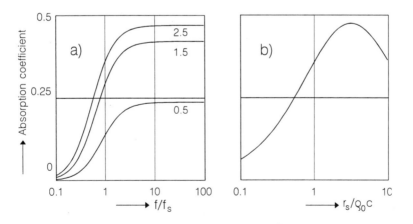

Figure 6.18 Absorption coefficient of a freely hanging curtain, random sound incidence: (a) as a function of frequency, parameter is $r_s/\rho_0 c$; (b) high-frequency limit of the absorption coefficient as a function of $r_s/\rho_0 c$.

becomes frequency independent. This limiting value is plotted in Fig. 6.18b as a function of the flow resistance $r_s$. It has a maximum $\alpha_{uni} = 0.446$ that occurs at $r_s = 3.136\rho_0 c$. This discussion shows that freely hanging curtains or large flags, etc., may considerably add to the absorption in a room and may well be used to control its reverberation.

The second object considered in this section is pseudostochastic diffusers. As mentioned near the end of Section 2.7, such devices show noticeable sound absorption even if they are manufactured from virtually loss-free materials.

The absorption of Schroeder diffusers was first observed in 1992 by Fujiwara and Miyajima.[10] By measurements carried out in the reverberation chamber, these authors found that diffusers of this type may have quite high absorption coefficients which cannot be attributed to poor surface finish. In Fig. 6.19 the absorption coefficient of a quadratic residue diffuser (QRD) with a design frequency of 285 Hz is plotted as a function of the frequency. It is negligible at low frequencies, but shows a marked rise well below the design frequency and remains at a relatively high level, although its value has several distinct maxima and minima.

This effect can be explained by the fact[11,12] that, according to the simplified theory presented in Section 2.7, the waves entering the troughs and reflected at their rear ends would cause abrupt phase jumps between adjacent openings, according to eqn (2.51). Physically, such a discontinuous

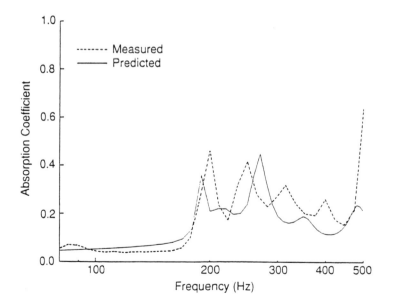

*Figure 6.19* Absorption coefficient of a quadratic residue diffuser (QRD) with $N = 7$ ( see Fig. 2.18), made of aluminium (after Fujiwara and Miyajama[10]).

phase distribution in the plane of the troughs cannot exist. In reality, any pressure differences in that plane will cause local air flows between the trough entrances which tend to equalise the pressure differences. These equalising flows may reach relatively high velocities which are not associated with far field radiation but which lead to high viscous and thermal losses in the channels. As pointed out by Mechel,[13] additional losses are probably due to the fact that the local flows are forced to go around the sharp edges of adjacent troughs.

At low frequencies all pressure differences in the diffuser plane will be almost completely levelled out. Hence the absorption can be described at least in a semi-quantitative manner from the average input admittance of the diffuser plane. (The electrical analogue would be a number of lossy capacitors connected in parallel and terminating a line.) The specific input admittance $\beta_n$ of one particular trough is the reciprocal of its input impedance according to eqn (2.21), multiplied by the characteristic impedance in the channel:

$$\beta_n = i \cdot \tan(kh_n) \tag{6.44}$$

It should be noted that in the present case the wave number $k$ is complex because it is to include the losses: $k \approx (\omega/c) - im/2$, where $m$ is the energy-related attenuation constant (see eqn (6.1)). For air under normal conditions, the contribution of the viscous and thermal losses of a channel with perfectly smooth walls in Neper/m is given by (see Ref. 3 of Chapter 4)

$$m = 3 \times 10^{-5} \frac{C}{S} \sqrt{f} \tag{6.45}$$

where $C$ denotes the circumference and $S$ the cross-sectional area of the channel, in m and m², respectively, and $f$ is the frequency in hertz. For a one-dimensional diffuser with the width $d$, the ratio $C/S = 1/d$.

The average specific admittance of the diffuser is

$$\langle \beta \rangle = \frac{1}{N} \sum_{0}^{N-1} \beta_n$$

The absorption coefficient is obtained from eqn (2.8) by inserting $\zeta = 1/\langle \beta \rangle$.

A more exact calculation of the absorption has to allow for non-constant sound pressures in the diffuser plane, as outlined in Section 2.7 (eqn (2.53)). More details are given in Refs 13 and 14.

As was shown by Fujiwara and Miyajima[10] and by Mechel,[13] surprisingly high absorptivities are not a peculiarity of pseudostochastic structures, as discussed above, but occur for any collection of wells or tubes with different lengths of openings which are close to each other. This leads us to the last 'absorbing' devices to be mentioned in this section, namely pipe organs, which show remarkable sound absorption, although no porous materials whatsoever are used in their construction. According to J. Meyer (see Ref. 6 of Chapter 1) the absorption coefficient of an organ, related to the area of its prospect, is as high as about 0.55–0.60 in the frequency range from 125 to 4000 Hz and has at least some influence on the reverberation time of a concert hall or a church.

## 6.9 Anechoic rooms

Certain types of acoustic measurements must be carried out in an environment which is free of reflections. This is true for the free field calibration of microphones, or for the determination of directional patterns of sound sources, etc. The same holds for psychoacoustic experiments. In all these cases the accuracy and the reliability of results would be impaired by the interference of the direct sound with sound components reflected from the boundaries.

One way to avoid reflections—except those from the ground—would be to perform such measurements or experiments in the open air. This has the disadvantage, however, that the experimenter depends on favourable weather conditions, which implies not only the absence of rain but also of wind. Furthermore, acoustic measurements in the open air can be affected by ambient noise.

A more convenient way is to use a so-called anechoic room or chamber, all the boundaries of which are treated in such a way that virtually no sound reflections are produced by them, at least in the frequency range of interest. How stringent the conditions are which have to be met by the acoustical treatment may be illustrated by a simple example. If all the boundaries of an enclosure have an absorption coefficient of 0.90, everybody would agree that the acoustics of this room is extremely 'dry' on account of its very low reverberation time. Nevertheless, the sound pressure level of a wave reflected from a wall would be only 10 dB lower than that of the incident wave! Therefore the usual requirement for the walls of an anechoic room is that the absorption coefficient is at least 0.99 for all angles of incidence. This requirement cannot be fulfilled with plane homogeneous layers of absorbent material; it can only be satisfied with a wall covering which achieves a stepwise or continuous transition from the air to a material with high internal energy losses.

In principle, this transition can be accomplished by a porous wall coating whose flow resistance increases in a well-defined way from the surface to the wall. It must be expected, however, that, at grazing sound incidence, the absorption of such a plane layer would be zero on account of total reflection: According to eqn (2.15), the reflection factor becomes $-1$ when the incidence angle $\theta$ approaches $90°$. Therefore it is more useful to achieve the desired transition by choosing a proper geometrical structure of the acoustical treatment than by varying the properties of the material. This can be achieved by forming pyramids or wedges of absorbent material which are mounted onto the walls. An incident sound wave then runs into channels with absorbent walls whose cross-sections steadily decrease in size, i.e. into reversed horns. The apertures at the front of these channels are well matched to the characteristic impedance of the air and thus no significant reflection will occur.

This is only true, however, as long as the length of the channels, i.e. the thickness of the lining, is at least about one-third of the acoustical wavelength. This condition can easily be fulfilled at high frequencies, but only with great expense at frequencies of 100 Hz or below. For this reason every anechoic room has a certain lower limiting frequency, usually defined as the frequency at which the absorption coefficient of its walls becomes less than 0.99.

As to the production of such a lining, it is easier to utilise wedges instead of pyramids. The wedges must be made of a material with suitable flow resistance and sufficient mechanical solidity, and the front edges of neighbouring wedges or packets of wedges must be arranged at right angles to each other (see Figs 6.20 and 6.21).

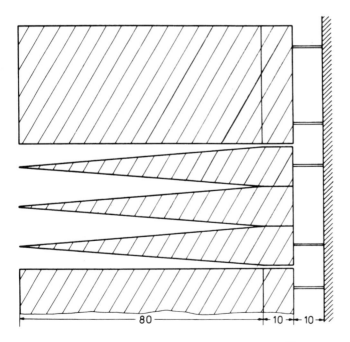

*Figure 6.20* Absorbing wall lining of an anechoic room (dimensions in cm).

*Figure 6.21* Anechoic room fitted with 65 loudspeakers for synthesising complex sound fields.

Since the floor, as well as the other walls, must be treated in the same way, a net of steel cables or plastic wires must be installed in order to give access to the space above the floor. The reflections from this net can be safely neglected at audio frequencies.

The lower limiting frequency of an anechoic room can be further reduced by combining the pyramids or wedges with cavity resonators which are located between the latter and the rigid wall. By choosing the apertures and the depths of the resonators carefully, the reflection can be suppressed at frequencies at which substantial reflection would occur without the resonators.

In this way, for a particular anechoic room,[15] a lower limiting frequency of 80 Hz was achieved by lining to a total depth of 1 m. The absorber material has a density of 150 kg/m$^3$ and a specific flow resistance of about $10^5$ kg/m$^{-3}$s$^{-1}$. It is fabricated in wedges of 13 cm $\times$ 40 cm base area and 80 cm length, which terminate in rectangular blocks of the same base area and 10 cm length. Between these blocks there are narrow gaps of 1 cm width which run into an air cushion of 10 cm depth between the absorber material and the concrete wall (see Fig. 6.20). The latter acts as a resonator volume, the necks of which are the gaps between the wedges. Three wedges with parallel edge are joined together in a packet; neighbouring packets are rotated by 90° with respect to each other. A view of the interior of this anechoic room is presented in Fig. 6.21.

Anechoic rooms are usually tested by observing the way in which the sound amplitude decreases when the distance from a sound source is increased. This decrease should take place according to a $1/r$ law to simulate perfect outdoor conditions. In practice, with increasing distance, deviations from this simple law become more and more apparent in the form of random fluctuations. Since these fluctuations are caused by wall reflections, they can be used to evaluate the average absorption coefficient of the walls. Several methods have been worked out to perform these measurements and to determine the wall absorption from their results.[16,17]

## References

1 Bass HE, Sutherland LC, Zuckerwar AJ, Blackstock DT, Hester DM. Atmospheric absorption of sound. Further developments. J Acoust Soc Am 1995; 97:680.

2 Fuchs HV, Zha X. Micro-perforated structures as sound absorbers—a review and outlook. Acta Acustica 2006; 92:139.

3 Kath U, Kuhl W. Messungen zur Schallabsorption von Polsterstühlen mit und ohne Personen. Acustica 1965; 15:127.

4 Beranek LL. Audience and seat absorption in large halls. J Acoust Soc Am 1960; 32:661.

5 Beranek LL. Concert and Opera Halls: How They Sound. Woodbury, New York: Acoustical Society of America, 1996.

6 Beranek LL, Hidaka T. Sound absorption in concert halls by seats, occupied and unoccupied, and by the hall's interior surfaces. J Acoust Soc Am 1998; 104:3169.

7 Schultz TJ, Watters BG. Propagation of sound across audience seating. J Acoust Soc Am 1964; 36:885.

8 Sessler GM, West JE. Sound transmission over theatre seats. J Acoust Soc Am 1964; 36:1725.

9 Mommertz E. Einige Messungen zur streifenden Schallausbreitung über Publikum und Gestühl. Acustica 1993; 79:42.

10 Fujiwara K, Miyajima T. Absorption characteristics of a practically constructed Schroeder diffuser of quadratic residue type. Appl Acoustics 1992; 35:149.

11 Kuttruff H. Sound absorption by pseudostochastic diffusers (Schroeder diffusers). Appl Acoustics 1994; 42:215.

12 Takahashi D. Sound absorption of a QRD. Proc Sabine Centennial Symposium, Cambridge, Mass. Woodbury, New York: Acoustical Society of America, 1994;149.

13 Mechel FP. The wide-angle diffuser—a wide-angle absorber? Acustica 1995; 81:379.

14 Fujiwara K. A study on the sound absorption of a quadratic-residue type diffuser. Acustica 1995; 81:370.

15 Meyer E, Kurtze G, Severin H, Tamm K. Ein neuer großer reflexionsfreier Raum für Schallwellen und kurze elektromagnetische Wellen. Acustica 1953; 3:409.
16 Diestel HG. Zur Schallausbreitung in reflexionsarmen Räumen. Acustica 1962; 12:113.
17 Delany ME, Bazley EN. The design and performance of free-field rooms. Proc Fourth Intern Congr on Acoustics, Budapest, 1971, Paper 24A6.

# Chapter 7

# Subjective room acoustics

The preceding chapters were devoted exclusively to the physical side of room acoustics, i.e. the objectively measurable properties of sound fields in a room and the circumstances which are responsible for their origin. We could be satisfied with this aspect if the only problems at stake were those of noise abatement by reverberation reduction and hence by reduction of the energy density, or if we only had to deal with problems of measuring techniques.

In most cases, however, the 'final consumer' of acoustics is the listener who wants to enjoy a concert, for example, or who attends a lecture or a theatre performance. This listener does not by any means require the reverberation time, at the various frequencies, to have certain values; neither does he insist that the sound energy at his seat should exhibit a certain directional distribution. Instead, he expects the room with its 'acoustics' to support the music being performed or to render speech easily intelligible (as far as this depends on acoustic properties).

The acoustical designer finds himself in a different situation. He must find ways to meet the expectations of the average or typical listener (whoever this is). To do this he needs knowledge of the relationship between properties of the physical sound field on the one hand and the listener's subjective impression (see Fig. 7.1). For this purpose, a set of physical parameters have been defined which are correlated more or less with certain aspects of the subjective listening impression. A second task is to find out in which way the physical sound field is determined by the architect's design, i.e. by the size and shape of a hall and the material of its boundary. This was the subject of the first chapters of this book. In this chapter we deal with the subjective aspects, i.e. with the left side of Fig. 7.1. With these considerations we clearly leave the region of purely physical fact and enter the realm of psychoacoustics.

The isolation of meaningful physical sound field parameters and the examination of their relevance have in the past been and still are the subject of numerous investigations—experimental investigations—since answers to these problems, which are not affected by the stigma of pure speculation, can only be obtained by experiments. Unfortunately the results of this research

*Figure 7.1* Relationship between architectural, physical and subjective aspects of room acoustics.

do not form an unequivocal picture, in contrast to what we are accustomed to in the purely physical branch of acoustics. This is ascribed to the lack of a generally agreed vocabulary to describe subjective impressions and to the very involved physiological properties of our hearing organ, including the manner in which hearing sensations are processed by our brain. It can also be attributed to listeners' hearing habits and, last but not least, to their personal aesthetic sensitivity—at least as far as musical productions are concerned. Another reason for our incomplete knowledge in this field is the great number of sound field components, which all may influence the subjective hearing impression. The experimental results which are available to date must therefore be considered in spite of the fact that many are unrelated or sometimes even inconsistent, and that every day new and surprising insights into psychoacoustic effects and their significance in room acoustics may be found.

There are basically two methods which are employed for investigating the subjective effects of complex sound fields: one method is to synthesise sound fields with well-defined properties and to have them judged by test subjects; the other method is to judge directly the acoustical qualities of completed halls, the objective properties of which are known from previous measurement. Each of these methods has its own merits and limitations, and each of them has augmented our knowledge in this field. The synthesis or laboratory simulation of sound fields permits easy and rapid variations of sound field parameters and the immediate comparison of different field configurations. However, it is not free from a somewhat artificial character in that it is impossible to simulate sound fields of real halls in their full complexity. Instead certain simplifications have to be made which restrict the application of this method to the investigation of particular aspects.

As with any psychoacoustic experiments, systematic listening tests with synthetic sound fields are very time-consuming, and the quality of their results depends a good deal on the experience and goodwill of the test subjects and on the instructions given to them prior to the tests.

On the other hand, the direct judgement of real halls leads to relatively reliable results if speech intelligibility is the only property in question, since the latter can be directly determined by articulation tests or special objective tests (see Section 7.4). The subjective assessment of the acoustics of concert halls or opera theatres, however, is affected by great uncertainties, one of which is the limited memory which impedes the direct comparison of different halls. Another one are differences in the listeners' qualification and experience. Furthermore, two halls differ invariably in more than one respect, which makes it difficult to correlate subjective opinions with one particular property such as reverberation. Nevertheless, much valuable knowledge has been collected by interviewing concert or opera goers.

The immediate comparison of aural impressions from different halls has been greatly improved by progress in sound reproduction and signal processing techniques that allow music samples played and recorded in different places to be reproduced with high fidelity. For this purpose it is not sufficient that all electro-acoustic components, such as microphones, recorders and loudspeakers, are of excellent quality. Of equal importance is that the system is capable of transmitting and reproducing all effects brought about by the complex structure of the original sound field. This is achieved by systems which exactly transplant the ear signals produced by the original sound field, to the ears of a remote listener.

As a first step, the signals in the original room must be recorded by a dummy head, i.e. by an artificial head with microphones built into the artificial ear channels. It is important that the dummy head, which usually includes the shoulders and part of the trunk, diffracts the arriving sound waves in a way which is representative for the majority of human listeners. Nowadays, several types of dummy heads are available which meet this requirement. An example is shown in Fig. 7.2. The signals recorded in this way are stored using a digital storage medium.

Equally important is the correct reproduction of the recorded and stored ear signals. In principle, high-quality headphones could be used for this purpose. Unfortunately, this kind of reproduction is often plagued by 'in-head-localisation', i.e. for some reason the listener has the impression that the sound source is located within the rear part of his head. The problem with loudspeaker reproduction, on the other hand, is that without particular preventative measures the right-hand loudspeaker will inevitably send a cross-talk signal to the left ear and vice versa. This can be avoided by a filter which 'foresees' this effect and adds proper cancellation signals to the input signals. This technique, nowadays referred to as 'cross-talk cancellation' (CTC), was invented and first demonstrated by Atal and Schroeder.[1] Figure 7.3 depicts the principle of CTC. A more detailed description of this reproduction technique may be found in Ref. 2.

In this way, a listener can immediately compare music motifs which have been recorded in different halls or at different places in one hall. Moreover, these techniques can be used to create realistic listening impressions from models of auditoria which are still under design. And finally, by using such a system the listening impressions in completely hypothetical enclosures can be demonstrated and compared with those in existing environments. As will

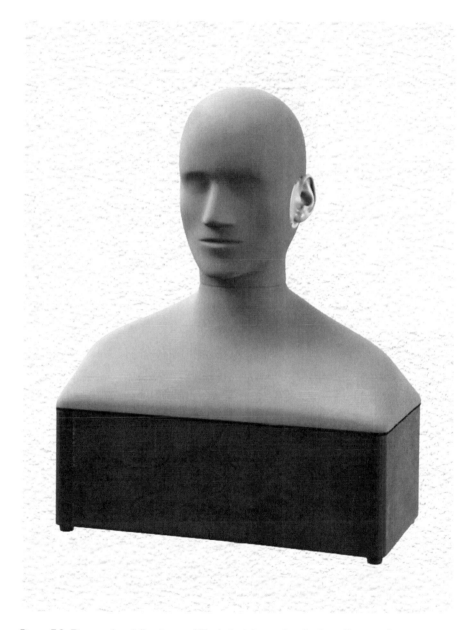

*Figure* 7.2  Dummy head (Institute of Technical Acoustics, Aachen, Germany).

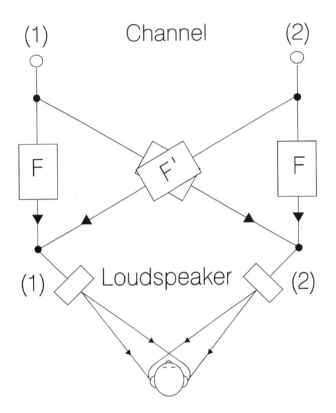

*Figure 7.3* Principle of cross-talk cancellation in loudspeaker reproduction of binaural signals. F and F' are digital filters.

be described in Section 9.7, the sound propagation in a room with given or assumed geometrical and acoustical data can be simulated and represented by binaural impulse responses. Once the impulse response for a particular listening position has been determined, a digital filter with exactly that response can be used to modify sound signals in the same way as the assumed room would modify them. Hence, these techniques are useful not only to learn more about the subjective effects of constructional features (reverberation, volume, ceiling height, etc.) but also as a valuable tool in the acoustical design of auditoria.

## 7.1 Some general remarks on reflections and echoes

In the following discussion we shall regard the sound transmission between two points of a room as formally represented by the impulse response of the transmission path. According to eqn (4.4$b$), this impulse response is composed of the direct sound and numerous repetitions of the original sound impulse caused by reflections of the sound signal from the boundary of the room. Each of these reflections is specified by its level and its time delay, both with respect to the direct sound. Since our hearing is sensitive to the direction of sound incidence as well, this description has to be completed by indicating the direction from which each reflection arrives at the receiving point. And finally there may be differences in spectrum since, as already mentioned in Section 4.1, the various components of the impulse response are not exact replicas of the original sound signal, strictly speaking, because of frequency-dependent wall reflectivities.

There are two experiences of the subjective effect of reflected portions of sound which are familiar to everyone: under certain conditions such a reflection can become a distinct 'echo'. In that case it is heard consciously as a repetition of the original signal. This can frequently be observed outdoors with sound reflections from the walls of houses or from the edge of forests. In closed rooms such experiences are less familiar, since the echoes in them are fortunately usually masked by the general reverberation of the room. Whether a reflection will become an echo or not depends on its delay with respect to the direct sound, on its relative strength, on the nature of the sound signal, and on the presence of other reflections which eventually mask the reflection under consideration.

The second common experience concerns our ability to localise sound sources in closed rooms. Although in a room which is not too heavily damped the sum of all reflected sound energies is mostly a multiple of the directly received energy, our hearing can usually localise the direction of the sound source without any difficulty. Obviously it is the sound signal to reach the listener first which subjectively determines the direction from which the sound comes. This observation is called—according to Cremer—the 'law of the first wave front'. In Section 7.3 we shall discuss the conditions under which it is valid.

In the following sections the subjective effects of sound fields with increasing complexity will be discussed. It is quite natural that the criteria of judgement become less and less detailed: in a sound field consisting of hundreds or thousands of reflections we cannot investigate the effects of each reflection separately.

Many of the experimental results to be reported on have been obtained with the use of synthetic sound fields, as mentioned earlier: in an anechoic chamber the reflections, as well as the direct sound, are 'simulated' by loudspeakers which have certain positions vis-à-vis the test subject; these positions correspond to the desired directional distribution. The differences in the strengths of the various reflections are achieved by attenuators in the electrical lines feeding the loudspeakers, whereas their mutual delay differences are produced by electrical delay units. When necessary or desired, reverberation with prescribed properties can be added to the signals. For this purpose signals are passed through a so-called reverberator, which is also most conveniently realised with digital circuits nowadays. (Other methods of reverberating signals will be described in Section 10.5.) A typical setup for psychoacoustic experiments related to room acoustics is depicted in Fig. 7.4; it allows simulation of the direct sound, two side wall reflections and one ceiling reflection. The reverberated signal is reproduced by four additional loudspeakers. A more flexible loudspeaker arrangement employed for investigations of this kind was shown in Fig. 6.21.

## 7.2 The perceptibility of reflections

In this and the next section we consider impulse responses with a very simple structure: they consist of the direct sound component and only a few or even one repetition of it. There are two questions which can be raised in this case, namely:

(1)  Under what condition is a reflection perceivable at all, without regard to the way in which its presence is manifested, and under what condition is it masked by the direct sound?
(2)  Under what condition does the presence of a reflection rate as a disturbance of the listening impression, as for instance an echo or a change of timbre ('colouration')?

In the present section we deal with the first question, postponing the discussion of the second one to the next section. Most of the reported results come from a research group of the Third Physical Institute at Göttingen (W Burgtorf, HK Oehlschlägel and HP Seraphim) and are published in *Acustica*, Vols 11–14.

We start with the hypothesis that there is a threshold level separating the levels at which a reflection is audible from those at which it is completely masked. This 'absolute threshold of perceptibility' or simply 'audibility threshold' is a function not only of the time delay with respect to the direct sound but also of the direction of its incidence, of the kind of signal and probably of other parameters. Through all the further discussions we assume the listener is looking into the direction of direct sound incidence.

To find this threshold a subject is presented with two alternate sound field configurations which differ in the presence or absence of a specified reflection. The test subject is asked to indicate solely whether he notices a difference or not. (One has to make sure, of course, that the test subjects do not know beforehand to which configuration they are listening at a given moment.) The answers of the subjects are evaluated statistically; the level at which 50% of the answers are positive is regarded as the threshold of absolute perceptibility.

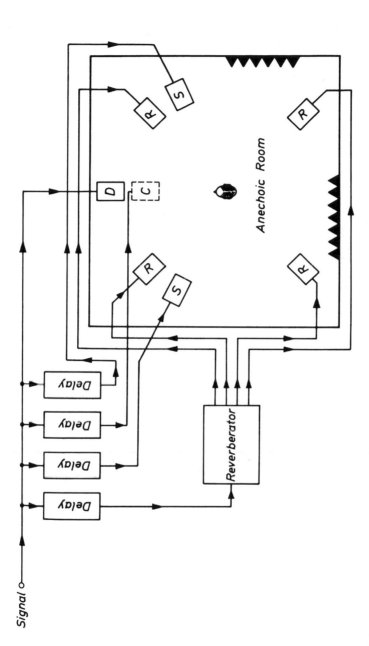

*Figure 7.4* Simulation of sound fields in an anechoic room. The loudspeakers are denoted D = direct sound, S = side wall reflections, C = ceiling reflection (elevated), R = reverberation (after Reichardt and Schmidt).

We start with an impulse response consisting of just two components: the direct sound and one reflection. For speech with a level of 70 dB, and for frontal incidence of the direct sound as well as of the reflected component, the audibility threshold is

$$\Delta L = -0.575t_0 - 6 \quad \text{decibels} \tag{7.1}$$

where $\Delta L$ is the pressure level of the reflected sound signal relative to the sound pressure of the direct sound and $t_0$ is its time delay in milliseconds. For an example take a reflection delayed by 60 ms with respect to the direct sound. According to eqn (7.1) it is audible even when its level is lower by 40 dB than that of the direct signal.

Figure 7.5 plots for three different signals (continuous speech, a short syllable and a white noise pulse with a duration of 50 ms), the audibility thresholds of a reflection delayed by 50 ms as a function of the angle under which the reflections arrive. (The investigated directions have been restricted to the horizontal plane.) It is evident to what extent the thresholds depend on the type of sound signal. In any case, however, the masking effect of the direct sound is most pronounced at equal directions of both components or, in other words, our hearing is more sensitive to reflections arriving from lateral directions than to those arriving from the front or the rear. It should be added that reflections arriving from above are also masked more effectively by the direct sound than are lateral reflections.

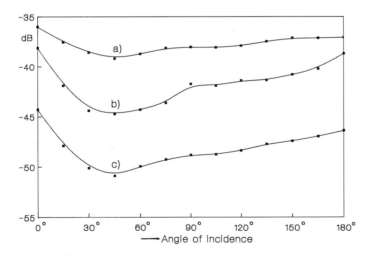

*Figure 7.5* Threshold of perception of a reflection with 50 ms delay, obtained with (a) continuous speech, (b) a short syllable, (c) noise pulses of 50 ms duration. Abscissa is the horizontal angle at which the reflection arrives. The level of the direct signal is 75 dB (after Burgtorf and Oehlschlägel).

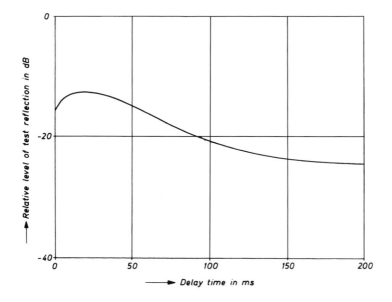

*Figure 7.6* Threshold of perception of a reflection as a function of its delay. The threshold is the average taken over six different music samples (frontal incidence of reflection) (after Schubert[3]).

If the sound signal is not speech but music, our hearing is generally much less sensitive to reflections. This is the general result of investigations carried out by Schubert,[3] who measured the threshold with various music motifs. One of his typical results is presented in Fig. 7.6, which plots the average threshold taken over six different music samples. With increasing delay time it falls much less rapidly than according to eqn (7.1); its maximum slope is about $-0.13$ dB/ms. As with speech, the threshold is noticeably lower for reflections arriving from lateral directions than with frontal incidence. Furthermore, added reverberation renders the detection of a reflection more difficult. Obviously reverberated sound components cause additional masking, at least with continuous sound signals.

For more than one reflection the number of parameters to be varied increases rapidly. Fortunately each additional reflection does not create a completely new situation for our hearing. This is demonstrated in Fig. 7.7, which shows the audibility threshold for a variable reflection which is added to a masking sound field consisting of the direct sound plus one, two, three or four reflections at fixed delay times. In this case all the reflections arrive from the same direction as the direct sound. The fixed reflections are indicated as vertical lines over the delay times belonging to them; their heights are a measure of the strength of the reflections. If all the reflections—the fixed

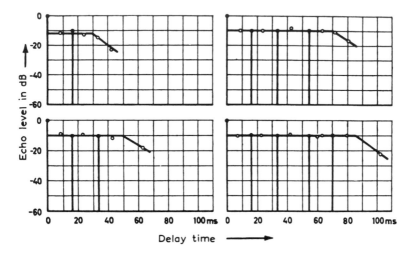

*Figure 7.7* Threshold of perception of a delayed reflection being added to a sound field consisting of direct sound plus one, two, three or four reflections at fixed delay times and relative levels as indicated by vertical lines. The test signal is speech. All components arriving from the front (after Seraphim).

ones as well as the variable one—and the direct sound arrive from different directions, the thresholds are different from those of Fig. 7.6 in that they immediately begin to fall and then jump back to the initial value at the delay time of one of the fixed reflections.

Apart from the absolute thresholds of perception, the subjective difference limen for reflections are also of interest. According to Reichardt and Schmidt, variations of the reflection level as small as about ±1.5 dB can be detected by our hearing if music is used as a test signal. In contrast, the auditive detection of differences in delay times is afflicted with great uncertainty.

## 7.3 Echoes and colouration

A reflection which is perceived at all does not necessarily reach the consciousness of a listener. At low levels it manifests itself only by an increase of loudness of the total sound signal, by a change in timbre, or by an increase of the apparent size of the sound source. But at higher levels a reflection can be heard as a separate event, i.e. as a repetition of the original sound signal. This effect is commonly known as an 'echo', as already mentioned in Section 7.1. But what outdoors usually appears as an interesting experience may be rather unpleasant in a concert hall or in a lecture room, in that it distracts the listeners' attention. In severe cases an echo may significantly reduce our enjoyment of music or impair the intelligibility of speech, since subsequent speech sounds or syllables are mixed up and the text is confused.

In the following the term 'echo' will be used for any sound reflection which is subjectively noticeable as a temporal or spatially separated repetition of the original sound signal, and we are discussing the conditions under which a reflection will become an echo. Thus we are taking up again the second question raised at the outset of the foregoing section.

From his outdoor experience the reader may know that the echo produced by sound reflection from a house front, etc., disappears when he approaches the reflecting wall and when his distance from it becomes less than about 10 m, although the wall still reflects the sound. Obviously it is the reduction of the delay time between the primary sound and its repetition which makes the echo vanish. This shows that our hearing has only a restricted ability to resolve succeeding acoustical events, a fact which is sometimes attributed to some kind of 'inertia' of hearing. Like the absolute threshold of perceptibility, however, the echo disturbance depends not only on the delay of the repetition but also on its relative strength, its direction, on the type of sound signal, on the presence of additional components in the impulse response and on other circumstances.

Systematic experiments to find the critical echo level of reflections are performed in much the same way as those investigating the threshold of absolute perceptibility, but with a different instruction given to the test subjects. It is clear that there is more ambiguity in fixing the critical echo levels than in establishing the absolute perception threshold since an event which is considered as disturbing by one person may be found tolerable by others.

Classical experiments of this kind were carried out as early as in 1950 by Haas[4] using continuous speech as a primary sound signal. This signal was presented to the test subjects with two loudspeakers: the input signal of one of them could be attenuated (or amplified) and delayed with respect to the other.

Figure 7.8 shows one of Haas' typical results. It plots the percentage of subjects who felt disturbed by an echo of given relative level as a function of the time delay between the undelayed signal (primary sound) and the delayed one (reflection). The numbers next to the curves indicate the level of the artificial reflection in dB relative to that of the primary sound. The rate of speech was 5.3 syllables per second; the listening room had a reverberation time of 0.8 s. At a delay time of 80 ms, for instance, only about 20% of the observers felt irritated by the presence of a reflection with a relative level of −3 dB, but the percentage was more than 80% when the level was +10 dB.

Analogue results have been obtained for different speaking rates or reverberation times of the listening room. They are summarised in Table 7.1. In these experiments both components of the sound field had equal levels.

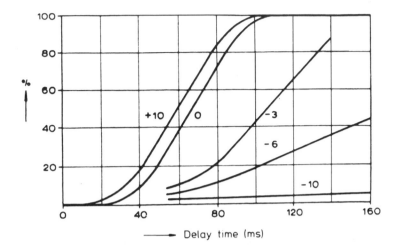

*Figure 7.8* Percentage of listeners disturbed by a delayed speech signal in a room with a reverberation time of 0.8 s. Speaking rate is 5.3 syllables per second. The relative echo levels (in dB) are indicated by numbers next to the curves (after Haas[4]).

*Table 7.1* Critical echo delays at equal levels of direct sound and reflection (after Haas[4])

| Reverberation time of listening room (s) | Speaking rate (syllables/s) | Critical delay time (ms) |
|---|---|---|
| 0 | 5.3 | 43 |
| 0.8 | 5.3 | 68 |
| 1.6 | 5.3 | 78 |
| 0.8 | 3.5 | 93 |
| 0.8 | 5.3 | 68 |
| 0.8 | 7.4 | 41 |

The numbers in the last column of the table denote the delay times at which the curves analogous to those of Fig. 7.8 cross the 50% line. It is evident that echoes are much more disturbing in surroundings with a low reverberation time than in more reverberant spaces.

Muncey et al.[5] have performed similar experiments for speech as well as for various kinds of music. The test signals were presented in a room which was virtually free of reverberation.

This is probably the reason why these authors found critical levels for speech which are well below those reported by Haas. Furthermore, their

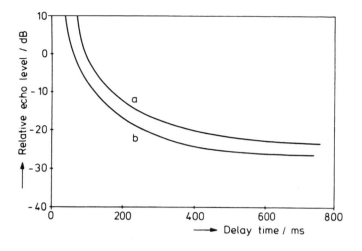

*Figure 7.9* Critical reflection level (re-direct sound level) as a function of delay time for (a) organ music and (b) string music (after Muncey et al.[5]). Note the different time scales in this figure and Figs 7.8 and 7.10.

investigations clearly showed that our hearing is less sensitive to echoes in music than in speech. The annoyance of echoes in very slow music, as for example organ music, is particularly low. In Fig. 7.9, the critical echo level (50% level) for fast string and organ music is plotted as a function of time delay.

A striking result of all these experiments can be seen in Fig. 7.8: if the relative level of the reflection is raised from 0 to +10 dB, there is only a small increase in the percentage of observers feeling disturbed by the reflected sound signal. Hence no disturbance is expected to occur for a reflection with time delay of, say, 20 ms even if its energy is 10 times the energy of the direct sound. This finding is frequently referred to as the 'Haas effect' and has important applications in the design of public address systems.

Careful investigations into this effect and of related phenomena have been carried out by Meyer and Schodder.[6] In order to restrict the range of possible judgements, the test subjects were not asked to indicate the level at which they were disturbed by an echo; instead they had to indicate the level at which they heard both the delayed signal and the undelayed one equally loudly. Since in these tests the undelayed signal reached the test subject from the front, and the delayed one from a lateral angle of 90°, the test subjects could also be asked to indicate the echo level at which the sound seemed to arrive from halfway between both directions. Both criteria of judgement led to the same results. One of them is shown in Fig. 7.10, where the critical level difference between primary sound and reflection is plotted as a function of

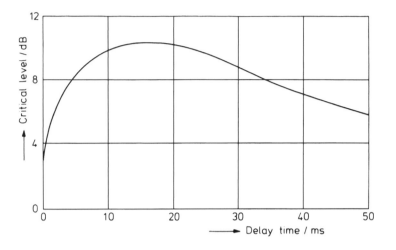

*Figure 7.10* Critical reflection level (relative to the direct sound level) as a function of delay time. At this level both components, the direct and the delayed, are heard equally loud (after Meyer and Schodder[6]).

the delay time. It virtually agrees with the test results obtained by Haas and renders them somewhat more precise. For our hearing sensation the primary sound determines the impression of direction even when the reflection—provided it has a suitable delay time—is stronger by up to about 10 dB. If the reflection is split up into several small reflections of equal strengths and with successive mutual delay times of 2.5 ms, leaving constant the total reflection energy, the curve shown in Fig. 7.10 is shifted upwards by another 2.5 dB. This result shows that many small reflections separated by short time intervals of the order of milliseconds cause about the same disturbance as one single reflection provided the total reflected energy and the (centre) delay time are the same for both configurations.

From these results we can draw the following practical conclusions. In room acoustics the law of the first wave front can be considered valid in general: exceptions, i.e. erroneous localisations and disturbing echoes, will occur only in special situations, as for example when most of the room boundaries, except for a few remote portions of wall, are lined with an absorbent material or when certain portions of wall are concavely curved and hence produce exceptionally strong reflections by focusing the sound.

The superposition of a strong isolated reflection onto the direct sound can cause another undesirable effect, especially with music, namely 'colouration': i.e. a characteristic change of timbre. The same is true for a succession of equidistant reflections.

If the impulse response $g_1(t)$ of a room consists only of the direct sound and one reflection which is weaker by a factor of $q$,

$$g_1(t) = \delta(t) + q\delta(t - t_0) \tag{7.2}$$

the corresponding squared absolute value of its Fourier transform is given according to eqn (1.48$a$) by

$$|G_1(f)|^2 = |1 + q\exp(2\pi i f t_0)|^2 = 1 + q^2 + 2q\cos(2\pi f t_0) \tag{7.3}$$

This is the squared transfer function of a comb filter with a ratio of maximum to minimum of $(1+q)^2/(1-q)^2$; the separation of adjacent maxima is $1/t_0$. An infinite and regular succession of reflections given by

$$g_2(t) = \sum_{n=0}^{\infty} q^n \delta(t - nt_0) \tag{7.4}$$

leads to the squared absolute spectrum

$$|G_2(f)|^2 = [1 + q^2 - 2q\cos(2\pi f t_0)]^{-1} \tag{7.5}$$

It has the same distance of the maxima and also the same maximum-to-minimum ratio as that in eqn (7.3). However, the peaks are much sharper here than with a single reflection, except for $q \ll 1$ (see Fig. 7.11).

Whether such a comb filter will produce audible colourations or not depends on the delay time $t_0$ and on the relative heights of the maxima.[7] The absolute threshold for audible colourations rises as the delay time or distance $t_0$ increases, i.e. when the distance $1/t_0$ between subsequent maxima on the frequency axis is smaller (see Fig. 7.12). This finding makes it understandable why the very pronounced but closely spaced irregularities of room frequency curves (see Fig. 3.7b) do not cause undesirable subjective effects.

If $t_0$ exceeds a certain value, say 25 ms or so, the regularity of impulse responses does not appear subjectively as colouration, i.e. of changes in the timbre of sound, but rather the sounds have a rough character, which means that we become aware of the regular repetitions of the signal as a phenomenon occurring in the time domain (echo or flutter echo). This is because our ear is not just a sort of frequency analyser but is also sensitive to the temporal structure of the sound signals. Or expressed more correctly: our hearing performs a short-time spectral analysis.

In Section 8.3, we shall discuss an objective criterion for the perceptibility of sound colouration or of a flutter echo which is based on thresholds of the kind shown in Fig. 7.12.

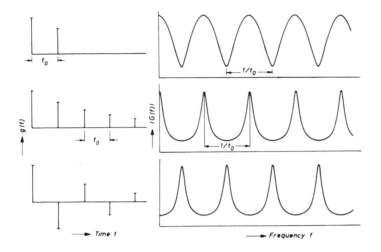

*Figure 7.11* Impulse responses (left) and absolute values of transfer function (right) of various comb filters ($q = 0.7$).

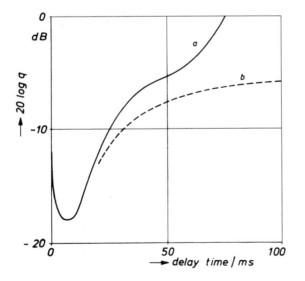

*Figure 7.12* Critical values of the factor $q$ resulting in just audible colouration of white noise passed through comb filter (a) according to eqn (7.3) and (b) according to eqn (7.5) (after Atal et al.[7]).

So far this discussion has been restricted to the somewhat artificial case that the sound field consists of the primary or direct sound followed by just one single repetition or a regular succession of repetitions of the sound signal. However, the impulse responses of most real rooms have a more complicated structure, and it is clear that the presence of numerous reflections must influence the way we perceive one of them in particular. On the other hand, from a practical point of view it would be desirable to have a criterion to indicate whether a certain peak in a measured impulse response or 'reflectogram' hints at an audible echo and should be removed by suitable constructive measures.

Such a criterion was proposed by Dietsch and Kraak[8] in 1986. It is based on the ratio

$$t_s(\tau) = \int_0^\tau |g(t)|^n t \, dt \bigg/ \int_0^\tau |g(t)|^n \, dt \tag{7.6}$$

with $g(t)$ denoting as before the impulse response of a room. Here $t_s(\tau)$ is a monotonous function of $\tau$ approaching a limiting value $t_s(\infty)$ as $\tau \to \infty$. This latter value is the first moment of $|g(t)|^n$, and the function $t_s(\tau)$ indicates its temporal build-up (Fig. 7.13). The quantity used for rating the strength of an echo is based upon the difference quotient of $t_s(\tau)$:

$$EC = \text{maximum of } \frac{\Delta t_s(\tau)}{\Delta \tau} \tag{7.7}$$

where $\Delta\tau$ can be adapted to the character of the sound signal. The dependence of the echo criterion EC on the directional distribution of the various reflections is accounted for by recording impulse responses with both microphones of a dummy head and adding their energies. By numerous subjective tests, carried out both with synthetic sound fields and in real halls, Dietsch and Kraak determined not only suitable values for the exponent $n$ and for $\Delta\tau$ but also the critical values $EC_{crit}$, which must not be exceeded to ensure that not more than 50% of the listeners will hear an echo (Table 7.2). (The 10% thresholds are slightly lower.) It should be noted that echo disturbances are mainly due to spectral components of somewhat elevated frequency. For practical purposes, however, it seems sufficient to employ test signals with a bandwidth of 1 or 2 octaves.

## 7.4 Early energy: definition, clarity index, speech transmission index

As explained in Section 4.2, the impulse response of most real rooms does not consist of four or six but of much more significant reflections, all of them contributing in a way to how a listener perceives a sound signal in the room.

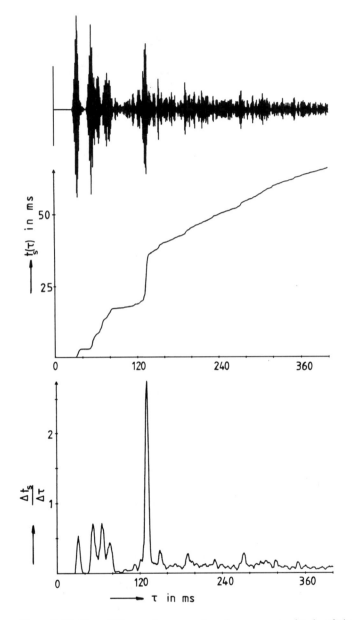

*Figure 7.13* First 400 ms of a room impulse response (top), of the associated build-up function $t_s(\tau)$ (middle) and of the difference quotient $\Delta t_s(\tau)/\Delta\tau$ (bottom, with $\Delta\tau = 5$ ms). EC is 2.75 in this example.

Table 7.2 Echo criterion of Dietsch and Kraak:[8] characteristic data

| Type of signal | n | $\Delta\tau$ (ms) | $EC_{crit}$ | Bandwidth of test signal (Hz) |
|---|---|---|---|---|
| Speech | 2/3 | 9 | 1.0 | 700–1400 |
| Music | 1 | 14 | 1.8 | 700–2800 |

The number of parameters characterising the response is therefore so large that their contribution cannot be investigated or described in detail, as has been done in the preceding sections. Instead the huge amount of information has to be condensed into certain parameters which summarize the situation in a subjectively meaningful way. This is mainly done by comparing the energies contained in different parts of the impulse response. Such a procedure is not just an expedient dictated by the limitations of time and facilities but is justified by the limited ability of our hearing to distinguish all the countless repeated sound signals.

In the preceding sections it was shown that a reflection is not perceived subjectively as something separate from the direct sound as long as its delay and its relative strength do not exceed certain limits. Its only effect is to make the sound source appear somewhat more extended and to increase the apparent loudness of the direct sound. Since these 'early reflections' give support to the sound source they are considered useful.

Reflections which arrive at the listener with longer delays are noticed as echoes in unfavourable cases; in favourable cases they contribute to the reverberation of the room. In principle, any reverberation impairs the intelligibility of speech because it blurs its time structure and mixes up the spectral characteristics of successive phonemes or syllables. Therefore, 'late reflections' are considered to be detrimental from the viewpoint of speech transmission. From our everyday experience with outdoor echoes, but even more precisely from Haas' results (see Fig. 7.8) and similar findings, it can be concluded that the critical delay time separating useful from detrimental reflections is somewhere in the range from 50 to 100 ms.

Most of the following criteria compare the energy conveyed in useful reflections, including that of the direct sound, with the energy contained in the remaining ones. To validate them it is necessary to determine the speech intelligibility directly. This can be done by articulation tests carried out in the following way. A sequence of meaningless syllables (so-called 'logatoms') is read aloud in the environment under test. To obtain representative results it is advisable to use phonetically balanced material (from so-called PB lists) for this purpose, i.e. sets of syllables in which initial consonants, vowels and final consonants are properly distributed. Listeners placed at various positions are asked to write down what they have heard.

The percentage of syllables which have been correctly understood is considered to be a relatively reliable measure of speech intelligibility, called 'syllable intelligibility'.

The earliest attempt to define an objective criterion of what may be called the distinctness of sound, derived from the impulse response, is due to Meyer and Thiele,[9] who named it 'definition' (originally 'Deutlichkeit'):

$$D = \left[ \int\limits_0^{50\,\text{ms}} [g(t)]^2\,dt \bigg/ \int\limits_0^{\infty} [g(t)]^2\,dt \right] \cdot 100\% \qquad (7.8)$$

Both integrals must include the direct sound. Obviously, $D$ will be 100% if the impulse response does not contain any components with delays in excess of 50 ms.

The relationship between $D$ and the syllable intelligibility is shown in Fig. 7.14. The plotted values of $D$ have been obtained by averaging over the frequency range 340–3500 Hz. Obviously, the 'definition' $D$ is a useful descriptor of speech intelligibility.

A quantity which is formally similar to 'definition' but intended to characterise the transparency of music in a concert hall is the 'clarity index' $C$ (originally 'Klarheitsmaß') as introduced by Reichardt et al.[10]

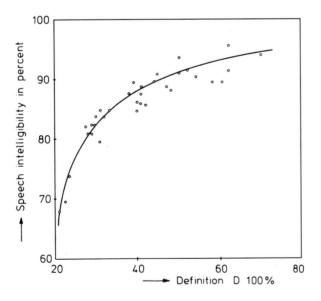

Figure 7.14 Relationship between syllable intelligibility and 'definition'.

It is defined by

$$C = 10 \log_{10} \left[ \int\limits_{0}^{80\,\text{ms}} [g(t)]^2 \, dt \middle/ \int\limits_{80\,\text{ms}}^{\infty} [g(t)]^2 \, dt \right] \quad \text{decibels} \qquad (7.9)$$

The higher limit of delay time (80 ms compared with 50 ms in eqn (7.8)) makes allowance for the fact that with music a reflection is less detectable than it is with speech signals. By subjective tests with synthetic sound fields these authors have determined the values of $C$ preferred for the presentation of various styles of orchestral music. They found that $C = 0$ dB indicates that the subjective clarity is sufficient even for fast musical passages, whereas a value of $C = -3$ dB seems to be still tolerable. Nowadays, $C$ (often referred to as C80 or $C_{80}$) is widely accepted as a useful criterion for the clarity and transparency of musical sounds in concert halls. According to an investigation of concert halls in Europe and the USA carried out by Gade[11] its typical range is from about $-5$ to $+3$ dB.

The assumption of a sharp delay limit separating useful from non-useful reflections is certainly a crude approximation to the way in which repetitions of sound signals are processed by our hearing. From a practical point of view it has the unfavourable effect that in critical cases a small change in the arrival time of a strong reflection may result in a significant change in $D$ or $C$. Therefore several authors have proposed a gradual transition from useful to detrimental reflections by calculating the useful energy with a continuous weighting function $a(t)$:

$$E_u = \int\limits_{0}^{\infty} a(t) \cdot [g(t)]^2 dt \qquad (7.10)$$

For a linear transition $a(t)$ is given by

$$a(t) = \begin{cases} 1 & \text{for } 0 \leq t < t_1 \\ \dfrac{t_2 - t}{t_2 - t_1} & \text{for } t_1 \leq t \leq t_2 \\ 0 & \text{for } t > t_2 \end{cases} \qquad (7.11)$$

A reasonable choice of $t_1$ and $t_2$ would be about 35 ms and 100 ms, for example.

No delay limit whatsoever is involved in the 'centre time' which was proposed and investigated by Kürer[12] and which is defined as the first moment of the squared impulse response:

$$t_s = \int\limits_{0}^{\infty} [g(t)]^2 \, t \, dt \middle/ \int\limits_{0}^{\infty} [g(t)]^2 \, dt \qquad (7.12)$$

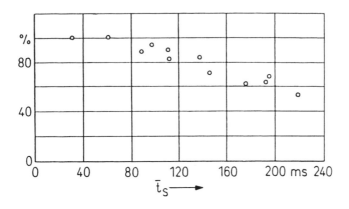

*Figure 7.15* Relationship between speech intelligibility and centre time $t_s$ (after Kürer[12]).

Obviously a reflection with given strength contributes more to $t_s$ the longer it is delayed with respect to the direct sound. High transparency or speech intelligibility is indicated by low values of the centre time $t_s$ and vice versa. The high (negative) correlation between measured values of $t_s$ and intelligibility scores is demonstrated in Fig. 7.15.

Quite a different approach to quantifying the speech intelligibility from objective sound field data is based on the modulation transfer function (MTF) already introduced in Section 5.5. It quantifies the smoothing effect of reverberation on the envelope of speech signals as well as that on their spectral components. For strictly exponential sound decay with a reverberation time $T = 6.91/\delta$ the complex MTF reads (see eqn (5.36c))

$$\underline{m} = \frac{1}{1 + i\Omega/2\delta} \tag{7.13}$$

where $\Omega$ is the angular frequency of the modulation. The absolute value of $\underline{m}(\Omega)$

$$m(\Omega) = \frac{1}{[1 + (\Omega T/13.8)^2]^{-1/2}}$$

(with $T = 6.91/\delta$) is plotted in Fig. 7.16. It shows that very slow variations of a signal's envelope are not levelled out to any noticeable extent, but very rapid fluctuations are almost completely eliminated by the reverberant trail. Usually, however, the sound decay does not follow a simple exponential law; hence, real MFTs deviate more or less from that shown in Fig. 7.16. Furthermore, the MTF depends on the spectral composition of the sound signal, for instance on its frequency, if a modulated sine signal is applied.

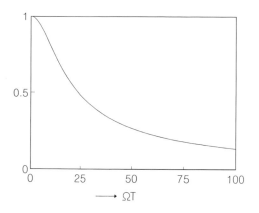

*Figure 7.16* Modulation transfer function of a room with exponential sound decay ($T =$ reverberation time, $\Omega =$ angular modulation frequency).

Houtgast and Steeneken[13] have developed a procedure to convert MTF data measured in 7 octave bands and at 14 modulation frequencies into one single figure of merit, which they called the 'speech transmission index' (STI). This conversion involves averaging over a certain range of modulation frequencies; furthermore, it makes allowance for the contributions of the various frequency bands to speech quality and also for the mutual masking between adjacent frequency bands occurring in our hearing organ. Finally, Houtgast and Steeneken[13] have shown in numerous experiments that the STI is very closely related to the results of articulation tests carried out with various types of speech signals (Fig. 7.17). Accordingly, to guarantee sufficient speech intelligibility the STI must be at least 0.5.

A simpler and less time-consuming version of this criterion is the 'RApid Speech Transmission Index' (RASTI).[14] It is obtained by applying only four modulation frequencies in the octave band centred at 500 Hz, and five modulation frequencies in the 2000 Hz octave band. The modulation frequencies range from 0.7 to 11.2 Hz. Each of the nine values of the modulation index $m$ is converted into an 'apparent signal-to-noise ratio':

$$(S/N)_{app} = 10 \log_{10} \left( \frac{m}{1-m} \right)$$

These figures are averaged after truncating those which exceed the range of $\pm 15$. The final parameter is obtained by forming the average $\overline{(S/N)}_{app}$ and normalising the result:

$$\text{RASTI} = \frac{1}{30} \left[ \overline{(S/N)}_{app} + 15 \right] \tag{7.14}$$

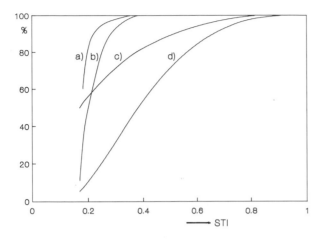

*Figure 7.17* Relationship between speech transmission index (STI) and speech intelligibility, obtained with: (a) numbers and spell alphabet; (b) short sentences;
(c) diagnostic rhyme test; and (d) logatoms (after Houtgast and Steeneken[13]).

*Table 7.3* Relation between speech
transmission quality and RASTI

| Quality score | RASTI |
| --- | --- |
| Bad | < 0.32 |
| Poor | 0.32–0.45 |
| Fair | 0.45–0.60 |
| Good | 0.60–0.75 |
| Excellent | > 0.75 |

For practical RASTI measurements both octave bands are emitted simultaneously, each with a complex power envelope containing five modulation
frequencies. Likewise, the automated analysis of the received sound signal is
performed in parallel. With these provisions it is possible to keep the duration
of one measurement as low as about 12 s. Table 7.3 shows the relationship
between five classes of speech quality and corresponding intervals of RASTI
values.

We conclude this section with two remarks which more or less apply to all
the above criteria. First, it is evident that they are highly correlated among
each other. If, for example, a particular impulse response is associated with
a short 'centre time' $t_s$, its evaluation according to eqn (7.8) will yield a
high value of 'definition' $D$ and vice versa. Therefore there is no point in

measuring many or all of them in order to collect as much information as possible.

Second, if the sound decay in the room under consideration would strictly obey an exponential law according to eqn (4.9) or (5.21), all the parameters defined above could be directly expressed by the reverberation time, as has already been done in eqn (7.14). Hence they would not yield any information beyond the reverberation time. In real situations, however, the exponential law is a useful but nevertheless crude approximation to a much more complicated decay process. Especially in its early portions an impulse response is far from being a smooth function of time. Furthermore, the pattern of reflections (see, for example, the upper part of Fig. 7.13) usually varies from one observation point to the other; accordingly, these parameters too may vary over a wide range within one hall and are quite sensitive to geometrical and acoustical details of a room. Therefore they are well suited to describe differences of listening conditions at different seats in a hall whereas the reverberation time does not significantly depend on the place where it has been measured.

## 7.5 Reverberation and reverberance

If we disregard all details of the impulse response of a room, we finally arrive at the general decay the sound energy undergoes after an impulse excitation or after a sound source has been stopped. As discussed in earlier chapters of this book, the duration of this decay is characterised by the reverberation time or decay time, at least if the energy decay obeys an exponential law in its gross appearance.

Historically, the outstanding role of reverberation was first recognized by Sabine (see Ref. 1 of Chapter 5) whose famous investigations, carried out during the last years of the 19th century, mark the origin of modern room acoustics. Sabine defined the reverberation time and developed several methods to measure it, and he was the first to formulate the laws of reverberation. Furthermore, he investigated the sound-absorbing power of numerous materials.

The reverberation time (or decay time) is still considered as the most important objective quantity in room acoustics, although it has been evident for some time that it characterises only one particular aspect of the sound field and needs to be supplemented by additional parameters if a full description of the prevailing listening conditions is to be obtained. This predominance of reverberation time has at least three reasons. First, it can be measured and predicted with reasonable accuracy and moderate expenditure. Secondly, the reverberation time of a room does not depend significantly on the observer's position in a room, a fact which is also underlined by the simple structure of the formulae by which it can be calculated from room data (see Section 5.3). Hence it is well suited to characterise the overall acoustic properties of a hall, neglecting details which may vary from one place to another. And, finally, abundant data on the reverberation time of existing halls are available nowadays, including their frequency dependence. They can be used as a yardstick to tell us which values are typical and which are not.

Before discussing the important question which reverberation times are desirable or optimal for the various types of rooms and halls, a remark on the just audible differences in reverberation time, i.e. on the difference threshold of reverberation time, may be in order. By presenting exponentially decaying noise impulses with variable decay times, bandwidths and centre frequencies to numerous test subjects, Seraphim[15] found that the relative difference limen for decay time is about 4%, at least in the most important range of decay times. Although these results were obtained under somewhat artificial conditions, they show at least that there is no point in giving reverberation times with a greater accuracy than about 0.05 or 0.1 s.

In principle, preferred ranges of the reverberation time can be determined by systematic listening tests, i.e. by presenting speech or music samples to a test audience. To lead to meaningful results such tests should be performed, strictly speaking, in environments (real enclosures or synthetic sound fields) which allow variations of the reverberation time under otherwise unchanged conditions. Just comparing sound recordings from different halls is not sufficient unless the answers of test persons are subject to a somewhat involved evaluation procedure (see Section 7.8).

A more empirical approach consists of collecting the reverberation times of halls which are generally considered as acoustically satisfactory or even excellent for the purpose they have to serve (lectures, drama theatre, operatic performances, orchestra or chamber music, etc.). It should be noticed, however, that subjective opinions on acoustical qualities and hence the conclusions drawn from them are afflicted with several factors of uncertainty, such as the question of who is able to utter meaningful acoustical criticism. Certainly musicians have the best opportunity of comparison, since many of them perform in different concert halls. On the other hand, musicians have a very special standpoint (meant literally as well as metaphorically) which does not necessarily agree with that of a listener. The same is true for acousticians and sound recording engineers, who have a professional attitude towards acoustical matters and may frequently concentrate their attention on special properties which are insignificant to the 'average' concert listeners. The latter, however, as for example the concert subscribers, often lack the opportunity to compare several concert halls or else they are not very critically minded in acoustical matters, or their opinion is influenced from the point of view of local patriotism. Furthermore, there are—and again this applies particularly to musical events and their appropriate surroundings—individual differences in taste which cannot be discussed in scientific terms.

And, finally, it is quite possible that there are certain trends of 'fashion' towards longer or shorter reverberation times. All these uncertainties make it understandable that it is impossible to specify one single optimum value of reverberation time for each room type or type of presentation; instead, only ranges of favourable values can be set up.

We start with rooms used only for speech, such as lecture rooms, congress halls, parliament, theatres for dramatic performances and so on. As mentioned earlier, no reverberation whatsoever is required for such rooms in principle, since any noticeable sound decay has the tendency to blur the syllables and thus to reduce speech intelligibility. On the other hand, a highly absorbing treatment of all walls and of the ceiling of a room would certainly remove virtually all the reverberation, but at the same time it would prevent the formation of useful reflections which increase the loudness of the perceived sounds and which are responsible for the relative ease with which communication is possible in enclosures as compared to outdoor communication. Furthermore, the lack of any audible reverberation in a closed space creates an unnatural and uncomfortable feeling, as can be observed when entering an anechoic room, for instance. Obviously one subconsciously expects to encounter some reverberation which bears a certain relation to the size of the room. Therefore the reverberation time in rooms of this kind should not fall short of say 0.5 s, except for very small rooms such as living rooms. For larger rooms such as drama theatres, values of about 1.2 s are still tolerable.

As is well known, low-frequency signal components contribute very little to speech intelligibility. Therefore it is advisable to provide for sufficient low-frequency absorption by applying suitably designed absorbers (resonance absorbers, see Section 6.4) to the boundary of such rooms in order to reduce the reverberation time and also the stationary sound level at low frequencies.

Now we shall turn to the reverberation times which can be considered to be optimum for concert halls. In order to discover these values, we depend completely on subjective opinions concerning existing halls, at least so long as there are no results available of systematic investigations with synthetic or simulated sound fields. As has been pointed out before, there is always some divergence in the opinions about a certain concert hall; furthermore, they are not always constant in time. Old concert halls particularly are often commented on enthusiastically, probably more than is justified by their real acoustical merits. (This is true especially for those halls which were destroyed by war or other catastrophes.) In spite of all these reservations, it is a matter of fact that certain concert halls enjoy a high reputation for acoustical reasons. This means, among other things, that at least their reverberation time does not give cause for complaint. On the whole it seems that the optimum values for occupied concert halls are in the range from about

1.6 to 2.1 s at mid-frequencies. Table 7.4 lists the reverberation times of several old and new concert halls, both for low frequencies (125 Hz) and for the medium-frequency range (500–1000 Hz).

At first glance it may seem curious that which is good for speech, namely a relatively short reverberation time, should be bad for music. This discrepancy can be resolved by bearing in mind that, when listening to speech, we are interested in perceiving each element of the sound signal, since this increases the ease with which we can understand what the speaker is saying. When listening to music, it would be rather disturbing to hear every detail, including the bowing noise of the string instruments or the air flow noise of flutes. Furthermore, it is impossible to achieve perfect synchronism among the various players of an orchestra, let alone differences in intonation. It is these inevitable imperfections, which are hidden or masked by reverberation. What is even more important, reverberation of sufficient strength improves the blending of musical sounds and increases their loudness and richness as well as the continuity of musical line. The importance of all these effects for musical enjoyment becomes obvious if one listens to music in an environment virtually free of reverberation, as for example to a military band or a light orchestra playing outdoors: the sounds are brittle and harsh, and it is obvious that it is of no advantage to be able to hear every detail. Furthermore, the loudness of music heard outdoors is reduced rapidly as the distance increases from the sound source.

But perhaps the most important reason why relatively long reverberation times are adequate for music is simply the fact that listeners are accustomed to hearing music in environments which happen to have reverberation times of the order of magnitude mentioned. This applies equally well to composers who unconsciously take into account the blending of sounds which is produced in concert halls of normal size.

As regards the frequency dependence of the reverberation time, it is generally considered tolerable, if not as favourable, to have an increase of the reverberation time towards lower frequencies, beginning at about 500 Hz (see Table 7.4). From the physical point of view, such an increase is quite natural since the sound absorption from the audience is generally lower at low frequencies than it is at medium and high frequencies (see, for instance, Tables 6.3 and 6.4). Concerning the subjective sensation, it is often believed that increasing the reverberation time towards low frequencies is responsible for what is called the 'warmth' of musical sounds. On the other hand, there are quite a number of concert halls with reverberation times which do not increase towards the low frequencies or which even have a slightly decreasing reverberation time and which are nevertheless considered to be excellent acoustically.

The optimum range of reverberation time as indicated above refers to the performance of orchestral and choral music. Smaller ensembles perform often in halls having seating capacities of 400–700 and reverberation times

Table 7.4 Reverberation time of concert halls, fully occupied

| Name and location | Volume ($m^3$) | Seating capacity | Year of completion (reconstruction) | $T_{125}$ | $T_{500-1000}$ | Source |
|---|---|---|---|---|---|---|
| Großer Musikvereinssaal, Vienna | 14 600 | 2000 | 1870 | 2.3 | 2.05 | Beranek |
| St. Andrew's Hall, Glasgow | 16 100 | 2130 | 1877 | 2.1 | 2.2 | Parkin et al. |
| Chiang Kai Shek Memorial, Taipei | 16 700 | 2077 | 1987 | 1.95 | 2.0 | Kuttruff |
| Symphony Hall, Boston | 18 800 | 2630 | 1900 | 2.2 | 1.8 | Beranek |
| Concertgebouw, Amsterdam | 19 000 | 2200 | 1887 | 2.3 | 2.2 | Geluk |
| Neues Gewandhaus, Leipzig | 21 000 | 1900 | 1884 (1981) | 2.0 | 2.0 | Fasold |
| Neue Philharmonie, Berlin | 24 500 | 2230 | 1963 | 2.4 | 1.95 | Cremer |
| Concert Hall De Doelen, Rotterdam | 27 000 | 2220 | 1979 | 2.3 | 2.2 | De Lange |

ranging from 1.5 to 1.7 s.[16] These shorter decay times are not only dictated by the smaller size of a chamber music hall but are more adequate to the character of this kind of music.

Even for orchestral music the optimum reverberation time depends on the style of music to be performed. This has been investigated by Kuhl[17] in a remarkable round robin experiment. In this experiment three different pieces of music were recorded in many concert halls and broadcasting studios with widely varying reverberation times. These recordings were later replayed to a great number of listeners—musicians as well as acousticians, music historians, recording engineers and other engineers, i.e. individuals who were competent in that field in one way or another. They were asked to indicate whether the reverberation times in the different recordings, whose origin they did not know, appeared too short or too long. The pieces of music played back were the first movement of Mozart's Jupiter Symphony (KV551), the fourth movement of Brahms' 4th Symphony (E minor) and the Danse Sacrale from Stravinsky's Le Sacre du Printemps. The final result was that a reverberation time of 1.5 s was considered to be most appropriate for the Mozart symphony as well as for the Stravinsky piece, whereas 2.1 s was felt to be most suitable for the Brahms symphony. In the first two pieces there was almost complete agreement in the listeners' opinions; in the Brahms symphony, however, there was considerable divergence of opinion.

These results should certainly not be overvalued since the conditions under which they have been obtained were far from ideal in that they were based on monophonic recordings, replayed in rooms with some reverberation. But they show clearly that no hall can offer optimum conditions for all types of music and explain the large range of 'optimum' reverberation times for concert halls.

In opera houses the listener should be able to enjoy the full sound of music as well as to understand the text, at least partially. Therefore one would expect that these somewhat contradictory requirements can be reconciled by a compromise as far as the reverberation time is concerned, and that consequently the optimum of the latter would be somewhere about 1.5 s. As a matter of fact, however, the reverberation times of well-renowned opera theatres scatter over a wide range (see Table 7.5). Traditional theatres have reverberation times close to 1 s only, whereas more modern ones show a definite trend towards longer values. One is tempted to explain these differences by a changed attitude of the listeners, who nowadays seem to give more preference to a full and smooth sound of music than to the intelligibility of the text, whereas earlier opera goers presumably just wanted to be entertained by the plot. This trend is supported by the tendency to perform operas in the original language and to display the translated text on a projection board. There is still another possible reason for the tendency towards longer reverberation times: Old theatres were designed in such a way as to

Table 7.5 Reverberation time of opera theatres, fully occupied

| Name and location | Volume ($m^3$) | Seating capacity (standees) | Year of completion (reconstruction) | $T_{125}$ | $T_{500-1000}$ | Source |
|---|---|---|---|---|---|---|
| La Scala, Milano | 10 000 | 2290 (400) | 1778 (1946) | 1.2 | 0.95 | Furrer |
| Covent Garden, London | 10 100 | 2180 (60) | 1858 | 1.2 | 1.1 | Parkin et al. |
| Festspielhaus, Bayreuth | 11 700 | 1800 | 1876 | 1.7 | 1.5 | Reichardt |
| National Theatre, Taipei | 11 200 | 1522 | 1987 | 1.6 | 1.4 | Kuttruff |
| Staatsoper, Wien | 11 600 | 1658 (560) | 1869 (1955) | 1.5 | 1.3 | Reichardt |
| Staatsoper, Dresden | 12 500 | 1290 | 1878 (1985) | 2.3 | 1.7 | Kraak |
| Neues Festspielhaus, Salzburg | 14 000 | 2158 | 1960 | 1.7 | 1.5 | Schwaiger |
| Metropolitan Opera House, New York | 30 500 | 3800 | 1966 | 2.25[a] | 1.8[a] | Jordan |

[a] With 80% occupancy.

seat as many spectators as possible, whereas the architects of more modern ones (including the Festspielhaus in Bayreuth, which was specially designed to stage Wagner's operas) tried to follow more or less elaborate acoustical concepts.

The rehearsal room of an orchestra cannot be expected to offer the same acoustical conditions as the hall where the orchestra performs. It has typically a volume of 1500–2500 $m^3$ and its reverberation time should not exceed about 1.2 s in order to ensure transparency of the produced sounds and to keep the sound level within tolerable limits. By providing for variable sound absorption in form of curtains, etc., musicians should be given the opportunity to do some experimentation in order to find the optimum conditions themselves.

The question of optimum reverberation times is even more difficult to answer if we turn to churches and other places of worship which cannot be considered merely under the heading of acoustics. It depends on the character of the service whether more emphasis is given to organ music and liturgical chants or to the sermon. In the first case longer reverberation times are to be preferred, but in the latter the reverberation time should certainly not exceed 2 s. Churches with still shorter reverberation times are not well accepted by the congregation for reasons which have nothing to do with acoustics. This shows that the churchgoers' acoustical expectations are not only influenced by rational arguments such as that of speech intelligibility but also by hearing habits.

As mentioned above, the reverberation time is a meaningful measure for the duration of the decay process as long as the latter is exponential, i.e. if the decay level decreases linearly with time. If, on the contrary, a logarithmic decay curve is bent and consequently each section of it has its own decay rate, the question arises as to which of these sections is most significant for the subjectively perceived 'reverberance' of a room.

To answer this question Atal et al.[18] passed speech and music samples through an artificial reverberator consisting of a combination of computer-simulated comb filters (see Section 10.5) with non-exponential decay characteristics. The signals modified in this way were presented with earphones to test subjects, who were asked to compare them with exponentially reverberated signals in order to find the subjectively relevant decay rate of the non-exponential decays. The results are plotted in Fig. 7.18. The abscissa represents the reverberation time corresponding to the initial slope (first 160 ms) of the non-exponential decay; the ordinate is the reverberation time of an exponential decay giving the same impression of reverberance, i.e. the subjectively effective reverberation time. Similar results were obtained with sound signals reverberated in concert halls.

These findings can be explained by the fact that the smoothing effect of reverberation on the irregular level fluctuations of continuous speech or music is mainly achieved by the initial portion of the decay process, while

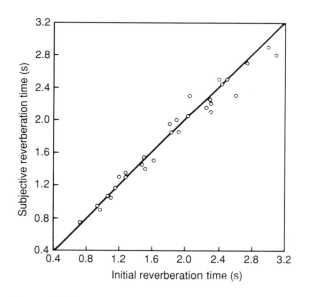

*Figure 7.18* Subjective reverberation time as a function of the initial reverberation time $T_{160}$ (after Atal et al.[18]).

its later portions add up to some general 'background' which is not felt subjectively as carrying much information on the signal. Only final or other isolated chords present the listener with the opportunity of hearing the complete decay process; but these chords occur too rarely for them to influence to any great degree the overall impression which a listener gains of the hall's reverberance.

Nowadays it has become common to characterise the rate of initial sound decay by the 'early decay time' (EDT) following a proposal of Jordan.[19] This is the time in which the first 10 dB fall of a decay process occurs, multiplied by a factor 6. It is mostly shorter than the Sabine reverberation time. Listening tests based on binaural impulse responses recorded in different concert halls confirmed that the perceived reverberance is closely related to EDT.[11]

The overall reverberation time does not show substantial variations with room shape. This is so because the decay process as a whole is made up of numerous reflections with different delays, strengths and wall portions where they originated. On the contrary, the 'early decay time' is strongly influenced by early reflections, and therefore depends noticeably on the measuring position; furthermore, it is sensitive to details of the room's geometry. In this respect it resembles to some extent the parameters discussed in the preceding sections.

## 7.6 Sound pressure level, strength factor

For a long time the stationary sound pressure level or energy density that a sound source produces in a hall was not considered as an acoustical quality criterion because it depends mostly on the power output of the source and the absorption area (or the reverberation time) of the room. In recent times, however, the general attitude towards the overall level has changed since high definition or clarity is of little use if the sound is too weak to be heard at a comfortable loudness. Moreover, the simple eqn (5.6), which relates the energy density to the absorption area and the source power P, is valid for diffuse sound fields only, and the field within a real hall deviates more or less from this ideal condition.

If the sound pressure level (SPL) in an enclosure is to merely reflect properties of the enclosure and not of the source, it must be measured by using a non-directional sound source and its power output must be accounted for by some suitable normalization. This can be done by subtracting the level $SPL_A$ which the sound source would produce in an anechoic room at a distance $r_A = 10$ m from the SPL, i.e. the level observed in the enclosure. The result is what is known as the 'strength factor' nowadays. Since the energy density at some place is proportional to the squared sound pressure of the direct sound plus that of all reflections, the strength factor is equal to the ratio:

$$G = 10 \log_{10} \left[ \int_0^\infty [g(t)]^2 \, dt \Big/ \int_0^\infty [g_A(t)]^2 \, dt \right] \quad dB \qquad (7.15)$$

where $g_A$ is the impulse response measured with the same sound source in an anechoic room at 10 m distance.

By inserting $A = 0.163V/T$ into eqn (5.6) and setting $w_A = P/4\pi r_A^2 c$ the strength factor which would be encountered in a diffuse sound field is

$$G_d = 10 \log_{10} \left( \frac{w}{w_A} \right) = 10 \log_{10} \left( \frac{T}{V} \right) + 45 \, dB \qquad (7.16)$$

On the other hand, Gade and Rindel[20] found from measurements in 21 Danish concert halls that the strength factor shows a linear decrease from the front to the rear of any hall which corresponds to 1.2–3.3 per distance doubling, and that its average in each hall falls short of the value predicted by eqn (7.16) by 2–3 dB. Furthermore, the steady-state level does not depend in a simple way on the reverberation time or on the geometrical data of the hall. From these results it may be concluded that the sound fields in real concert halls are not diffuse, and that the strength factor is indeed a useful figure of merit.

## 7.7 Spaciousness of sound fields

The preceding discussions of this chapter predominantly referred to the temporal structure of the impulse response of a room and to the auditive sensations associated with them. A subjective effect not mentioned so far, which is nevertheless of utmost importance, at least for concert halls, is the acoustical 'sensation of space' which a listener usually experiences in a room. It is caused by the fact that the sound in a closed room reaches the listener from quite different directions and that our hearing, although not able to locate these directions separately, processes them into an overall impression, namely the mentioned sensation or feeling of space.

It is quite evident that this sensation is not achieved just by reverberation. If music recorded in a reverberant room is replayed through a single loudspeaker in a non-reverberant environment, it never suggests acoustically the illusion of being in a room of some size, no matter if the reverberation time is long or short. Likewise, if the music is replayed through several loudspeakers which are placed at equal distance but in different directions seen from the listener, and which are fed by identical signals, the listener will not feel more enveloped by the sound. Instead, all the sound seems to arrive from a single imaginary sound source, a so-called 'phantom source' which can easily be located and which seems, at best, somewhat more extended than a single loudspeaker. The same effect occurs if a great number of loudspeakers in an anechoic room are arranged in a hemisphere (see Fig. 6.21) and are connected to the same signal source. A listener at the centre of the hemisphere does not perceive a 'spacious' or 'subjectively diffuse' sound field but instead he perceives a phantom sound source immediately overhead. Even the usual two-channel stereophony employing two similar loudspeaker signals, which differ in a certain way from each other, cannot provide a full acoustical impression of space since the apparent directions of sound incidence remain restricted to the region between both loudspeakers.

The 'sensation of space' has attracted the interest of acoustic researchers for many years, but only since the late 1960s has real progress been made in finding the cause of this subjective property of sound fields. The different authors used expressions like 'spatial responsiveness', 'spatial impression', 'ambience', 'apparent source width', 'subjective diffusion', 'Räumlichkeit', 'spaciousness', 'listener envelopment' and others to circumscribe this sensation. Assuming that all these verbal descriptions signify the same thing, we prefer the term 'spaciousness' or 'spatial impression' in the following.

For a long time it was common belief among acousticians that spaciousness was a direct function of the uniformity of the directional distribution (see Section 4.3) in a sound field: the more uniform this distribution, the higher the degree of spaciousness. This belief originated from the fact that the ceiling and walls of many famous concert halls are highly structured by cofferings, niches, pillars, statuettes and other projections which supposedly reflect the sound in a diffuse manner rather than specularly.

It was the introduction of synthetic sound fields as a research tool which led to the insight that the uniformity of the stationary directional distribution is not the primary cause of spaciousness. According to Damaske,[21] spatial impression can be created with quite a few synthetic reflections provided they reach the listener from lateral directions, and the signals they carry are mutually incoherent (as is usually true in a large hall). In a recent publication[22] which summarizes and amends his earlier investigations, Damaske proved that the objective directional distribution has not much to do with the subjective feeling of space for which he coined the term 'diffuseness' in order to distinguish it from 'diffusion' or 'diffusivity', which is a physical sound field property. According to this work, diffuseness is caused by early reflections with delays of up to, say, 50 ms.

That only the lateral reflections (i.e. not reflections from the front, from overhead or from the rear) contribute to spaciousness has been observed by several earlier authors, and many subsequent publications have confirmed this finding. A very extensive investigation into spaciousness was made by Barron.[23] He found that the contribution of a reflection to spaciousness is proportional to its energy and to $\cos\theta$, where $\theta$ is the angle between the axis through a listener's ears and the direction of sound incidence, provided the delay is in the range from 5 to 80 ms. Furthermore, this contribution is independent of other reflections and of the presence or absence of reverberation. Based on these results the 'lateral energy fraction' (LEF)[24] was proposed as an objective measure for the spatial impression:

$$LEF = \int_{5\,ms}^{80\,ms} [g(t)\cos\theta]^2\,dt \left/ \int_{0}^{80\,ms} [g(t)]^2\,dt \right. \tag{7.17}$$

(The replacement of $\cos\theta$ with $\cos^2\theta$ in the numerator of this expression is a concession to the simplicity of measurement.) The LEF is usually averaged over four octave bands of 125 Hz, 250 Hz, 500 Hz and 1000 Hz mid-frequency. This accounts for the fact that the low- and mid-frequency components contribute most to the sensation of spaciousness. In large halls, the lateral energy fraction may vary from 0 to about 0.5.

Another way of characterising the laterality of reflected sounds is based upon the fact that sound impinging on a listener's head from its vertical symmetry plane will produce equal sound pressures at both his ears, whereas a sound wave from outside the symmetry plane will produce different ear signals. Generally, the similarity or dissimilarity of two signals is measured by their cross-correlation function or by the correlation coefficient as defined in eqn (1.41). In the following we apply this equation to the impulse responses $g_r$ and $g_l$ measured at the right and the left ear. Since these signals are transient, time averaging would be meaningless and is therefore replaced by integration over the time. Then the correlation coefficient reads:

$$\varphi_{rl}(\tau) = \int_{t_1}^{t_2} g_r(t)g_l(t+\tau)\,dt \left/ \left\{ \int_{t_1}^{t_2} [g_r(t)]^2\,dt \int_{t_1}^{t_2} [g_l(t)]^2\,dt \right\}^{1/2} \right. \qquad (7.18)$$

Usually the limits $t_1$ and $t_2$ are set 0 and 100 ms in order to restrict the integration over the arrival times of the 'early reflections'; hence, $\varphi_{rl}$ is a 'short time' correlation factor. The maximum of its absolute value within the range $|\tau| < 1$ ms is called the 'interaural cross-correlation' (IACC):[25]

$$\text{IACC} = \max\{|\varphi_{rl}|\} \qquad \text{within} - 1 \text{ ms } < \tau < 1 \text{ ms} \qquad (7.19)$$

Of course, this quantity is negatively correlated with the spatial impression, and high values of the IACC mark a low degree of spaciousness and vice versa.

It should be mentioned that several versions of the IACC are in use which differ in the integration limits in eqn (7.18), or in the way the impulse responses are frequency-filtered before processing them according to eqn (7.18). Beranek[26] prefers the use of an IACC version which comprises three octave bands with mid-frequencies 0.5, 1 and 2 kHz. This limitation takes into consideration that at low frequencies the difference between $g_r$ and $g_l$ is negligibly small. Beranek reports that the values of this quantity are as low as 0.3 in excellent concert halls.

Laterality of the early reflections is certainly the most decisive factor on which the impression of spaciousness depends but it is not the only one. Several researchers have observed that the spatial impression increases monotonically with the listening level. Furthermore, there seems to be some agreement nowadays that spatial impression is not a 'one-dimensional' sensation, but consists of at least two different components which are more or less independent. The most prominent of these are the 'apparent source width' (ASW) and 'listener envelopment' (LEV).[27] If this view is adopted it is reasonable to attribute Damaske's and Barron's results to the ASW and hence to consider the objective quantities LEF and IACC as predictors of this particular aspect of spaciousness.

This leaves us with the question of which objective parameters of the sound field are responsible for the remaining component, namely the sense of 'listener envelopment'.

This question was the subject of careful listening experiments performed by Bradley and Soulodre[27] using a synthetic sound field similar to that shown in Fig. 7.4. With this system the reverberation time, the early-to-late energy ratio expressed by the clarity index $C_{80}$ (see eqn (7.9), the $A$-weighted sound pressure level, and the angular distribution of the reverberated signals could be independently varied, whereas the IACC and the LEF were kept constant through all combinations.

It turned out that the angular distribution had the largest effect on spaciousness. The wider it was, the higher is the listener envelopment. Another strong influence on the perceived LEV was that of the overall sound level, while the relative strength of the early energy ($C_{80}$) and the reverberation time was found to be less significant. A correlation analysis of the collected data revealed that the best objective predictor for listener envelopment is the strength (see eqn (7.15)) of the late lateral reflections

$$LG_{80}^{\infty} = 10 \log_{10} \left\{ \int_{80\text{ms}}^{\infty} [g(t) \cos \theta]^2 \, dt \Big/ \int_0^{\infty} [g_A(t)]^2 \, dt \right\} \quad \text{decibels}$$

(7.20)

averaged over four octave bands with mid-frequencies 125, 250, 500 and 1000 Hz, thus accounting for the fact that high-frequency components do not contribute much to the sense of listener envelopment. The range from missing to maximum envelopment corresponds to a range in $LG_{80}^{\infty}$ from $-20$ to 2 dB.

We return to the measures of ASW: namely, the LEF and the IACC. Both quantities are not very highly correlated to each other, which may be a consequence of the different frequency ranges in which they are usually determined. Another puzzling finding is that these parameters show considerable fluctuations with the position of the receiver (microphones or dummy head), as has been revealed by careful experiments carried out by de Vries et al.[28] These fluctuations are caused by interference of different components of the wave field and are not accompanied by corresponding variations of the ASW. Even within one seat width, the fluctuations of the parameter are quite high although it is inconceivable that a listener in a concert hall will perceive any change in spatial impression when he moves his head by a few centimetres. This shows us that still more research on significant spaciousness parameters is needed.

## 7.8 Assessment of concert hall acoustics

Although nowadays quite a number of parameters are at the acoustician's disposal to quantify the listening conditions in a concert hall (or particular aspects of them), the situation is still unsatisfactory in that important questions remain unanswered. Do these parameters yield a complete description of the acoustics of a hall? Are they independent from each other?

Which relative weight is to be given to each of them? Is it possible, for instance, to compensate insufficient reverberation by a large amount of early lateral energy?

Conventional attempts to correlate the acoustical quality of a concert hall with an objective measure or a set of them have not been very satisfactory because they have concentrated on one particular aspect only or, as for instance Beranek's elaborate rating system,[29] relied on plausible but unproven assumptions. Since about 1970, however, researchers have tried to get a complete picture of the factors which contribute to good acoustics, including their relative significance, by employing modern psychometric methods.

The basic idea of this approach is first to detect the number and significance of independent scales of auditive perception and then, as a second step, to find out the physical parameters which show the highest correlation to these scales.

The standard method of collecting the material for such an analysis is to record music samples in different concert halls, or at different places in one hall, using a dummy head. In a second step, these are presented to test subjects, either by headphones or by loudspeakers in an anechoic room. In the latter case it is advantageous to employ the cross-talk cancellation (CTC) techniques described in the introduction to this chapter. The subjects are asked to assess in some way the differences between subsequent listening impressions. This can be done by assigning scores to the presented signals using a number of bipolar rating scales with verbally labelled extremes such as 'dull–brilliant', 'cold–warm', etc. Another possibility is having the subjects indicate which of two presentations they prefer, and to construct a preference scale on the basis of these judgements.

Subsequently, these data, collected from many listeners, are subjected to a mathematical procedure called factor analysis. In the following an attempt is made to give at least an idea of this technique. A more detailed explanation of factor analysis may be found in Ref. 3 of Chapter 4.

As a first step, the test data are arranged in the columns (objects) and rows (attributes) of a matrix $Z$ from which the so-called 'correlation matrix' $C$ is derived:

$$C = \frac{1}{N-1}ZZ' \tag{7.21}$$

$Z'$ is the transposed matrix, i.e. the matrix $Z$ with lines and columns interchanged, and $N$ is the number of 'objects'(for instance, halls or different locations in a hall).The matrix $C$ is quadratic, with $n \times n$ elements, where $n$ denotes the number of attributes. Next, the matrix $C$ is orthogonalized, which means it is transformed into a matrix $D$, the only elements of which are the ordered eigenvalues of $\mathbf{C}$ situated on its main diagonal while all the other elements are zero. This is achieved by finding a suitable transformation matrix $\mathbf{W}$:

$$\mathbf{WCW}^{-1} = \mathbf{D} = \begin{pmatrix} \lambda_1 & 0 & 0 \cdots \cdots 0 \\ 0 & \lambda_2 & 0 \cdots \cdots 0 \\ \cdots \cdots \cdots \cdots \\ 0 & 0 & 0 \cdots \cdots \lambda_n \end{pmatrix} \tag{7.22}$$

$\mathbf{W}^{-1}$ is the inverse of $\mathbf{W}$. At the same time $\mathbf{W}$ transforms the space of original attributes (or preferences) into a perceptual space, the coordinate axes of which are the 'factors', i.e. independent aspects of perception. Usually, the number of significant factors is three or four. The meaning of these factors or scales is principally unknown beforehand, but sometimes they can be circumscribed vaguely by verbal labels such as 'resonance' or 'proximity'.

Suppose the perceptual space consists of just two factors. Then the perceptual space is a plane with rectangular coordinates $F_1$ and $F_2$, as shown in Fig. 7.19. If the preceding factor analysis was based on preference tests, the individual preference scale of each subject can be represented as a vector in this plane. Figure 7.19 shows just one of them. According to the angle $\alpha$ which it includes with the axis $F_1$, this particular listener gives in his preference judgement a weight of $\cos \alpha$ to the factor $F_1$ and of $\sin \alpha$ to the factor $F_2$. The projections of the 'concert hall points' A, B, C, etc., on this vector

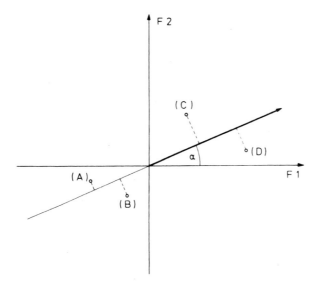

*Figure 7.19* Two-dimensional perceptual space with factors $F_1$ and $F_2$. The vector represents the preference scale of one particular subject. The points A, B, C, etc., indicate different concert halls or seats.

indicate this listener's personal preference rating of these halls. Similarly, if bipolar rating scales have been used, these scales can be presented as vectors in such a diagram.

A careful study using bipolar rating scales has been carried out by Wilkens[30] and his coworkers. They collected 30 test samples by accompanying the Berlin Philharmonic Orchestra on a concert tour where it played the same programme (Mozart, Bartok, Brahms) in six different halls. Wilkens used headphones for the reproduction of his samples and found three factors to be relevant, which he labelled as follows:

$F_1$: strength and extension of the source (47%)
$F_2$: clearness or distinctness (28%)
$F_3$: timbre (14%)

The numbers in parentheses denote the relative significance of the three factors. They add up to 89% only, which means that there are still more (however, less significant) factors. It is interesting to note that there seem to be two groups of listeners: one group which prefers loud sounds (high values of $F_1$) and one giving more preference to distinct sounds (high values of $F_2$). (A similar division of the subjects into two groups with different preferences has been found by Barron.[31] The members of the first group prefer 'intimacy' while the other group give preference to 'reverberance') In the course of this research project, physical sound field parameters have been measured at the same positions in which the sound recordings have been made. A subsequent correlation analysis revealed that $F_1$ is highly correlated with the strength factor $G$ from eqn (7.15), whereas $F_2$ shows high (negative) correlation with Kürer's 'centre time' $t_s$ defined in eqn (7.12). Finally, the factor $F_3$ seems to be related to the frequency dependence of the 'early decay time'.

Siebrasse[32] and his co-workers collected their test samples in a different way. They replayed a 'dry', i.e. reverberation-free motif of Mozart's Jupiter Symphony stereophonically from the stages of 25 European concert halls and re-recorded them with a dummy head placed at various positions. The samples prepared in this way were presented to the test subjects in the free field at constant level employing a CTC system mentioned in the Introduction to this chapter (see also Ref. 33). The subjects were asked to judge preference between pairs of presentations. Application of factor analysis indicated four factors $F_1$–$F_4$ with relative significance of 45%, 16%, 12% and 7%, respectively. In Fig. 7.20, which is analogous to Fig. 7.19, part of Siebrasse's results are plotted in the plane of factors $F_1$ and $F_2$. The vectors representing the individual preference scales have different lengths since they have nonvanishing components also in $F_3$- and $F_4$-directions. The fact that they all are directed towards the right side leads to the conclusion that $F_1$ is a 'consensus factor', whereas the components in the $F_2$-direction reflect differences in the listener's personal taste. This holds even more with regard to the factors $F_3$ and $F_4$.

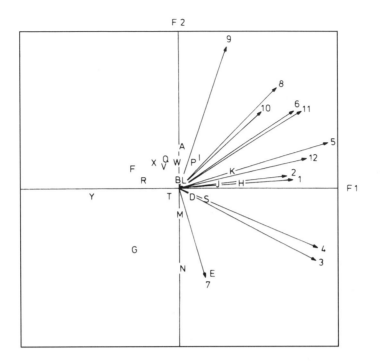

*Figure 7.20* Individual preference scales of 12 test subjects (arrows) and 22 concert halls (points A, B, ... ,Y) represented in the $F_1 - F_2$ plane of a four-dimensional perceptual space (from Schroeder et al.[33]).

In order to find out the objective sound field parameter with the highest correlation to the consensus factor $F_1$, Gottlob[34] analysed the impulse responses observed at the same places where the music samples had been recorded. The reverberation time proved to be important only when it deviated substantially from an optimum range centred on 2 s. If halls with unfavourable reverberation times are excluded, $F_1$ shows high correlation with several parameters, of which the IACC (see Section 7.7) seems to be of particular interest, since it is virtually independent of the reverberation time. Another highly correlated quantity is the width of the concert hall. Both the IACC and the width of the hall are negatively correlated with $F_1$, i.e. narrow concert halls are generally preferred. This again proves the importance of early lateral reflections, which are particularly strong in narrow halls but which lead to a low value of the IACC.

The fact that both investigations show such poor agreement may have several explanations, such as the differences in the way the test samples were collected and presented to the subjects, different choice of halls and different judging schemes. It is not surprising, however, that in the second study the strength or loudness of the sound signal does not show up since here all test samples were presented to the subjects at equal loudness.

A different approach to arrive at a consistent rating system for concert halls was proposed by Ando,[35] who simulated sound fields consisting of a direct sound, two distinct reflections and subsequent electronically generated reverberation (see Section 10.5). The latter was radiated by four loudspeakers which were distributed along a circle and fed with mutually delayed signals. The electrical signals were computer controlled in such a way that four independent parameters could be varied: namely, the overall listening level, the 'initial time delay gap' $\Delta t_1$ (i.e. the delay of the first reflection with respect to the direct sound), the reverberation time $T$ and the IACC. Ando summarised the results of his extended preference tests in a figure-of-merit which he called the 'total subjective preference':

$$S_a = -\sum_{i=1}^{4} \alpha_i |X_i|^{3/2} \tag{7.23}$$

with

$$X_1 = 20 \log_{10}\left(\frac{p}{p_p}\right), \qquad \alpha_1 = \begin{cases} 0.07 \text{ for } X_1 \geq 0 \\ 0.04 \text{ for } X_1 < 0 \end{cases}$$

where $p$ is the root-mean-square of the sound pressure in arbitrary units,

$$X_2 = \log_{10}\left(\frac{\Delta t_1}{(\Delta t_1)_p}\right), \qquad \alpha_2 = \begin{cases} 1.42 \text{ for } X_2 \geq 0 \\ 1.11 \text{ for } X_2 < 0 \end{cases}$$

$$X_3 = \log_{10}\left(\frac{T}{T_p}\right), \qquad \alpha_3 = \begin{cases} 0.45 + 0.74A \text{ for } X_3 \geq 0 \\ 2.36 - 0.42A \text{ for } X_3 < 0 \end{cases}$$

$$X_4 = IACC \qquad \alpha_4 = 1.45$$

$A$ is the square root of the energy contained in all reflections relative to the energy of the direct sound.

The quantities $X_1$, $X_2$ and $X_3$ are deviations of the A-weighted sound pressure level, of the 'initial time delay gap' $\Delta t_1$ and of the reverberation time $T$ from certain standard values, the so-called 'most preferred values' marked with the subscript p. The weighting factors $\alpha_i$ depend on the sign of the deviation. Here $p_p$ corresponds to a level of 79 dB(A), whereas $(\Delta t_1)_p = (1 - \log_{10} A)\tau_e$ and $T_p = 23\tau_e$. The quantity $\tau_e$ is the effective duration of the autocorrelation function of the sound signal (see Section 1.7). The most preferred value of the IACC is zero. Ando's 'total subjective preference'

$S_a$ vanishes if all the parameters assume their most preferred values; negative values of $S_a$ indicate certain acoustical deficiencies.

Beranek[26] has modified this rating scheme by adding two further components, —'warmth' (equal to low-frequency divided by mid-frequency reverberation time) and an index characterising the surface diffusivity of a concert hall, determined by visual inspection. He applied it to his collection of data on concert halls and opera theatres. On the other hand, Beranek attributed each of these concert halls to one of six categories of acoustical quality, ranging from 'fair' (category C) to 'superior' (category A+), by evaluating numerous interviews with musicians and music critics and found satisfactory agreement between these subjective categories and the rating numbers derived from the objective data.

In a way, Ando's approach seems to answer the questions raised at the beginning of this section. However, it does not fully agree with the reported results of factor analysis from which the initial time delay gap did not emerge as a significant acoustical criterion. Furthermore, it takes it for granted that there exists an optimum acoustical environment regardless of differences in listeners' taste. Consequently, if all designers of concert halls decided to follow the guidelines of this system, the result would be halls not only equally good but of similar acoustics. One may doubt whether this is a desirable goal of room acoustical efforts, since differences in listening environments are just as enjoyable as different architectural solutions or different musical interpretations.

We have to admit that the insights which have been reported in the preceding sections do not combine to form a well-rounded picture of concert hall acoustics; they are not free of inconsistencies and even contradictions, and hence are to be considered as preliminary only. Nevertheless, it is obvious that important insights have been gained in what is relevant for good concert hall acoustics. These insights along with the auralisation techniques to be described in Section 9.7 will certainly help acoustical designers to create concert halls with satisfactory or even excellent listening conditions and to avoid serious mistakes.

# References

1 Atal BS, Schroeder MR. Gravesaner Blätter 1966; 27/28:124.
2 Neu G, Mommertz E, Schmitz A. Untersuchungen zur richtungstreuen Schallwiedergabe bei Darbietung von kopfbezogenen Aufnahmen über zwei Lautsprecher. Acustica 1992; 76:183; Urbach G, Mommertz E, Schmitz A. Acustica 1992; 77:153.
3 Schubert P. Die Wahrnehmbarkeit von Rückwürfen bei Musik. Zeitschr Hochfrequenztechn Elektroakust 1969; 78:230.
4 Haas H. Über den Einfluß eines Einfachechos auf die Hörsamkeit von Sprache. Acustica 1951; 149.

5 Muncey RW, Nickson AFB, Dubout P. The acceptability of speech and music with a single artificial echo. Acustica 1953; 3:168.

6 Meyer E, Schodder GR. Über den Einfluß von Schallrückwürfen auf Richtungslokalisation und Lautstärke bei Sprache. Nachr Akad Wissensch Göttingen Math-Phys Kl 1952; No. 6:31.

7 Atal BS, Schroeder MR, Kuttruff H. Perception of coloration in filtered Gaussian noise—short-time spectral analysis by the ear. Proc Fourth Intern Congr Acoustics, Copenhagen, 1962, Paper H31.

8 Dietsch L, Kraak W. Ein objektives Kriterium zur Erfassung von Echostörungen bei Musik- und Sprachdarbietungen. Acustica 1986; 60:205.

9 Meyer E, Thiele R. Raumakustische Untersuchungen in zahlreichen Konzertsälen und Rundfunkstudios unter Anwendung neuerer Meßverfahren. Acustica 1956; 6:425.

10 Reichardt W, Abdel Alim O, Schmidt W. Abhängigkeit der Grenzen zwischen brauchbarer und unbrauchbarer Durchsichtigkeit von der Art des Musikmotivs, der Nachhallzeit und der Nachhalleinsatzzeit. Appl Acoustics 1974; 7:243.

11 Gade AC. Acoustic properties of concert halls in the US and in Europe. Proc Sabine Centennial Symposium, Cambridge, Mass. Woodbury, New York: Acoustical Society of America, 1994; 191.

12 Kürer R. Zur Gewinnung von Einzahlkriterien bei Impulsmessung in der Raumakustik. Acustica 1969; 21:370; Einfaches Messverfahren zur Bestimmung der "Schwerpunktszeit" raumakustischer Impulsantworten. Proc Seventh Intern Congr on Acoustics, Budapest, 1971, Paper 23 A5.

13 Houtgast T, Steeneken HJM. The modulation transfer function in room acoustics as a predictor of speech intelligibility. Acustica 1973; 28:66; Steeneken HJM, Houtgast T. A physical method for measuring speech-transmission quality. J Acoust Soc Am 1980; 67:318.

14 Houtgast T, Steencken HJM. A multi-language evaluation of the RASTI-method for estimating speech intelligibility. Acustica 1984; 54:186.

15 Seraphim HP. Untersuchungen über die Unterschiedsschwelle exponentiellen Abklingens von Rauschbandimpulsen. Acustica 1958; 8:280.

16 Hidaka T, Nishihara N. Objective evaluation of chamber-music halls in Europe and Japan. J Acoust Soc Am 2004; 116:357.

17 Kuhl W. Über Versuche zur Ermittlung der günstigsten Nachhallzeit großer Musikstudios. Acustica 1954; 4:618.

18 Atal BS, Schroeder MR, Sessler GM. Subjective reverberation time and its relation to sound decay. Proc Fifth Intern Congr on Acoustics, Liege, 1965, Paper G32.

19 Jordan VL. Acoustical criteria for auditoriums and their relation to model techniques. J Acoust Soc Am 1970; 47:408.

20 Gade AC, Rindel JH. Die Abstandsabhängigkeit von Schallpegeln in Konzertsälen, Fortschr. d. Akustik – DAGA '85, Bad Honnef, DPG-GmbH, 1985, p 435.

21 Damaske P. Subjektive Untersuchung von Schallfeldern Acustica 1967/68; 19:199.

22 Damaske P. Acoustics and Hearing. Berlin: Springer-Verlag, 2008.

23 Barron M. The subjective effects of first reflections in concert halls. J Sound Vibr 1971; 15:475.

24 Barron M, Marshall AH. Spatial impression due to early lateral reflections in concert halls. J Sound Vibr 1981; 77(2):211.

25 Damaske P, Ando Y. Interaural crosscorrelation for multichannel loudspeaker reproduction. Acustica 1972; 27:232.

26 Beranek LL. Concert and Opera Halls: How They Sound. Woodbury, New York: Acoust Soc of America, 1996.

27 Bradley JS, Soulodre GA. Objective measures of listener envelopment. J Acoust Soc Am 1995; 98:2590.

28 de Vries D, Hulsebos EM, Baal J. Spatial fluctuations in measures for spaciousness. J Acoust Soc Am 2001; 110:947.

29 Beranek LL. Music, Acoustics and Architecture. New York: John Wiley, 1962.

30 Wilkens H, Plenge G. The correlation between subjective and objective data of concert halls. In: Mackenzie R (ed.). Auditorium Acoustics. London: Applied Science, 1974; Wilkens H. Eine mehrdimensionale Beschreibung subjektiver Beurteilungen der Konzertsaalakustik. Acustica 1977; 38:10.

31 Barron M. Subjective study of British symphony concert halls. Acustica 1988; 66:1.

32 Siebrasse KF. Vergleichende subjektive Untersuchungen zur Akustik von Konzertsälen. Dissertation, University of Göttingen, 1973.

33 Schroeder MR, Gottlob D, Siebrasse KF. Comparative study of European concert halls: correlation of subjective preference with geometric and acoustic parameters. J Acoust Soc Am 1974; 56:1195.

34 Gottlob D. Vergleich objektiver Parameter mit Ergebnissen subjektiver Untersuchungen an Konzertsälen. Dissertation, University of Göttingen, 1973.

35 Ando Y. Calculation of subjective preference at each seat in a concert hall. J Acoust Soc Am 1983; 74:873.

# Chapter 8

# Measuring techniques in room acoustics

The starting point of modern room acoustics is marked by attempts to define physical sound field parameters which can be considered as the objective counterpart to the various components of the subjective acoustic impression that a listener has when sitting in a hall and attending a presentation. As shown in Fig. 7.1, these parameters or a set of them can be thought of as a link between the subjective world and the realm of architecture which yields geometrical and other room data. As we have seen from Chapter 7, quite a number of such parameters have been established. It is the object of this chapter to describe how such quantities can be measured and which kind of equipment is required for this purpose.

Room acoustical measurements are not only necessary for research purposes but are also a valuable diagnostic tool. Suppose there are recurrent complaints about the acoustics of a particular hall. Then measurements are needed to reveal and to characterize the deficiencies and to find their cause. This holds not only if the complaining people are concert or opera goers but also to the same extent for performers, in particular for musicians. It is sometimes difficult to translate the comments of musicians, who often exaggerate their point into the language of acousticians. On the other hand, their opinions must be taken seriously; a concert hall is not very likely to get a good reputation if the performing artists are continuously dissatisfied. Another typical situation is that the performance of an electroacoustic reinforcement system is not satisfactory in its interaction with the room in which it is operated. Then the examination of the sound field indicates whether the location or the directivities of the loudspeakers should be improved, or whether it is rather a matter of readjusting the equaliser (is there is any). And finally, it may be advisable to take at least some measurements in a new or refurbished hall prior to its opening or reopening when minor corrections are still possible.

Other acoustical measurements which are not directly related to subjective impressions concern the acoustic properties of materials, especially of the absorption of materials for walls and ceiling, of seats, etc. Knowledge of such data is essential for any planning in room acoustics. Many of them can be found in published collections, but often the acoustical consultant is faced with new products or with specially designed wall linings for which no absorption data are available. This holds even more with regard to the scattering efficiency of acoustically 'rough' surfaces (see Section 8.8).

## 8.1 General remarks on instrumentation

Viewed from our present state of the art, the equipment which Sabine[1] had at his disposal for his famous investigation of reverberation appears quite modest: he excited the room under test with a few organ pipes and used his ear as a measuring instrument together with a simple stop watch. With the development and introduction of the electrical amplifier in the 1920s, almost all measuring techniques became electrical. In acoustics, the ear was replaced with a microphone and the fall in sound level was observed with an electromechanical level recorder. Furthermore, electrical filters became available, and the mechanical sound source was often replaced with a loudspeaker. More recently, the introduction of the digital computer has triggered off a second revolution in measuring techniques: all kinds of signal processing (filtering, storing and evaluation of measured signals as well as the presentation of results) can now be done with digital equipment which is generally more powerful, precise and flexible and less expensive than the more traditional equipment.

For field measurement of the reverberation time, the most convenient way to excite the room is still by firing a pistol which produces sufficient sound power even if there is some background noise. The same holds for wooden hand clappers or bursting balloons which yield strong excitation of a room, especially at low frequencies. However, for the generation of reproducible excitation signals the use of an electrical loudspeaker, driven by a power amplifier, is indispensable. This method enables the experimenter to produce a great variety of well-defined and reproducible excitation signals and thus to apply more sophisticated measuring schemes in order to improve the signal-to-noise ratio and to suppress non-linear distortions. This holds in particular for the precise measurement of room impulse responses.

For reverberation measurements, no strict requirements concerning the uniformity of radiation must be met since the various sound components will anyway be mixed during the decay process. This is different when if comes to measuring the correct impulse response of a room, or when the directional distribution of sound energy is to be observed. Then any directionality of the sound source should be avoided, otherwise the result of the measurement would depend not only on the position of the source but also on its orientation. One way to achieve omnidirectional sound radiation, at least at lower frequencies, is to employ 12 or 20 equal and equally fed loudspeaker systems mounted on the faces of a regular polyhedron (dodecahedron or icosahedron). A dodecahedron loudspeaker is shown in Fig. 8.1.

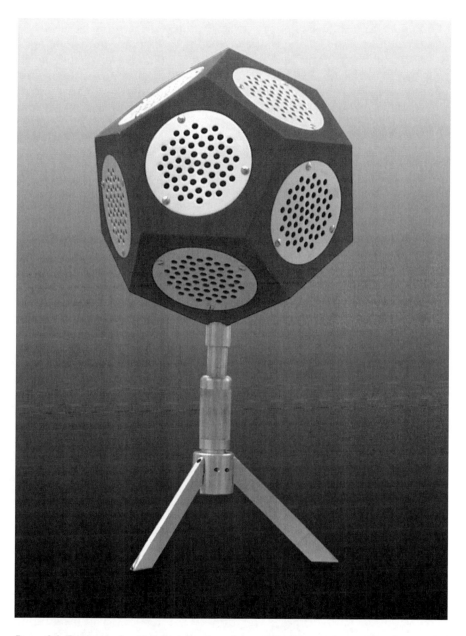

*Figure 8.1* Dodecahedron loudspeaker for measurements in room acoustics. (Institute of Technical Acoustics, Aachen)

It should be noted that even this kind of source shows some directionality at elevated frequencies. For this reason two polyhedral loudspeaker systems of different size are sometimes employed. Another useful device is the driver of a powerful horn loudspeaker with the horn replaced by a tube, the open end of which radiates the sound uniformly provided that its diameter is small compared with the wavelength. The resonances of the tube can be suppressed by inserting some damping material into it. Almost omnidirectional radiation of powerful acoustical wide-band impulses can also be achieved by specially designed electrical spark gaps.

To pick up the sound in the enclosure, pressure-sensitive microphones with omnidirectional characteristics are most commonly used. For certain measurements, however, a gradient receiver, i.e. figure-of-eight microphones, or a dummy head must be employed. Receivers with still higher directivity are needed to determine the directional distribution of sound energy at a certain position. Examples of directional microphones are an array of equal microphones, or a standard microphone mounted in the focus of a concave mirror, or the 'line microphone', which is the combination of a usual pressure microphone with a slotted tube. Although the experimentor should have a certain idea of the sensitivity of his microphones, absolute calibration is not required since virtually all measurements in room acoustics are relative sound measurements. To observe the sound intensity in a wave field, both the sound pressure and the particle velocity have to be measured, according to eqn (1.26). An indirect method of measuring particle velocity employs two pressure microphones with equal sensitivity separated by some distance $d$. If $d$ is very small compared to the acoustical wavelength, the sound pressure at both microphones differs by

$$\Delta p \approx \left(\frac{\partial p}{\partial x}\right) \cdot d = -\rho_0 \frac{\partial v_x}{\partial t} \tag{8.1}$$

i.e. it is proportional to the time derivative of the velocity component parallel to the line connecting both microphones. A direct measurement of the particle velocity is possible by means of a hot-wire anemometer. Probes of this kind developed for acoustic measurements are commercially available. To determine the complete intensity vector all components of the particle velocity must be measured, either by changing the orientation of the velocity probe or by using a sensor consisting of three velocity probes.[2]

After proper pre-amplification, the output signal of the microphone may be stored until further evaluation in the laboratory. This can be achieved with a magnetic tape recorder, preferably a digital recorder. The most convenient method, however, is to feed the microphone signal immediately into a portable digital computer where it is stored in the computer's hard disk or directly processed to yield the parameters which one is interested in. This requires, of course, that the computer is fitted out with an analog-to-digital converter which covers a sufficiently wide dynamic range.

## 8.2 Measurement of the impulse response

According to system theory all properties of a linear transmission system are contained in its impulse response or, alternatively, in its transfer function, which is the Fourier transform of the impulse response. Since a room can be considered as an acoustical transmission system, the impulse response yields a complete description of the changes a sound signal undergoes when it travels from one point in a room to another, and almost all of the parameters we discussed in the preceding chapter can be derived from it, at least in principle. Parameters related to spatial or directional effects can be based upon the 'binaural impulse response' picked up at both ears of a listener or a dummy head. From these remarks it is clear that the determination of impulse responses is one of the most fundamental tasks in experimental room acoustics. It requires high-quality standards for all measuring components, which must be free of linear or non-linear distortions including phase shifts. By its very definition the impulse response of a system is the signal obtained at the system's output after its excitation by a vanishingly short impulse (yet with non-vanishing energy), i.e. by a Dirac or delta impulse (see Section 1.4). Since we are interested only in frequencies below, say, 10 kHz, this signal can be approximated by a short impulse of arbitrary shape, the duration of which is smaller than about 20 $\mu$s. However, it should be kept in mind that all loudspeakers show at least linear distortions. This means a loudspeaker transforms a Dirac impulse $\delta(t)$ applied to its input terminal into a signal $g_{LS}(t)$. The response $g'(t)$ of the room to this signal is its own impulse response $g(t)$ convolved with $g_{LS}(t)$ after eqn (1.46) or (1.46a):

$$g'(t) = \int\limits_{-\infty}^{\infty} g(t')g_{LS}(t - t')\,dt' = g(t) * g_{LS}(t) \tag{8.2}$$

The linear distortion caused by the loudspeaker can be eliminated by 'deconvolution', i.e. by undoing the convolution shown in eqn (8.2): Let $G(f)$ and $G_{LS}(f)$ denote the Fourier transforms of $g(t)$ and $g_{LS}(t)$, respectively. Then the Fourier transform of eqn (8.2) reads

$$G'(f) = G(f) \cdot G_{LS}(f) \tag{8.3}$$

Dividing $G'(f)$ with $G_{LS}$ yields $G(f)$, the transfer function of the room, from which the corrected room impulse response $g(t)$ can be recovered by inverse Fourier transformation (see eqn (1.33a)). It is practical to avoid the influence of the loudspeaker by pre-emphasis, i.e. by passing the test signal through a filter with the transfer function $[G_{LS}(f)]^{-1}$ prior to feeding it to the loudspeaker.

In practice, however, impulse responses of a room measured with this straightforward scheme are more or less impaired by the omnipresent background noise: for instance, by traffic noise intruding into the room from the outside, or by noise from technical equipment, etc. One way to overcome this difficulty is by repeating the measurement several or many times and averaging the results. If $N$ is the number of measurements, the total energy of the collected impulse responses grows in proportion to $N^2$ while the energy of the resulting noise grows only proportionally to $N$. Hence the signal-to-noise ratio, expressed in decibels, is increased by $10 \cdot \log_{10} N$ decibels.

A less time-consuming method employs test signals which are stretched in time and hence can carry more energy than a short impulse. Suppose the system under test, i.e. the room, is excited by an arbitrary signal $s(t)$. Principally, the impulse response can be recovered from the output signal by applying the same recipe as before, namely by 'deconvolution'. The signal modified by transmission in the room is

$$s'(t) = \int_{-\infty}^{\infty} s(t')g(t-t')\,dt' = \int_{-\infty}^{\infty} g(t')s(t-t')\,dt' \qquad (8.4)$$

or expressed in the frequency domain

$$S'(f) = S(f) \cdot G(f) \qquad (8.5)$$

The transfer function $G(f)$ is obtained as $S'/S$ provided $S$ is non-zero within the whole frequency range of interest. Figure 8.2 shows the principle of this method, which is known as dual-channel analysis. If $s(t)$ is a random signal, for instance white noise, averaging over several or many single measurements is required to arrive at a reliable result. The execution of several Fourier transformations offers no difficulties using current

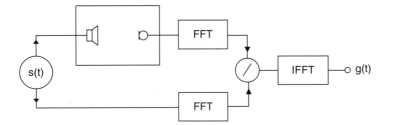

*Figure 8.2* Principle of dual-channel analysis. (FFT = Fast Fourier Transform, IFFT = Inverse FFT)

computer technology. If the signal $s(t)$ is known beforehand, its spectrum $S(f)$ must be calculated only once and can be used in all subsequent measurements.

Particularly well-suited for this technique are excitation signals with flat power spectra. Their autocorrelation function is a Dirac impulse (or a regular sequence of Dirac impulses if $s(t)$ is periodic). The frequency spectrum of such a signal $s(t)$ is of the all-pass type: $S(f) = (\exp[i\varphi(f)]$. Accordingly, $S(f)^{-1} = (\exp[-i\varphi(f)]) = S^*(f)$, where the asterisk indicates the conjugate-complex. Hence we obtain from eqn (8.5)

$$G(f) = S'(f) \cdot S^*(f) \qquad (8.6)$$

which corresponds to

$$g(t) = s'(t) * s(-t) \qquad (8.7)$$

in the time domain since $s(-t)$ is the inverse Fourier transform of $S^*(f)$. This means that the measured signal is passed through a filter, the impulse response of which is the time-inversed excitation signal. This techniques is also known as 'matched filtering'.

One signal of this kind is the Dirac impulse, as already discussed. Other test signals with the required property are trains of equally spaced Dirac impulses, the signs of which alternate according to a particular pattern. The most popular type of such binary signals are 'maximum length sequences'. Their application to room acoustics was first described by Alrutz and Schroeder.[3]

Maximum length sequences are periodic pseudorandom signals. The number of elements per period is

$$L = 2^n - 1$$

where $n$ is a positive integer. They can be generated by a digital $n$-step shift register with the outputs of certain stages fed back to the input.[4]

Let $s_k$ (with $k = 0, 1, \ldots, L - 1$) be a maximum length sequence with the length $L$, for instance for $n = 3$:

$$-1, +1, +1, -1, +1, -1, -1$$

Because of its periodicity we have $s_{k+L} = s_k$, and the sequence has the general property

$$\sum_{k=o}^{L-1} s_k = -1 \qquad (8.8)$$

Since $s_k$ is a discrete function, its autocorrelation function is defined by

$$(\phi_{ss})_m = \sum_{k=0}^{L-1} s_k s_{k+m} \tag{8.9}$$

where the subscript $k+m$ has to be taken modulo $L$, the periodicity of the sequence. It is given by (see Fig. 8.3):

$$(\phi_{ss})_m = \begin{cases} L & \text{for } m = 0 \\ -1 & \text{for } m \neq 0 \end{cases} \tag{8.10}$$

Now suppose the room under consideration is excited with a close succession of equidistant Dirac impulses, with their signs corresponding to the sequence $s_k$. The mutual distance of subsequent impulses is $\Delta t$. The received signal $s'(t)$ is sampled at the same rate. Its samples are given by

$$s'_k = \sum_{j=0}^{L-1} s_j g_{k-j} = \sum_{j=0}^{L-1} g_j s_{k-j} \tag{8.11}$$

which is the discrete version of eqn (8.4). In accordance with the sampling theorem, which is fundamental in digital signal processing, the sampling interval $\Delta t$ should be shorter than half the period of the highest frequency $f_m$ encountered in $g(t)$, i.e. $\Delta t < 1/2f_m$.

a)

b)

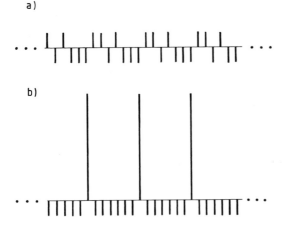

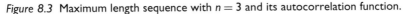

Figure 8.3 Maximum length sequence with $n = 3$ and its autocorrelation function.

To calculate the impulse response $g_k'$ from the measured values $s_k'$ we multiply eqn (8.11) (second version) on both sides with $s_{k-n}$ and sum up all products:

$$\sum_{k=0}^{L-1} s_k' s_{k-n} = \sum_{k=0}^{L-1} s_{k-n} \sum_{j=0}^{L-1} g_j s_{k-j} = \sum_{j=0}^{L-1} g_j \sum_{k=0}^{L-1} s_{k-j} s_{k-n}$$

The last expression is obtained by interchanging the order of summation. According to eqn (8.9), the last sum on the right is $(\phi_{ss})_{n-j}$. Thus, we can insert eqn (8.10) and arrive, after a little algebra, at:

$$g_n = \frac{1}{L+1} \left[ \sum_{k=0}^{L-1} s_k' s_{k-n} - \bar{s} \right] \tag{8.12}$$

with

$$\bar{s} = \sum_{k=0}^{L-1} s_k' = - \sum_{k=0}^{L-1} g_k$$

The main advantage of maximum length sequences is the possibility of retrieving the impulse response without leaving the time domain. This is because the elements $s_k$ can be arranged in a cyclic matrix, which in turn is transformed into a Hadamard matrix by simple reordering the elements.[5] Hadamard matrices have particular symmetry and recursive properties which permit carrying out the calculations required by eqn (8.12) in a very time-efficient way because the only operations needed are additions and subtractions.

In the practical application the room is not excited with a train of Dirac impulses as assumed before but with a close succession of rectangular short impulses of duration $\Delta t$ which feed more energy to the room. This implies a linear distortion which must be compensated, either afterward or by applying proper pre-emphasis to the excitation signal. In order to avoid time-aliasing, i.e. overlap of different portions of the room impulse response, the period of the sequence must exceed the length of the impulse response to be measured. (In practice, the impulse response ends when it drops below the level of the background noise.) Suppose the sampling frequency $1/\Delta t$ has the usual value of 44.1 kHz and we choose $n = 18$; then the period of the sequence $L$ has the duration of about 6 seconds, which is sufficient for most purposes.

Finally, we describe another class of excitation signals that are particularly well-suited for acquiring the impulse response of rooms. These are the sine

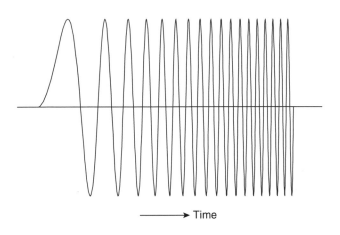

→ Time

*Figure 8.4* Sine sweep signal.

sweep signals, i.e. signals with continuously varying frequency. The simplest example of such signals is the linear sine sweep:

$$s(t) = A \sin(bt^2) \tag{8.13}$$

(see Fig. 8.4) with the instantaneous frequency

$$f(t) = \frac{1}{2\pi} \frac{d(bt^2)}{dt} = \frac{1}{\pi} bt \tag{8.14}$$

It is obvious that the frequency spectrum of this signal $s(t)$ has constant absolute value. Hence the impulse response of the room can be retrieved immediately by 'matched filtering' according to eqn (8.7), i.e. by applying the output signal of the microphone to a filter with the impulse response $s(-t)$. It is easy to see how matched filtering works: the spectral components received last are processed first and vice versa. However, in view of modern computer technology it is more practical to perform this operation in the frequency domain shown in Fig. 8.5. Furthermore, the duration of the sine sweep must be limited, for instance, by defining the start and end frequencies, $f_1$ and $f_2$, respectively. Then the constant $b$ can be determined from these frequencies and the duration $t_s$ of the sweep:

$$b = \pi \frac{f_2 - f_1}{t_s} \tag{8.15}$$

Truncating the sine sweep at finite frequencies introduces some ripple at the extremities of the amplitude spectrum which can be minimalized by

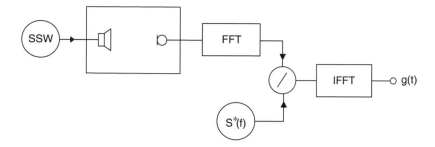

*Figure 8.5* Measurement of room impulse responses with sine sweep signals. (SSW sine sweep signal with spectrum S(f))

providing for a soft switch-on and switch-off at the frequencies $f_1$ and $f_2$. The constant $b$ must be chosen small enough to make the duration $t_s$ of the sweep sufficiently long.

Besides the linear sine sweep represented by eqn (8.13), the logarithmic sine sweep is also in use:

$$s(t) = A \sin \left[ \exp(bt) \right] \tag{8.16}$$

with the instantaneous frequency

$$f(t) = b \cdot \exp(bt) \tag{8.17}$$

Its power spectrum is not constant but drops proportionally to $1/f$, corresponding to 3 dB per octave. One of the assets of this signal is that it gives more emphasis to the low-frequency range. Of course, the influence of this spectrum must be eliminated, for instance by pre-emphasis. In any case, the duration of the sweep must be a little longer than that of the impulse response to be measured.

Careful comparisons of the various methods of measuring the impulse response of a room have been published by Müller and Massarani[6] and by Stan et al.[7] It seems that sine sweeps as excitation signals are more advantageous because of the high attainable signal-to-noise ratio and the possibility of eliminating all distortion products. The longer processing times are no longer of concern because of today's powerful digital processors. However, when it comes to measurements in occupied rooms or in other noisy environments maximum length sequences may still be superior to sweep techniques.

## 8.3 Examination of the impulse response

As mentioned before, the impulse response of a room or, more precisely, of a particular transmission path within a room, is the most characteristic objective feature of its 'acoustics'. Some of the information it contains can be found by direct inspection of a 'reflectogram', by which term we mean the graphical representation of the impulse response. Furthermore, certain parameters which have been discussed in the preceding chapter may be extracted from the impulse response. (The measurement of reverberation time will be postponed to a separate section below.) In any case some further processing of the measured impulse response is useful or necessary.

From the visual inspection of a reflectogram the experienced acoustician may learn quite a bit about the acoustical merits and faults of the place for which it has been measured. One important question is, for instance, to what extent the direct sound will be supported by shortly delayed reflections, and how these are distributed in time. Furthermore, strong and isolated peaks with long delays which hint at the danger of echoes are easily detected.

The examination of a reflectogram can be facilitated—especially that of a band-limited reflectogram—by removing insignificant details beforehand. In principle, this can be effected by rectifying and smoothing the impulse response. This process, however, introduces some arbitrariness into the obtained reflectogram with regard to the applied time constant: if it is too short, the smoothing effect may be insufficient; if it is too long, important details of the reflectogram will be suppressed. One way to avoid this uncertainty is to apply a mathematically well-defined procedure to the impulse response, namely to form its 'envelope'. Let $s(t)$ denote any signal, then its envelope is defined as

$$e(t) = \sqrt{\left([s(t)]^2 + [\check{s}(t)]^2\right)} \qquad (8.18)$$

Here $\check{s}(t)$ denotes the Hilbert transform of $s(t)$:

$$\check{s}(t) = \frac{1}{\pi} \int_{-\infty}^{\infty} \frac{s(t - t')}{t'} \, dt' \qquad (8.19)$$

Probably the most convenient way to calculate the Hilbert transform is by exploiting its spectral properties. Let $S(f)$ denote the Fourier transform of $s(t)$, then the Fourier transform of $\check{s}(t)$ is

$$\check{S}(f) = -iS(f) \, \text{sign} \, (f) = \begin{cases} -iS(f) & \text{for } f > 0 \\ iS(f) & \text{for } f < 0 \end{cases} \qquad (8.20)$$

Hence, a function $s(t)$ may be Hilbert-transformed by computing its spectral function $S$, modifying it according to eqn (8.20) and transforming the result back into the time domain. There are also efficient methods to compute the Hilbert transform in the time domain.

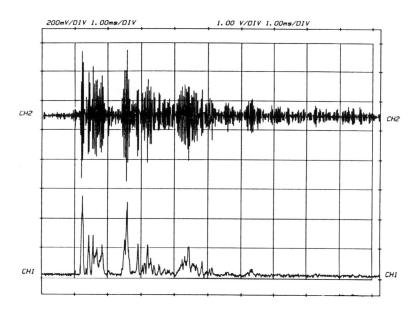

200mV/DIV 1.00ms/DIV          1.00 V/DIV 1.00ms/DIV

CH2                                                        CH2

CH1                                                        CH1

*Figure 8.6* Room impulse response (upper part) and its squared envelope (lower part). Total range of abscissa is 400 ms.

Figure 8.6 illustrates how an experimental reflectogram is modified by forming its squared envelope $[e(t)]^2$. It is obvious that the significant features are shown much more clearly now.

A reflectogram may be further modified by smoothing its envelope in order to simulate the integrating properties of our hearing. For this purpose the envelope $e(t)$ (or the rectified impulse response $|g(t)|$) is convolved with $\exp(-t/\tau)$. This corresponds to applying the signal $|g(t)|$ or $e(t)$ to a simple RC network with the time constant $\tau = RC$. A reasonable choice of the time constant is 25 ms.

Figure 8.7 shows three example of measured 'reflectograms', obtained at several places in a lecture room which was excited by short tone bursts with a centre frequency of 3000 Hz and a duration of about 1 ms; the frequency bandwidth was about 500 Hz. The lower trace of each registration shows the result of the smoothing described above. The total length of an oscillogram corresponds to a time interval of 190 ms. The uppermost reflectogram was taken at a place close to the sound source; consequently, the direct sound is relatively strong. The most outstanding feature in the lowest oscillogram is the strong reflection delayed by about 40 ms with regard to the direct sound. It is not heard as an echo since it still lies within the integration time of our ear (see Section 7.3).

Although just looking at a reflectogram or its modification is very suggestive it does not permit a safe decision as to whether a particular reflection

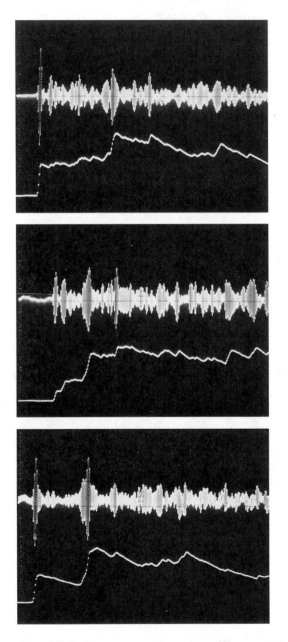

*Figure 8.7* Reflectograms measured at different positions in a lecture room (centre frequency = 3000 Hz, impulse duration about 1 ms). Upper traces: original response of the room to the excitation signal. Lower traces: same, after rectifying and smoothing with time constant $\tau = 25$ ms.

will be heard as an echo or not. This can only be achieved by applying an objective echo criterion, for instance that discussed in Section 7.3. About the same holds for a periodic train of reflections which can give rise to quite undesirable subjective effects even when the periodic components are buried among other nonperiodically distributed reflections. Usually the periodic components are caused by repeated reflections of sound rays between parallel walls, or generally in rooms with a very regular shape, as for instance in rooms which are circular or regularly polygonal in cross-section. At relatively short repetition times they are perceived as colouration, at least under certain conditions. But even a single dominating reflection may cause audible colouration, especially of music, since the corresponding transmission function has a regular structure or substructure (see Section 7.3).

The most adequate technique for testing the randomness or pseudorandomness of an impulse response is autocorrelation analysis. In this procedure all the irregularly distributed components are swept together into a single central peak (see Fig. 8.8), whereas the remaining reflections will form side components or even satellite maxima in the autocorrelogram. Since $g(t)$ is a transient signal, the autocorrelation function we are looking for is

$$\phi_{gg}(\tau) = \int\limits_{-\infty}^{\infty} g(t')g(t'+\tau)\,dt' = \int\limits_{-\infty}^{\infty} g(-t')g(\tau-t')\,dt' \qquad (8.21)$$

The second expression is just the convolution of the impulse response with its time-inverse replica and tells us that the autocorrelation function can be obtained in real time by exciting the room with its stored and time-inversed impulse response. In this way the functions shown in Fig. 8.8 have been obtained.

In order to decide whether or not a certain side maximum indicates audible colouration, we form the 'weighted autocorrelation function':[8]

$$\phi'_{gg}(\tau) = b(\tau) \cdot \phi_{gg}(\tau) \qquad (8.22)$$

The weighting function $b(\tau)$ can be calculated from the thresholds, represented in Fig. 7.12, and is shown in Fig. 8.9.

Let us denote by $\tau_0$ the argument at which the side maximum, which is next to the central maximum (at $t = 0$), appears. Then we have to expect audible colouration if

$$\phi'_{gg}(\tau_0) > 0.06 \cdot \phi'_{gg}(0) \qquad (8.23)$$

no matter if this side maximum is caused by a single strong reflection or by a periodic succession of reflections.

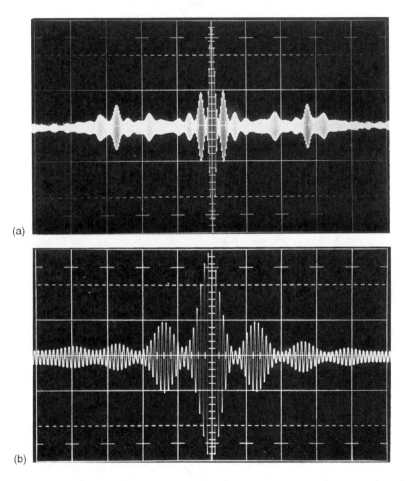

(a)

(b)

*Figure 8.8* Autocorrelation function of a reflectogram, measured in a reverberation chamber. The room was excited with a filtered impulse of about 1 ms duration and a centre frequency of 2000 Hz: (a) abscissa unit = 20 ms; (b) same as (a) but abscissa, unit = 5 ms.

The temporal structure of a room's impulse response determines not only the shape of the autocorrelation function obtained in this room but also its modulation transfer function (MTF), which was introduced in Section 5.5. Indeed, as was shown by Schroeder,[9] the complex MTF for white noise as a primary sound signal is related to the impulse response by

$$\underline{m}(\Omega) = \int\limits_{0}^{\infty} [g(t)]^2 \exp{(i\Omega t)}\,dt \bigg/ \int\limits_{0}^{\infty} [g(t)]^2 \, dt \qquad (8.24)$$

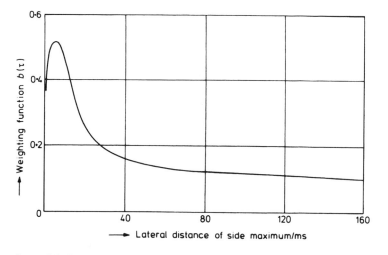

*Figure 8.9* Detecting colouration: weighting function for autocorrelation functions.

*Table 8.1* Objective quality criteria

| Name of criterion | Symbol | Defined by equation |
|---|---|---|
| Definition ('Deutlichkeit') | $D$ | 7.8 |
| Clarity index ('Klarheitsmaß') | $C, C_{80}$ | 7.9 |
| Centre time | $t_s$ | 7.12 |
| Echo criterion (Dietsch and Kraak) | EC | 7.6, 7.7 |
| Speech transmission index | STI | Ref. 13 of Chapter 7 |
| Reverberation time and early decay time | $T$, EDT | (see Section 8.4) |
| Strength factor | $G$ | 7.15 |
| Lateral energy fraction | LEF | 7.17 |
| Late lateral energy | $LG_{80}^{\infty}$ | 7.20 |
| Interaural cross-correlation | IACC | 7.18, 7.19 |

This means that the complex modulation transfer is the Fourier transform of the squared room impulse response divided by the integral over the squared response. Of course, this formula applies to an impulse response $g'(t)$ which has been confined to a suitable frequency band by bandpass filtering.

It goes without saying that all the parameters introduced in Chapter 7 can be evaluated from impulse responses by suitable operations. Table 8.1 lists these criteria along with the equations by which they are defined and which can be used to compute them.

The experimental determination of the 'early lateral energy fraction' (LEF) and the 'late lateral energy' requires the use of a gradient microphone (figure-of-eight microphone) with its direction of minimum sensitivity directed towards the sound source. For measuring the 'early lateral energy fraction', an additional non-directional microphone placed at the same position as the gradient microphone is needed. In contrast, the normalising term of $LG_{80}^{\infty}$, namely the denominator in eqn (7.20), is independent of the measuring position in the room and must be determined just once for a given sound source. The 'interaural cross-correlation' (IACC) is obtained by cross-correlating the impulse responses describing the sound transmission from the sound source to both ears of a human head. If such measurements are carried out only occasionally, the responses can be obtained with two small microphones fixed in the entrance of both ear channels of a person whose only function is to scatter the sound waves properly. For routine work it is certainly more convenient to replace the human head by a dummy head with built-in microphones.

## 8.4 Measurement of reverberation

For reasons which have been discussed earlier, Sabine's reverberation time is still the best known and most important quantity in room acoustics. This fact is the justification for describing the measurement of this important parameter in a separate section which includes that of its younger relative, the 'early decay time' (EDT).

Although both the reverberation time of a room as well as the early decay time at a particular location in it can be derived from the corresponding impulse response, we start by describing the more traditional method, the principle of which agrees with Sabine's measuring procedure, apart from the equipment. Since the reverberation time is usually determined by evaluation of decay curves, the first step is recording decay curves over a sufficiently wide range of sound level.

The standard equipment for this measurement (which can be modified in many ways) is depicted schematically in Fig. 8.10. A loudspeaker LS, driven by a signal generator, excites the room to steady-state conditions. The output voltage of the microphone M is fed to an amplifier and filter F, and then to a logarithmic recorder LR, whose deflection is calibrated in decibels. At a given

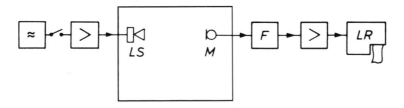

*Figure 8.10* Measurement of sound decay and reverberation time.

moment the excitation is interrupted by a switch, and at the same time the recorder is triggered and starts to record the decay process.

The signal from the generator is either a frequency-modulated sinusoidal signal whose momentary frequency covers a narrow range or it is random noise filtered by an octave or a third octave filter. The range of mid-frequencies, for which reverberation time measurements are usually taken, extends from about 50 to 10 000 Hz; most frequently, however, the range from 100 to 5000 Hz is considered. As mentioned in Section 8.1, excitation by a pistol shot is often a practical alternative. Pure sinusoidal tones are used only occasionally, as for example to excite individual modes in the range well below the Schroeder frequency, eqn (3.32).

The loudspeaker, or more generally the sound source, is usually placed at the location of the source when the room is in its normal use: for instance, on the stage. This applies not only to reverberation measurements but also to other measurements. Because of the validity of the reciprocity principle, however (see Section 3.1), the location of sound source and microphone can be interchanged without altering the results, in cases where this is practical and provided that the sound source and the microphone have no directionality. In any case, it is important that the distance between the sound source and the microphone is much larger than the diffuse-field distance given by eqn (5.39), otherwise the direct sound would have an undue influence on the shape of the decay curve.

If the sound field were completely diffuse, the decay should be the same throughout the room. Since this ideal condition is hardly ever met in normal rooms, it is advisable to carry out several measurements for each frequency band at different microphone positions. This is even more important if the quantity to be evaluated is the early decay time, which may vary considerably from one place to the next within one hall.

The microphone is followed by a filter—usually an octave or third octave filter—mainly to improve the signal-to-noise ratio, i.e. to reduce the disturbing effects of noise produced in the hall itself as well as that of the microphone and amplifier noise. If the room is excited by a pistol shot or another wide-band impulse, it is this filter which defines the frequency band and hence yields the frequency dependence of the reverberation time.

For recording the decay curves the conventional electromechanical level recorder has been replaced nowadays by the digital computer, which converts the sound pressure amplitude of the received signal into the instantaneous level. Usually, the level $L(t)$ in the experimental decay curves does not fall in a strictly linear way but contains random fluctuations which are due, as explained in Section 3.6, to complicated interferences between decaying normal modes. If these fluctuations are not too strong, it is easy to approximate the decay within the desired section by a straight line. In many cases it is sufficient to do this with just a ruler and pencil. Any arbitrariness of evaluation is avoided, however, if a 'least square fit' is carried out. Let $t_1$ and $t_2$ denote

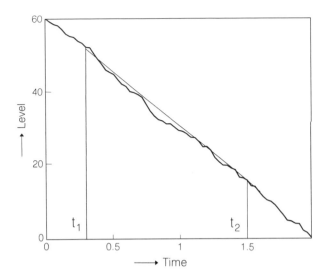

*Figure 8.11* Logarithmic decay curve and its approximation within the limits $t_1$ and $t_2$ by a straight line.

the interval in which the decay curve is to be approximated (see Fig. 8.11), then the following integrations must be performed:

$$L_1 = \int_{t_1}^{t_2} L(t)\,dt, \qquad L_2 = \int_{t_1}^{t_2} L(t)t\,dt \tag{8.25}$$

With the further abbreviations $I_1 = t_2 - t_1$, $I_2 = (t_2^2 - t_1^2)/2$ and $I_3 = (t_2^3 - t_1^3)/3$, the mean slope of the straight line in the given interval is given by:

$$-\frac{\Delta L}{\Delta t} = \frac{I_1 L_2 - I_2 L_1}{I_1 I_3 - I_2^2} \tag{8.26}$$

This procedure is particularly useful when it comes to the 'early decay time'. In any case, the slope of the decay curve is related to the reverberation time by

$$T = 60 \cdot \left(\frac{\Delta L}{\Delta t}\right)^{-1} \tag{8.27}$$

The quasi-random fluctuations of the decay level and the associated uncertainties about the true shape of a decay curve can be avoided, in principle, by averaging over a great number of individual reverberation curves, each of which was obtained by random noise excitation of the room. Fortunately the same result is arrived at with an elegant method, called 'backward integration', which was proposed and first applied by Schroeder.[10] It is based on the following relationship between the ensemble average $\langle h^2(t) \rangle$ of all possible decay curves (for a certain place and bandwidth of exciting noise) and the corresponding impulse response $g(t)$:

$$\langle h^2(t) \rangle = \int_t^\infty [g(x)]^2 \mathrm{d}x = \int_0^\infty [g(x)]^2 \mathrm{d}x - \int_0^t [g(x)]^2 \mathrm{d}x \qquad (8.28)$$

To prove this relationship let us suppose that the room is excited by white noise $n(t)$, which is switched off at the time $t = 0$. According to eqn (1.46), the sound decay is given by

$$h(t) = \int_t^\infty g(x) \cdot n(t - x) \, \mathrm{d}x \qquad \text{for } t > 0$$

Squaring this latter expression yields a double integral, which after averaging reads

$$\langle h^2(t) \rangle = \int g(x) \, \mathrm{d}x \int g(y) \langle n(t - x) n(t - y) \rangle \, \mathrm{d}y$$

The brackets $\langle \rangle$ on the right-hand side indicate ensemble averaging, strictly speaking, which however is equivalent to temporal averaging. Hence

$$\langle n(t - x) n(t - y) \rangle = \phi_{nn}(x - y)$$

is the autocorrelation function of white noise which is a Dirac function. Thus, the double integral is reduced to the single integral of eqn (8.28). This derivation is valid no matter whether the impulse response is that measured for the full frequency range or for only of a part of it.

The merits of this method may be underlined by the examples shown in Fig. 8.12. The upper decay curves have been measured with the traditional method, i.e. with random noise excitation, according to Fig. 8.10. They exhibit strong fluctuations of the decaying level which do not reflect any acoustical properties of the transmission path and hence of the room, but are due to the random character of the exciting signal; if one of these recordings were repeated, it would differ from the preceding one in many details. In contrast, the lower curves, obtained from different impulse responses by backward integration according to eqn (8.28), are free of such confusing

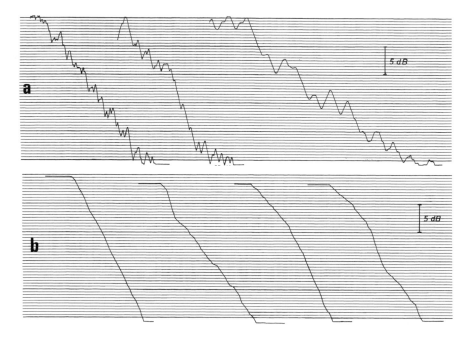

*Figure 8.12* Experimental reverberation curves: (a) conventional procedure according to Fig. 8.10; (b) recorded with backward integration according to eqn (8.28).

fluctuations and hence contain only significant information. Repeated measurements for one situation yield identical results, which is not surprising since these decay curves are unique functions of impulse responses. It is clear that from such registrations the reverberation time can be determined with much greater accuracy than from decay curves recorded in the traditional way. This is even more true for the EDT. Furthermore, any characteristic deviations of the sound decay from exponential behaviour are much more obvious.

For the practical execution, the second version of eqn (8.28) is more useful in that it yields the decaying quantity in real time. This means that the integral of the full impulse response must be measured and stored beforehand. Of course, the upper limit $\infty$ must be replaced with a finite value, named $t_\infty$, which, however, is not uncritical because there is always some acoustical or electrical background noise. Its effect on basically linear decay curves corresponding to a reverberation time $T = 2$ s is demonstrated by Fig. 8.13. If the limit $t_\infty$ is too long, the microphone will pick up too much noise, which causes the characteristic tail at the lower end of the decay curve. Too short an integration time will lead to an early downward bend of the curve, which is also awkward. Obviously, there exists an optimum for $t_\infty$ which depends on the relative noise level and the decay time.

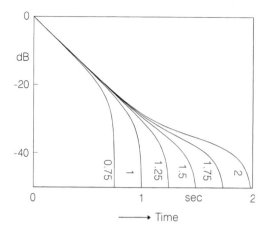

*Figure 8.13* Effect of background noise on decay curves processed with backward integration. The noise level is −41.4 dB. Parameter: upper integration limit $t_\infty$ (in seconds).

To determine the reverberation time after eqn (8.27), the slope of the decay curve is frequently evaluated in the level range from −5 to −35 dB relative to the initial level. This procedure is intended to improve the comparability and reproducibility of reverberation times in such cases where the fall in level does not occur linearly. It is doubtful, however, whether the evaluation of an average slope from curves which are noticeably bent is very meaningful, or whether the evaluation should rather be restricted to their initial parts, i.e. to the EDT which is anyway a more reliable indicator of the subjective impression of reverberance. The same argument applies if the absorption coefficient of a test material is to be determined from reverberation measurements (Section 8.7) since it is not some average slope but the initial slope which is related to the average damping constant of all excited normal modes (see eqn (3.49a)).

It should be mentioned that the absorption of a room and hence its reverberation time could be obtained, at least in principle, from the steady-state sound level or energy density according to eqn (5.37). Likewise, the modulation transfer function could be used to determine the reverberation time. In practice, however, these methods do not offer any advantages compare with those described above, since they are more time consuming and less accurate.

## 8.5 Diffusion

To this day it is not clear whether sound field diffusion is an acoustical quality parameter in its own right, or whether it is just the condition which ensures the validity of the simple reverberation formulae which are in common use. This is because the direct measurement of diffusion is difficult and time consuming. Accordingly, not many data on diffusion, collected in concert or other large halls, are available. Since 'sound field diffusion' are kind of magic words in room acoustics, this book would remain incomplete without a section on diffusion measurement.

The straightforward way to measure sound field diffusion is certainly to determine the directional distribution of sound energy flow. As mentioned in Section 4.3, this distribution is characterised by a function $I(\varphi, \vartheta)$, provided the field is stationary. It can be determined by scanning all directions with a directional microphone of sufficiently high angular resolution. Let $\Gamma(\varphi, \vartheta)$ be the directional factor, i.e. the relative sensitivity of the microphone as a function of angles $\varphi$ and $\vartheta$ in a suitably chosen polar coordinate system; then the squared output voltage of the microphone is proportional to

$$I'(\varphi, \vartheta) = \iint_{\Omega} I(\varphi', \vartheta') \left| \Gamma\left(\varphi - \varphi', \vartheta - \vartheta'\right) \right|^2 d\Omega' \qquad (8.29)$$

i.e. it is the two-dimensional convolution of $I(\varphi, \vartheta)$ and $|\Gamma(\varphi, \vartheta)|^2$; $d\Omega'$ is the solid angle element.

A measure of the isotropy of the sound field is the 'directional diffusion' introduced by Thiele (see Ref. 9 of Chapter 7). Let $\langle I' \rangle$ be the measured intensity averaged over all directions and

$$m = \frac{1}{4\pi \langle I' \rangle} \iint |I' - \langle I' \rangle| \, d\Omega \qquad (8.30)$$

the average of the absolute deviation from it. Let $m_0$ be the same quantity obtained in an anechoic chamber. Then the directional diffusion is defined as

$$d = \left(1 - \frac{m}{m_0}\right) \cdot 100\% \qquad (8.31)$$

Dividing $m$ by $m_0$ effects a certain normalisation and, consequently, $d = 100\%$ in a perfectly diffuse sound field, whereas in the sound field consisting of one single plane wave the directional diffusion becomes zero. This simple procedure, however, does not completely eliminate the influence of the microphone characteristics. Therefore, for experimental $d$ values are only comparable when they have been gained with similar microphones. (The correct way to 'clean' the experimental data would be to perform a two-dimensional deconvolution.)

Directional distributions measured in this way in many concert halls and broadcasting studios have been published by Meyer and Thiele (see Ref. 9 of

Chapter 7). As a directional microphone these authors used a concave metal mirror (see Fig. 8.21) with a pressure microphone arranged in its focus. The diameter of the mirror was 1.20 m, which led to a half width of its directional characteristics (angular distance between 3 dB points) of about $8.6^0$ at 2000 Hz. The quantity $d$, as determined from the directional distribution, varied between 35% and 75% without showing a clear tendency.

A more manageable device has been employed by Tachibana et al[11] in the course of a survey of auditoriums in Europe and Japan comprising 20 large halls. To detect the directional distribution of low-order reflections they applied a method which was developed by Yamasaki and Itow.[12] The sensor these authors used consists of four ¼inch pressure microphones which define a Cartesian coordinate system, with one of the microphones at the origin while the remaining ones are placed on the three axis of the imagined coordinate system; their distances from the origin are 50 mm, sometimes only 33 mm. The room is excited by an impulse whose width is 5 $\mu$s. To improve the signal-to-noise ratio, the impulse responses picked up simultaneously by the microphones are averaged over up to 256 shots. Suppose the impulse responses contain just the direct sound and one reflection. The arrival times of both components are slightly different in the four final responses, and from these differences one obtains the direction of the incident reflections. Together with the absolute delay this leads to the location of the image source which caused this reflection. The arrival times are determined from the peaks of the six short-time correlation coefficients

$$\varphi_{ik}(\tau) = \int\limits_{t_a}^{t_a+\Delta} g_i(t)g_k(t+\tau)\,dt \bigg/ \left\{ \int\limits_{t_a}^{t_a+\Delta} [g_i(t)]^2\,dt \int\limits_{t_a}^{t_a+\Delta} [g_k(t)]^2\,dt \right\}^{1/2}$$

with $i, k = 1, 2, 3$ and $i \neq k$. $\Delta$ denotes a suitably chosen time interval. Figure 8.14 presents the impulse responses, the spatial distributions of image sources and the corresponding directional distributions of the sound for three rooms, namely a living room, the Boston Symphony Hall and a cathedral (Münster in Freiburg, Germany). The lengths of the 'rays' in the directional diagrams are proportional to the level of the received sound. (Note both the pattern of image sources and the directional distributions are projected into the floor plane.)

As an alternative Yamasaki and Itow used the aforementioned four-microphone probe as an intensity sensor, according to eqn (8.1), with the goal of identifing image sources. To this end they filtered the impulse responses with a suitable band-pass filter and applied a variable time window to them with a width of 10 ms. From the output of each microphone pair, the corresponding component of the short-time intensity was obtained. These components determine the magnitude and the direction of the intensity vector.

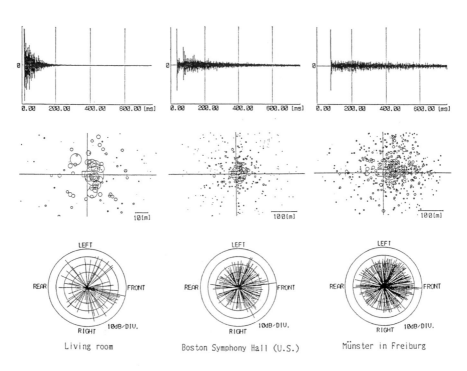

*Figure 8.14* Impulse response (top), image sources (middle) and directional distribution of reflected sound (bottom) for three different enclosures: living room (left), Boston Symphony Hall (middle) and Münster in Freiburg/Germany (right). The pattern of image sources and the directional distributions are projected into the ground plane. (After Yamasaki and Itow.[12] Courtesy J Acoust Soc Japan.)

A procedure similar to that of Yamasaki and Itow was developed by Sekiguchi et al.[13] These authors used a probe consisting of four ¼inch microphones forming the corners of a regular tetrahedron with a side length of 17 cm. For room excitation they used 2 kHz tone bursts with a duration of 80 $\mu$s; to improve the signal-to-noise ratio, 32 or 64 impulse responses were averaged. Because of the relatively large dimensions of the probe, the authors achieved sufficiently high accuracy without performing the time-consuming correlation operations described above.

It should be noted that the application of both methods is restricted to the early part of a room impulse response, which consists of a few well-separated reflections. In the reverberant part of the response the density of reflections is so high that no individual image sources can be identified.

If one is not interested in all the details of the directional distribution but only in a measure for its uniformity, more indirect methods can be applied. One of these measures is the correlation coefficient of two sound pressures, defined in eqn (1.41):

$$\Psi = \frac{\overline{p_1 \cdot p_2}}{\sqrt{\overline{p_1^2} \cdot \overline{p_2^2}}}$$

(8.32)

To calculate the correlation coefficient in a diffuse sound field we assume that the room is excited by random noise with a very small bandwidth. The sound field can be considered to be composed of plane waves with equal amplitudes and randomly distributed phase angles $\psi_n$. The sound pressure due to one such wave at points 1 and 2 at distance $x$ (see Fig. 8.15) is

$$p_1(t) = A\cos(\omega t - \psi_n) \quad \text{and} \quad p_2(t) = A\cos(\omega t - \psi_n - kx\cos\vartheta_n)$$

The angle $\vartheta_n$ characterizes the direction of the incident sound wave. Because of the periodicity of $p_1$ and $p_2$, the averaging process indicated by the over-bars in eqn (8.32) should be extended over one period or an integer multiple of it. It is easy to see that

$$\overline{p_1^2} = \overline{p_2^2} = \frac{A^2}{2}$$

The time average of the product of both pressures is

$$\overline{p_1 \cdot p_2} = \frac{A^2}{2}\cos(kx\cos\vartheta_n)$$

Inserting these expressions into eqn (8.32) leads to $\Psi(x, \vartheta_n) = \cos(kx\cos\vartheta_n)$ which has to be averaged with constant weight (corresponding to complete diffusion) over all possible directions of incidence. This yields

$$\Psi(x) = \frac{\sin kx}{kx}$$

(8.33a)

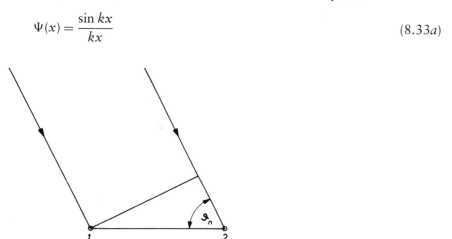

Figure 8.15 Derivation of eqns (8.33a–8.33c).

If, however, the directions of incident sound waves are not uniformly distributed over the entire solid angle, but only in a plane containing both points, we obtain, instead of eqn (8.33a),

$$\Psi(x) = J_0(kx) \qquad (8.33b)$$

where $J_0$ = Bessel function of order zero.

If the connection between both points is perpendicular to the plane of two-dimensional diffusion, the result is

$$\Psi(x) = 1 \qquad (8.33c)$$

The functions given by eqns (8.33a)–(8.33c) are plotted in Fig. 8.16. The most important is curve a; each deviation of the measured correlation coefficient $\Psi$ from this curve hints at a lack of diffusion.

The derivation presented above is strictly valid only for signals with vanishing frequency bandwidths. Practically, however, its result can be applied with sufficient accuracy to signals with bandwidths of up to a third octave. For wider frequency bands an additional frequency averaging of eqn (8.33a) is necessary.

The correlation coefficient $\Psi$ is the cross-correlation function $\phi_{p_1 p_2}$ at $\tau = 0$ (see eqn (1.40)), divided by the root-mean-square of $p_1$ and $p_2$. It can

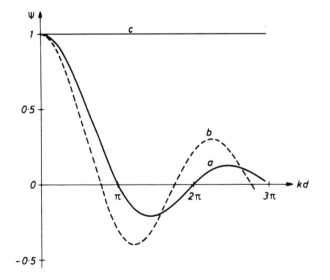

*Figure 8.16* Dependence of the correlation coefficient $\Psi$ on the distance $x$ between two observation points: (a) in a three-dimensional diffuse field; (b) in a two-dimensional diffuse field with the measuring line 1–2 (in Fig. 8.15) within its plane; (c) same as before, but the measuring line perpendicular to the plane of diffusion.

be determined with an obvious modification of the playback techniques (see eqn (8.21). First the impulse response of the system is measured at point 1 and is stored in the memory of the computer after time inversion. Then the latter signal excites the system for a second time; the cross-correlation function which we are looking for is picked up at point 2 as a function of real time. To catch its maximum it is advisable to observe the cross-correlation function in the vicinity of $\tau = 0$ in order to allow for delays in the signal paths.

An even simpler method is to measure the squared sound pressure amplitude in front of a sufficiently rigid wall as a function of the distance as discussed in Section 2.5. In fact, eqn (8.33a) agrees—apart from a factor 2 in the argument—with the second term of eqn (2.38) which describes the pressure fluctuations in front of a rigid wall at random sound incidence. This similarity is not accidental, since these fluctuations are caused by interference of the incident and the reflected waves which become less distinct with increasing distance from the wall according to the decreasing coherence of those waves. On the other hand, it is just the correlation factor of eqn (8.32) which characterises the degree of coherence of two signals. The additional factor of 2 in the argument of eqn (2.38) is due to the fact that the distance of both observation points here is equivalent to the distance of the point from its image, the rigid wall being considered as a mirror.

An alternative way to check the degree of sound field diffusion which is not restricted to a particular kind of room excitation can be arrived at by exploiting the fact that in a perfectly diffuse field the sound intensity is zero. Hence a useful descriptor of diffusion is the quantity

$$q_{\mathrm{d}} = 1 - \frac{c\,|\mathbf{I}|}{w} \tag{8.34}$$

The magnitude of the intensity vector is calculated from its three Cartesian components:

$$|\mathbf{I}|^2 = I_x^2 + I_y^2 + I_z^2$$

which are determined by using a three-component intensity probe (see Section 8.1).

## 8.6 Sound absorption—tube methods

The knowledge of sound absorption of typical building materials is indispensable for all tasks related to room acoustical design: i.e. for the prediction of reverberation times in the planning phase of auditoria and other rooms, for model experiments, for acoustical computer simulation of environments and for other purposes.

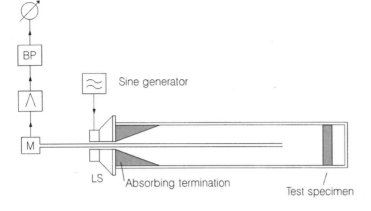

*Figure 8.17* Conventional impedance tube, schematic.

Basically, there are two standard methods of measuring absorption coefficients. In one of them, the test field is a plane wave which is mostly enclosed in a rigid tube, while the other one requires a diffuse field in a so-called reverberation chamber. In this section the first method will be described. It is restricted to the examination of small, locally reacting materials with a plane or nearly plane surface, and also to normal wave incidence onto the test specimen.

A typical set-up, known as 'Kundt's tube' or 'impedance tube', is shown in Fig. 8.17. It has rigid walls and a rectangular or circular cross-section. At one of its ends there is a loudspeaker which generates a sinusoidal sound signal. This signal travels along the tube as a plane wave towards the test specimen, which terminates at the other end of the tube and which must be mounted in the same way as it is to be used in practice (e.g. at some distance in front of a rigid wall). To reduce tube resonances it may be useful (although not essential for the principle of the method) to place an absorbing termination in front of the loudspeaker. The test sample reflects the incident wave more or less, leading to the formation of a partially standing wave, as described in Section 2.2. The sound pressure maxima and minima of this wave are measured by a movable probe microphone that must be small enough not to distort the sound field to any great extent. As an alternative, a miniature microphone mounted on the tip of a thin movable rod may be employed as well.

The tube should be long enough to permit the formation of at least one maximum and one minimum of the pressure distribution at the lowest frequency of interest. Its lateral dimensions have to be chosen in such a way that at the highest measuring frequency they are still smaller than a certain fraction of the wavelength $\lambda_{min}$. Or, more exactly, the following requirements must be met:

$$\text{Dimension of the wider side} < 0.5\lambda_{\min} \quad \text{(rectangular tubes)}$$
$$\text{Diameter} < 0.586\lambda_{\min} \quad \text{(circular tubes)} \quad (8.35)$$

Otherwise, higher-order wave types may occur with non-constant lateral pressure distributions and with different and frequency-dependent wave velocities. On the other hand, the cross-section of the tube must not be too small, since otherwise the wave attenuation due to losses at the wall would become too high. Generally, at least two tubes of different dimensions are needed in order to cover the frequency range from about 100 to 5000 Hz.

For the determination of the absorption coefficient it is sufficient to measure the maximum and the minimum values of the sound pressure amplitudes, i.e. the pressures in the nodes and the antinodes of the standing wave. According to eqns (2.10) and (2.1), the absolute value of the reflection factor and the absorption coefficient are obtained by

$$|R| = \frac{\hat{p}_{\max} - \hat{p}_{\min}}{\hat{p}_{\max} + \hat{p}_{\min}} \tag{8.36}$$

$$\alpha = \frac{4\hat{p}_{\max} \cdot \hat{p}_{\min}}{(\hat{p}_{\max} + \hat{p}_{\min})^2} \tag{8.37}$$

If possible, the maxima and minima closest to the test specimen should be used for the evaluation of $R$ and $\alpha$ since these values are influenced least by the attenuation of the waves. It is possible, however, to eliminate this influence by interpolation or by calculation, but in most cases it is hardly worthwhile doing this.

The absorption coefficient is not the only quantity which can be obtained by probing the standing wave; additional information can be derived from the location of the pressure maximum (pressure node) which is next to the test specimen. According to eqn (2.9a), the condition for the occurrence of a pressure node is $\cos(2kx + \chi) = -1$. If $x_{\min}$ is the distance of the nearest pressure node from the surface of the sample, this condition yields the phase angle $\chi$ of the reflection factor

$$\chi = \pi\left(1 - \frac{4x_{\min}}{\lambda}\right) \tag{8.38}$$

Once the complex reflection factor is known, the wall impedance $Z$ or the specific impedance $\zeta$ of the material under test can be obtained by eqns (2.6) and (2.2a), or by applying a graphical representation of these relations, known as a 'Smith chart' (see Fig. 8.18). Furthermore, the specific impedance

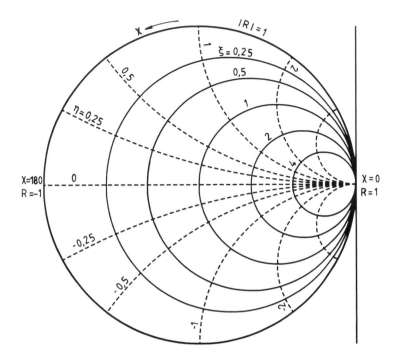

*Figure 8.18* Smith chart: circles of constant real part $\xi$ (solid circles) and imaginary parts $\eta$ of the specific impedance $\zeta = Z/\rho_0 c$, represented in the complex R plane.

of the sample can be used to determine its absorption coefficient $\alpha_{\text{uni}}$ for random sound incidence, either from Fig. 2.12 or by applying eqn (2.42). In many practical situations this latter absorption coefficient is more relevant than that for normal sound incidence. However, the result of this procedure will be correct only if the material under test can be assumed to be of the 'locally reacting' type (see Section 2.3).

Several attempts have been made to replace the somewhat involved and time-consuming standing wave method by faster and more modern procedures. A typical arrangement is sketched in Fig. 8.19. The movable probe is replaced with two fixed microphones which are mounted flush into the wall of the tube. Let $S(f)$ denote the spectrum of the stationary sound signal emitted by the loudspeaker, for instance of random noise. Then the spectra of the sound signals received at both microphone positions are

$$S_1(f) = S(f)\left[\exp(ikd) + R(f)\exp(-ikd)\right]$$

and

$$S_2(f) = S(f)\left\{\exp\left[ik(d+\Delta)\right] + R(f)\exp\left[-ik(d+\Delta)\right]\right\}$$

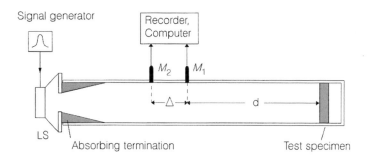

*Figure 8.19* Impedance tube with two fixed microphones.

As shown in Fig. 8.19, $d$ is the distance of microphone $M_1$ from the surface of the sample under test, and $\Delta$ denotes the distance between both microphones. From these equations, the complex reflection factor is obtained as

$$R(f) = \exp\left[ik(2d+\Delta)\right] \frac{S_2 - S_1 \exp(ik\Delta)}{S_1 - S_2 \exp(ik\Delta)} = \exp(2ikd) \frac{\exp(ik\Delta) - H_{12}}{H_{12} - \exp(-ik\Delta)}$$

$$(8.39)$$

with $H_{12} = S_2(f)/S_1(f)$ denoting the transfer function between both microphone positions. Critical are those frequencies for which $|\exp(ik\Delta)|$ is close to unity, i.e. the distance $\Delta$ is about an integer multiple of half the wavelength. In such regions the accuracy of measurement is not satisfactory. This problem can be circumvented by providing for a third microphone position. Of course, the relative sensitivities of all microphones must be taken into account, or the same microphone is used to measure $S_1$ and $S_2$ in succession.

If a short impulse, idealised as a Dirac impulse $\delta(t)$, is used as a test signal, the measurement can be carried out with one microphone since the short signals produced by the incident and the reflected wave can be separated by proper time windows. The sound signal received by the microphone $M_1$ is, apart from an insignificant time delay,

$$s'(t) = \delta(t) + r(t) * \delta(t - 2d/c) = \delta(t) + r(t - 2d/c)$$

$$(8.40)$$

where $r(t)$ is the 'reflection response' of the test material, the inverse Fourier transform of the reflection factor $R$ (see Section 4.1). Equation (8.40) represents the impulse response of the arrangement. The Fourier transform of this equation reads

$$S'(f) = 1 + R(f) \cdot \exp(-2ikd)$$

$$(8.41)$$

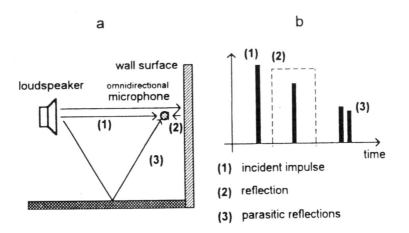

*Figure 8.20* In-situ measurement of acoustical wall properties: (a) experimental set-up; (b) sequence of reflected signals. Reflection 2 is selected by applying a suitable time window (after Mommertz[14]).

from which the complex reflection factor is easily obtained. If needed, the signal-to-noise ratio can be improved by replacing the test impulse by a time-stretched test signal with constant amplitude spectrum: e.g a sine sweep, as described in Section 8.2.

In order to separate safely the reflected signal from the primary one, the latter must be sufficiently short, and the distance $d$ of the microphone from the sample must be large enough. The same holds for any reflections from the loudspeaker. This may lead to impractically long tubes. An alternative is to omit the tube, as depicted in Fig. 8.20.

This one-microphone techniques can be used for in-situ measurement of acoustical wall and ceiling properties in existing enclosures. In this case, however, the waves are not plane but spherical. For this reason the results may be not too exact because the reflection of spherical waves is different from that of plane waves. Furthermore, the $1/r$ law of spherical wave propagation has to be accounted for by proper correction terms in eqn (8.41). Further refinements of this useful method are described in Ref. 14.

## 8.7 Sound absorption—reverberation chamber

The reverberation method of absorption measurement is superior to the impedance tube method in several respects. First of all, the measurement is performed with a diffuse sound field, i.e. under conditions which are much more realistic than those encountered in a one-dimensional waveguide. Secondly, there are no limitations concerning the type and construction of the absorber, and the test specimen can be set up in much the same way in which the material is to be used in the particular practical application. Hence the reverberation method is well suited for measuring the absorption coefficient of almost any type of wall linings and of ceilings. And finally the absorption of discrete objects which cannot be characterized by an absorption coefficient can be determined in a reverberation chamber. This concerns, for instance, any kind of chairs, empty or with persons seated on them.

A reverberation chamber is a small room with a volume of at least 100 m³, better still 200–300 m³, whose walls are as smooth and rigid as possible. The absorption coefficient $\alpha_0$ of the bare walls, which should be uniform in construction and finish, is determined by measuring the reverberation time of the empty chamber:

$$T_0 = 0.161 \frac{V}{S\alpha_0} \tag{8.42}$$

($V$ = volume in m³, $S$ = wall area in m²). Then a certain amount of the material to be tested is introduced into the chamber; it goes without saying that the test sample must be mounted in the same way as in the intended application. The test specimen reduces the reverberation time from $T_0$ to $T$, given by

$$T = 0.161 \frac{V}{S_s \alpha + (S - S_s)\,\alpha_0} \tag{8.43}$$

From this equation the absorption coefficient $\alpha$ of the test sample is easily calculated ($S_s$ is the area of the test sample).

The air absorption term $4mV$ appearing in eqn (5.25) can be neglected, since it is contained in the absorption of the empty chamber as well as in that of the chamber containing the test material and therefore will almost cancel out. Its effect is small anyway because of the small chamber volume.

In the case of discrete objects, the product $S_s\alpha$ in the denominator of eqn (8.43) is replaced by the total absorption area of the objects.

The measurement of reverberation has been described in detail in Section 8.4, and therefore no further discussion on this point is necessary. To increase the accuracy of the decay measurement it is recommended to use several different source and microphone positions and to average the results. Usually, these measurements are performed with frequency bands of third octave bandwidth.

As mentioned earlier the application of the Sabine equation produces systematic errors in that it overrates the absorption coefficients; sometimes the determined value of $\alpha$ is larger than unity. An example is shown in Fig. 6.13 (dashed line). Such inconsistencies could be avoided by using other, more exact decay formulae, for instance Eyring's formula (5.24) without the term $4\,mV$. (Another reason for absorption coefficients exceeding unity is edge diffraction, as will be discussed below.) Although these facts have been known for a long time, the international standardization of this measuring procedure continues to recommend the use of eqn (8.43).[15] The coefficient determined in this way is sometimes given the name 'Sabine absorption coefficient' in order to distinguish it from the 'statistical absorption coefficient' or the random incidence absorption coefficient $\alpha_{uni}$ after eqn (2.41).

The advantages of the reverberation method as mentioned at the beginning of this section are offset by considerable uncertainty as regards its reliability and the accuracy of the results obtained with it. In fact, several round robin tests, in which the same specimen of an absorbing material has been tested in different laboratories, have revealed a remarkable disagreement in the results. This must certainly be attributed to different degrees of sound field diffusion established in the various chambers and shows that increased attention must be paid to the methods of enforcing sufficient diffusion.

A first step towards a high degree of sound field diffusion is to design the reverberation chamber without parallel pairs of walls and thus avoid sound waves which can be reflected repeatedly between vertical walls without being influenced by the floor and the ceiling.

Among all further methods to achieve a diffuse sound field the introduction of volume scatterers, as described in Section 5.1, seems to be most adequate for reverberation chambers, since an existing arrangement of scatterers can easily be changed if it does not prove satisfactory. Practically, such scatterers can be realised as bent shells of wood, plastics or metal which are suspended from the ceiling by cables in an irregular arrangement (see, for instance, Fig. 8.21). If necessary, bending resonances of these shells can be damped by applying thin layers of lossy material onto them. It should be noted, however, that too many diffusers may also affect the validity of the usual reverberation formulae and that therefore the density of scatterers has a certain optimum.[16] If $H$ is the distance of the test specimen from the wall (or ceiling) opposite to it, this optimum range is about

$$0.5 < \langle n \rangle Q_s H < 2 \tag{8.44}$$

with $\langle n \rangle$ and $Q_s$ denoting the average density and scattering cross-section of the diffusers, respectively. This condition has also been proven experimentally. For not too low frequencies the scattering cross-section $Q_s$ is roughly half the geometrical area of one side of a shell (see Section 5.1).

Another source of systematic errors is the so-called 'edge effect'. If an absorbing area has free edges, i.e. edges not adjacent to a perpendicular wall, it will usually absorb more sound energy per second than in proportion to its geometrical area, the difference being caused by diffraction of sound into the absorbing area. It can be reduced although not eliminated by covering the free edges of the test specimen with a reflective frame. Such a frame made of reflective panels is mandatory for measuring the absorption of chairs or of

*Figure 8.21*  Reverberation chamber fitted out with 25 diffusers of Perspex (volume 324 m³; dimensions of one shell 1.54 m × 1.28 m).

a seated audience. Formally, this effect can be accounted for by introducing an 'effective absorption coefficient':[17]

$$\alpha_{\text{eff}} = \alpha_\infty + \beta L' \tag{8.45}$$

where $\alpha_\infty$ is the absorption coefficient of the unbounded test material and $L'$ denotes the total length of free edges divided by the area of the actual sample. The factor $\beta$ depends on the frequency and the type of material. It may be as high as 0.2 metres or more and can be determined experimentally using test pieces of different sizes and shapes. In rare cases, $\beta$ may even turn out slightly negative. A comprehensive treatment of the edge effect can be found in Ref. 1 of Chapter 2.

In principle, the edge effect can be avoided by covering one wall of the reverberation chamber completely with the material to be tested, since then there will be no free edges. However, the adjacent rigid or nearly rigid walls cause another, although less serious, edge effect, sometimes referred to as 'Waterhouse effect'.[18] According to eqn (2.38) (see also Fig. 2.11), the square of the sound pressure amplitude in front of a rigid wall exceeds its value far from the wall, and the same holds for the energy absorbed per unit time and area by the test specimen which perpendicularly adjoins that wall. This effect can be corrected for by replacing the geometrical area $S$ of the test specimen with

$$S_{\text{eff}} = S + \frac{1}{8}L''\lambda \tag{8.46}$$

where $\lambda$ is the wavelength corresponding to the centre frequency of the selected frequency band and $L''$ is the length of the edges adjacent to perpendicular walls.

Finally, a remark may be appropriate on the frequency range in which a given reverberation chamber can be used. If the linear chamber dimensions are in the range of a few wavelengths only, then statistical reverberation theories are no longer applicable to the decay process and hence to the process of sound absorption. Likewise, a diffuse sound field cannot be established when the number and density of eigenfrequencies are small. Suppose we require that at least ten normal modes will be excited by a sound signal of third-octave bandwidth. This is the case if the mid-frequency of the band exceeds a limiting frequency $f_l$, which can be found from eqn (3.20):

$$f_l \approx \frac{500}{\sqrt[3]{V}} \tag{8.47}$$

where the room volume $V$ has to be expressed in m³ and the frequency in Hz.

## 8.8 Scattering coefficient

According to the discussion at the end of Section 4.5, most real boundaries reflect incident sound waves partially specularly, i.e. like a mirror in optics, and partially in a diffuse manner. The direct way to measure the diffuse reflectivity of a surface is to irradiate a test specimen with a sound wave under a certain angle of incidence and to record the sound reflected (or scattered) into the various directions by swivelling a microphone at fixed distance around the specimen. In many cases, of course, this measurement must be carried out with scale models of the surfaces. Its result depends on the direction of irradiation and—like everything in acoustics—on the frequency. It can be represented as a scattering diagram or a collection of scattering diagrams in which the scattered sound components can be separated from the specular one. The scattering characteristics from a particular object, namely an irregularly structured ceiling, have been shown in Fig. 2.16.

Very often one is interested not so much in scattering diagrams but in a single figure which characterises the diffuse reflectivity of a wall. For this purpose, in Section 4.5 the total reflected energy was split up into the fractions $s$ and $1 - s$. The quantity defined by

$$s = 1 - \frac{I_{spec}}{I_0(1 - \alpha)} \tag{8.48}$$

is named the scattering coefficient, where $I_0$ and $I_{spec}$ denote the intensities of the incident and the specular component of the reflected wave, respectively; $\alpha$ is the absorption coefficient of the test specimen.

Of course, the scattering coefficient can be determined from a measured scattering diagram.[19] In the following, however, our goal is to describe more indirect procedures for measuring this quantity which circumvent the time-consuming measurement of scattering diagrams. Vorländer and Mommertz[20] have developed two methods which are intended for application in the model scale. Both of them employ a turntable which carries the test specimen.

The free-field method is carried out with a source–microphone arrangement, as shown in Fig. 8.22. The complex reflection factor consists of two parts, one due to the specular reflection and the other one representing the scattered part. Suppose we have determined the reflection factors

$$R_i = R_{spec} + S_i$$

for $n$ different orientations of the turntable. When they are added, the energy contained in the specular component increases proportional to $n^2$, while that corresponding to the scattered component increases in proportion to $n$ only. If the number $n$ of measurements is sufficiently large the latter part

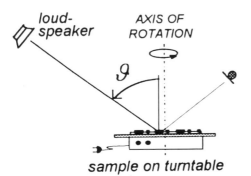

sample on turntable

*Figure 8.22* Experimental set-up for measuring the angle-dependent scattering coefficient.

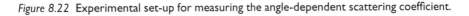

can be neglected, leaving

$$R_{\text{spec}} \approx \frac{1}{n} \sum_{i=1}^{n} R_i \qquad \text{for } n \gg 1$$

and, according to eqn (8.48) with $I_{\text{spec}}/I_0 = |R_{\text{spec}}|^2$ :

$$s(1 - \alpha) = 1 - |R_{\text{spec}}|^2 \tag{8.49}$$

the absorption coefficient is given by

$$\alpha \approx 1 - \frac{1}{n} \sum_{i=1}^{n} |R_i|^2 \tag{8.50}$$

The second method is based on the same idea. But now the turntable with the test specimen is placed in an otherwise empty reverberation chamber. Accordingly the scattering coefficient obtain with it is valid for random sound incidence. Again we observe $n$ impulse responses $g_1(t), g_2(t), \ldots g_n(t)$ at slightly different positions of the sample. Each of them consists of an invariable part $g_0$ due to the specular component, and a part $g'$ which differs from one measurement to the other; thus $g_i(t) = g_0(t) + g'_i(t)$. Hence the sum of all these impulse responses is

$$h(t) = \sum_{i=1}^{n} g_i(t) = ng_0(t) + \sum_{i=1}^{n} g'_i(t)$$

The square of this sum has the expectation value

$$\langle h^2 \rangle = n^2 \cdot g_0^2 + n \cdot \langle g'^2 \rangle \tag{8.51}$$

Both terms decay with different decay rates, since the first term is subject to an additional reduction by the scattering process, which converts 'specular energy' into 'diffuse energy'. Hence, another version of eqn (8.51) reads:

$$\langle h^2 \rangle \propto \quad n^2 \cdot \exp(-2\delta_1 t) + n \cdot \exp(-2\delta_2 t) \tag{8.52}$$

The damping constant $\delta_2$ is determined by the sound absorption of the reverberation chamber:

$$\delta_2 = \frac{cA}{8V} = \frac{c}{8V}\left[(S - S_\mathrm{s})\alpha_0 + S_\mathrm{s}\alpha\right] \tag{8.53a}$$

where $V$ denotes the volume of the chamber and $S$ is the area of its boundary with the absorption coefficient $\alpha_0$; $S_\mathrm{s}$ is the area of the sample. On the other hand, the first term of eqn (8.50) decays with the larger decay constant which contains the losses due to scattering

$$\delta_1 = \delta_2 + s(1 - \alpha)\frac{c}{8V} \tag{8.53b}$$

Figure 8.23 shows several logarithmic decay curves measured with this method. They are double-sloped as was to be expected. The parameter is $n$, the number of individual decays from which the averaged decay has been formed. With increasing $n$, the initial slope corresponding to the average decay constant (see eqn (3.49a))

$$\bar{\delta} = \frac{n\delta_1 + \delta_2}{n + 1} \approx \delta_1 \quad \text{if } n \gg 1 \tag{8.54}$$

becomes more prominent. If the number $n$ of decays is high enough, both decay constants $\delta_1$ and $\delta_2$ can be evaluated with sufficient accuracy, particularly if the backward integration technique (see Section 8.4) is applied.

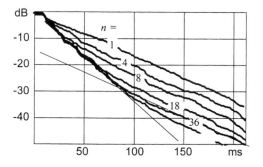

*Figure 8.23* Average decay curves (model measurements). Here $n$ is the number of individual decays from which the average has been formed. (From Vorländer and Mommertz.[20])

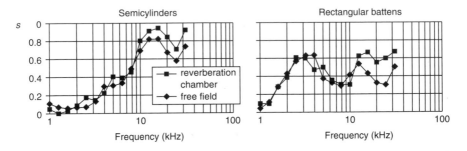

*Figure 8.24* Random incidence scattering coefficients *s* of irregular arrangements of battens on a plane panel as a function of frequency: ♦———♦ free field method; ■———■ reverberation method. Left side: battens with quadratic cross-section, side length = 2 cm. Right side: with semicircular cross-section, diameter = 2 cm (after Mommertz and Vorländer [21]).

From their difference the scattering coefficient for random sound incidence *s* is readily obtained, using eqns (8.53).

Figure 8.24 represents the 'random incidence scattering coefficient' of two surfaces obtained with the direct method (averaged over all directions of incidence) and with the reverberation method. The test objects were battens with quadratic or semicylindrical cross-section (side length or diameter 2 cm) irregularly mounted on a plane panel. The agreement of both results is obvious. The small differences at high frequencies are probably caused by the quadratic shape of the sample.

More data on scattering surfaces can be found in Ref. 22 and in Ref. 12 of the next chapter.

## References

1 Sabine WC. Collected Papers on Acoustics. Dover, 1964 (first published in 1922).
2 Fahy F. Sound Intensity, 2nd edn. London: E & FN Spon, 1995.
3 Alrutz H, Schroeder MR. A fast Hadamard transform method for the evaluation of measurements using pseudorandom test signals. Proc. 11th Intern Congr on Acoustics, Paris, 1983, 6:235.
4 MacWilliams FJ, Sloane, NJA. Pseudo-random sequences and arrays. Proc IEEE 1976; 84:1715.
5 Borish J, Angell JB. An efficient algorithm for measuring the impulse response using pseudorandom noise. J Audio Eng Soc 1983; 31:478.
6 Müller S, Massarani P. Transfer-function measurement with sweeps. J Audio Eng Soc 2001; 49:443.
7 Stan GB, Embrechts JJ, Archambeau D. Comparison of different impulse response measurement techniques. J Audio Eng Soc 2002; 50:249, etc.
8 Bilsen FA. Thresholds of perception of pitch. Conclusions concerning coloration in room acoustics and correlation in the hearing organ. Acustica 1967; 19:27.

9 Schroeder MR. Modulation transfer functions: definition and measurement. Acustica 1981; 49:179.

10 Schroeder MR. New method of measuring reverberation time. J Acoust Soc Am 1965; 37:409.

11 Tachibana H, Yamasaki Y, Morimoto M, et al. Acoustic survey of auditoriums in Europe and Japan. J Acoust Soc Japan (E) 1989; 10:73.

12 Yamasaki Y, Itow T. Measurement of spatial information in sound fields by closely located four point microphone method. J Acoust Soc Japan E 1989; 10:101.

13 Sekiguchi K, Kimura S, Hanyuu T. Analysis of sound field on spatial information using a four-channel microphone system based on a regular tetrahedron peak point method. Appl Acoustics 1992; 37:305.

14 Mommertz E. Angle-dependent in-situ measurements of reflection coefficients using a subtraction technique. Appl Acoustics 1995; 46:251.

15 International Standardization Organization. Measurement of sound absorption in a reverberation room. ISO 354/2000.

16 Kuttruff H. Sound decay in reverberation chambers with diffusing elements. J Acoust Soc Am 1981; 69:1716.

17 de Brujin A. A mathematical analysis concerning the edge effect of sound absorbing materials. Acustica 1973; 28:33.

18 Waterhouse RV. Interference patterns in reverberant sound fields. J Acoust Soc Am 1955; 27:247.

19 Mommertz E. Determination of scattering coefficients from the reflection directivity of architectural surfaces. Appl Acoustics 2000; 60:201.

20 Vorländer M, Mommertz E. Definition and measurement of random-incidence scattering coefficients. Appl Acoustics 2000; 60:187.

21 Mommertz E, Vorländer M. Measurement of the scattering coefficient of surfaces in the reverberation chamber and in the free field. Proc 15th Intern Congr on Acoustics, Vol. II, Trondheim, 1995, p 577.

22 Cox TJ, D'Antonio P. Acoustic Absorbers and Diffusers. London: Spon Press, 2004.

# Chapter 9

# Design considerations and design procedures

The purpose of this chapter is to describe and to discuss some practical aspects of room acoustics, namely the acoustical design of auditoria in which some kind of performance (lectures, music, theatre, etc.) is to be presented to an audience, or of spaces in which the reduction of noise levels is the main interest. Its contents are not just an extension of fundamental laws and scientific insights towards the practical world, nor are they a collection of guidelines and rules deduced from them. In fact, the reader should be aware that the art of room acoustical design is only partially based on theoretical considerations, and that it cannot be learned from this or any other book but that successful work in this field requires considerable practical experience. On the other hand, mere experience without at least some insight into the physics of sound fields and without certain knowledge of psychoacoustic facts is of little worth, or is even dangerous in that it may lead to unacceptable generalisations.

Usually the practical work of an acoustic consultant starts with drawings being presented to him which show the details of a hall or some other room which is at the planning stage or under construction, or even one which is already in existence and in full use. First of all, he must ascertain the purpose for which the hall is to be used, i.e. which type of performances or presentations are to take place in it. This is more difficult than appears at first sight, as the economic necessities sometimes clash with the original ideas of the owner or the architects. Second, he must gain some idea of the objective structure of the sound field to be expected, for instance the values of the parameters characterising the acoustical behaviour of the room. Third, he must decide whether or not the result of his investigations favours the intended use of the room; and finally, if necessary, he must work out proposals for changes or measures which are aimed at improving the acoustics, keeping in mind that these may be very costly or may substantially modify the architect's original ideas and therefore have to be given very careful consideration.

In order to solve these tasks there is so far no generally accepted procedure which would lead with absolute certainty to a good result. Perhaps it is too much to expect there ever to be the possibility of such a 'recipe', since one project is usually different from the next due to the efforts of architects and owners to create something quite new and original in each theatre or concert hall.

Nevertheless, a few standard methods of acoustical design have evolved which have proven useful and which can be applied in virtually every case. The importance which the acoustic consultant will attribute to one or the other, the practical consequences which he will draw from his examination, whether he favours reverberation calculations more than geometrical considerations or vice versa—all this is left entirely to him, to his skill and to his experience. It is a fact, however, that an excellent result requires close and trustful cooperation with the architect—and a certain amount of luck too.

As we have seen in preceding chapters, there are a few objective sound field properties which are beyond question regarding their importance for what we call good or poor acoustics of a hall: namely, the strength of the direct sound, the temporal and directional distribution of the early sound energy and the duration of reverberation processes. These properties depend on constructional data, in particular on the

(a)  shape of the room
(b)  volume of the room
(c)  number of seats and their arrangement
(d)  materials of walls, ceiling, floor, seats, etc.

While the reverberation time is determined by factors (b)–(d) and not significantly by (a), the room shape influences strongly the number, directions, delays and strengths of the early reflections received at a given position or seat. The strength of the direct sound depends on the distances to be covered, and also on the arrangement of the audience.

In the following discussion we shall start with the last point: i.e. the factors that determine the strength of the direct sound in a hall.

## 9.1 Direct sound

The direct sound signal travelling from the sound source to a listener along a straight line is not influenced at all by the walls or the ceiling of a room. Nevertheless, its strength depends on the geometrical data of the hall: i.e. on the (average) length of paths which it has to travel and on the height at which it propagates over the audience until it reaches a particular listener.

Of course, the direct sound intensity under otherwise constant conditions is higher the closer the listener is seated to the sound source. Different plans of halls can be compared in this respect by a dimensionless figure of merit, which is the average distance of all listeners from the sound source divided by the square root of the area occupied by audience. For illustration, Fig. 9.1 presents a few types of floor plans; the numbers indicate this normalised

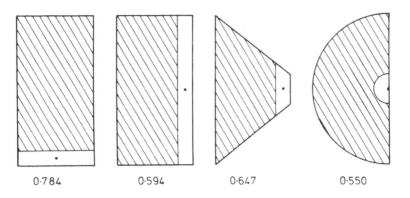

0·784            0·594            0·647            0·550

*Figure 9.1* Normalised average distance from listeners to source for various room shapes.

average distance. The audience areas are shaded and the sound source is marked by a point.

It is seen that a long rectangular room with the sound source on its short side seats the listeners relatively far from the source, whereas a room with a semicircular floor plan provides particularly short direct sound paths. For this reason, many large lecture halls, theatres and session halls of parliaments are of this type. Likewise, most ancient amphitheatres have been given this shape by their builders. (The same figure holds for any circular sector, i.e. also for the full circle.) However, for a closed room this shape has certain acoustical risks in that it concentrates the sound reflected from the rear wall toward certain regions. Generally, considerations of this sort should not be given too much weight since they are only concerned with one aspect of acoustics, which may conflict with other ones.

Attenuation of the direct sound due to grazing propagation over the heads of the audience (see Section 6.7) can be reduced or avoided by sloping the audience area upwardly instead of arranging the seats on a horizontal floor. This holds also for the attenuation of side or front wall reflections. As is easily seen by comparing Figs 9.2a and 9.2b, a constant slope is less favourable than an increasing ascent of the audience area because in the former case the angle of incidence and hence the attenuation shows a stronger dependence on the distance from the source. The optimum slope (which is optimal as well with respect to the listener's visual contact with the stage) is reached if all sound rays originating from the sound source S strike the audience area at the same incidence angle $\vartheta$. The mathematical expression for this condition is

$$r(\varphi) = r_0 \exp(\varphi \cdot \cot \gamma) \tag{9.1}$$

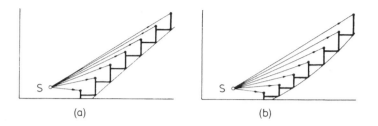

Figure 9.2 Reducing the attenuation of direct sound by sloping the seating area: (a) constant slope; (b) increasing slope.

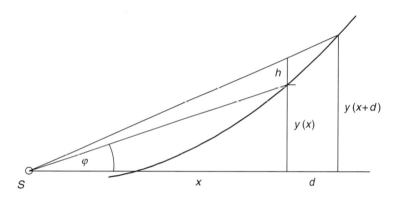

Figure 9.3 Notations in eqns (9.1) and (9.2). S = sound source, h = clearance.

In this formula, which describes what is called a logarithmic spiral, $r(\varphi)$ is the length of the sound ray leaving the source under an elevation angle $\varphi$, and the constant $r_0$ is the length of the sound ray at $\varphi = 0$ (see Fig. 9.3). The angle $\gamma = (\pi/2) - \vartheta$ is named the 'grazing' angle. For design purposes eqn (9.1) is not well-suited; however, it can be simplified by setting $\cot \gamma \approx 1/\gamma$ and $r \approx x$. With these approximations, the height $y$ of a point can be expressed as a function of its horizontal distance $x$:

$$y(x) \approx \gamma x \ln \left( \frac{x}{r_0} \right) \tag{9.1a}$$

Another important figure is the clearance $h$, i.e. the vertical distance between a ray arriving at a particular seating row (a 'sight-line') and the corresponding point of the preceding row. Let $d$ denote the row distance; then the clearance for a seating arrangement after eqn (9.1a) is

$$h \approx \gamma d \left( 1 - \frac{d}{2x} \right) \tag{9.2}$$

According to this formula, $h$ does not vary strongly with distance $x$, apart from the rows next to the sound source. For very distant seats it approaches the constant value $\gamma d$. Thus, for $d = 1$ metre and $\gamma = 5°$ this limiting value is 8.7 cm. To achieve a significant effect the clearance should not be smaller. Of course, higher values are more favourable. However, a gradually increasing slope of the seating area has certain practical disadvantages. They can be circumvented by approximating the sloping function of eqn (9.1) by a few sections with constant but increasing slope, and thus subdividing the audience area in a few sections with uniform seating rake within each of them.

Front seats on galleries or balconies are generally well supplied with direct sound since they do not suffer at all from sound attenuation due to listeners sitting immediately in front. This is one of the reasons why seats on galleries or in elevated boxes are often known for excellent listening conditions.

## 9.2 Examination of the room shape

As already mentioned, the delay times, the strengths and the directions of the reflections—and in particular of early reflections—are determined by the position and the orientation of reflecting areas, i.e. by the shape of a room. Since these properties of reflected sound portions are to a high degree responsible for good or poor acoustics, it is indispensable to examine the shape of a room carefully in order to get a survey on the reflections produced by the enclosure.

A simple way to obtain this survey is to trace the paths of sound rays which emerge from an assumed sound source, using drawings of the room under consideration. In most cases the assumption of specular reflections is more or less justified.

The sound rays reflected from a wall portion are found very easily if the enclosure is made up of plane boundaries; then one can take advantage of the concept of image sound sources as described in Chapter 4. This procedure, however, is feasible for first-order or at best for second-order reflections only, which is often sufficient.

For curved walls the method of image sources cannot be applied. In this case we have to determine the tangential plane (or the normal) of a considered boundary element and find the reflection from this plane. A schematic example of sound ray construction is presented in Fig. 9.4.

At any event the construction of sound paths gives us some ideas on the distribution of the reflected sound energy, i.e. whether it is concentrated in a limited region, and where a focal point or a caustic is to be expected. If a sufficiently large portion of the wall or the ceiling has a circular shape in the sectional drawing, or can be approximated by a circle, the location of the focus associated with it may be found from eqn (4.17). Furthermore, the

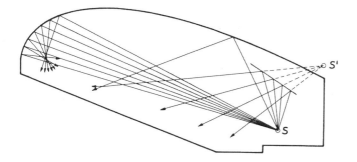

*Figure 9.4* Construction of sound rays paths in the longitudinal section of a hypothetical hall.

directions of incidence at various seats can be seen immediately, whereas the delay time between a reflection with respect to the direct sound is determined from the difference in path lengths after dividing the latter by the sound velocity.

The decision whether a particular reflection will be perceivable at all, whether it will contribute to speech intelligibility, to 'clarity' or to 'spaciousness', or whether it will be heard as a disturbing echo requires knowledge of its relative intensity (see Chapter 7). If the reflection occurs at a plane boundary with dimensions large compared to the acoustical wavelengths, the $1/r$ law of spherical wave propagation can be applied. Let $r_0$ and $r_i$ be the path lengths of the direct sound ray and of a particular reflection, measured from the sound source to the listener, then

$$L_0 - L_i = 20 \cdot \log_{10}(r_i/r_o) \quad \text{dB} \tag{9.3}$$

is the difference between the level of the direct sound and that of the reflection. Equations (2.44) or (2.45) are employed to decide whether a particular wall portion, a balcony face or suspended reflector can be regarded as an efficient reflector. If the reflecting boundary has an absorption coefficient $\alpha$, the level of a first-order reflection is reduced by another $10 \cdot \log(1/\alpha)$ decibels. Irregularities on walls and ceiling can be neglected as long as their dimensions are small compared to the wavelength; this requirement may impose restrictions on the frequency range for which the results obtained with the formula above are valid. The intensities or pressure levels of reflections from a curved wall section can be estimated by comparing the density of the reflected rays in the observation point with the ray density which would be observed if that wall section were plane. For spherical or cylindrical wall portions the ratio of the reflected and the incident energy can be calculated from eqn (4.19), which is equivalent to

$$\Delta L = 10 \cdot \log_{10} \left| \frac{1 + x/a}{1 - x/a} \right|^n \tag{9.4}$$

As in eqn (4.19) $a$ and $x$ are the distances of the source and the observation point from the reflecting wall portion, respectively; $n$ is unity if the curved boundary is cylindrical and for a spherical wall portion $n = 2$.

The techniques of ray tracing with pencil and ruler apply only to sound paths which are situated in the plane of the drawing to hand. Sound paths in different planes can be constructed by applying the methods of constructive geometry. This, of course, involves considerably more time and labour, and it is questionable whether this effort is worthwhile considering the rather qualitative character of the information gained by it. For rooms of more complicated geometry it is much more practical to apply computerised ray tracing techniques (see Section 9.6).

So far we have described methods to investigate the effects of a given enclosure upon sound reflections. Beyond that, there are some general conclusions which can be drawn from geometrical considerations, and also from experiences with existing halls. They are summarised briefly below.

If a room is to be used for speech, it would be desirable to provide for high 'definition' or short 'centre time' (see Section 7.4). Hence, the direct sound should be supported by many strong reflections with delay times not exceeding about 50 ms. Reflecting areas (wall portions, screens) placed very close to the sound source are generally favourable since they collect a great deal of the emitted sound energy and reflect it in the direction of the audience. For the same reason it is wrong to have heavy curtains of fabric behind the speaker. On the contrary, the speaker should be surrounded by hard and properly orientated surfaces, which can even be in the form of portable screens, for instance. Similarly, reflecting surfaces above the speaker have a favourable effect. If the ceiling over the speaker is too high to produce strong and early reflections, the installation of suspended and suitably tilted reflectors should be taken into consideration (see Fig. 9.4). An old and familiar example of a special sound reflector is the canopy above the pulpits in churches. The acoustical advantage of these canopies can be observed very clearly when it is removed during modern restoration.

Unfortunately these principles can only be applied to a limited extent to theatres, where such measures could in fact be particularly useful. This is because the stage is the realm of the stage designer, of the stage manager and of the actors; in short, of people who sometimes complain bitterly about the acoustics but who are not ready to sacrifice one iota of their artistic intentions in favour of acoustical requirements. It is all the more important to shape the wall and ceiling portions which are close to the stage in such a way as to direct the incident sound immediately onto the audience.

In conference rooms, school classrooms, lecture halls, etc., at least the front and central parts of the ceiling should be made reflecting, since, in most cases, the ceiling is low enough to produce reflections which support the direct sound. Absorbent materials required for the reduction of the reverberation time can thus only be mounted on more remote ceiling portions (and on the rear wall, of course).

When it comes to the design of concert halls the economic use of sound energy is not of foremost interest as it is in a lecture hall or a drama theatre except for very large halls. Therefore the shape of a concert hall should not project as much sound energy as possible immediately towards the audience. For this would result in a high fraction of early energy and—in severe cases—to subjective masking of the sound decay in the hall and hence too low reverberance even if the objective reverberation time has correct values. In fact, many concert halls which are famous for their acoustics are of the shoe-box type (Vienna Musikvereinssal, Boston Symphony Hall.)

This does not mean that early sound reflections can be disregarded at all. As we have seen in Section 7.7, it is the energy reaching the listener from lateral directions which is responsible for the 'spatial impression' or 'spaciousness' in a concert hall. For this reason particular attention must be given to the design of the side walls, especially to their distance and to the angle which they include with the longitudinal axis of the plan.

This may be illustrated by Fig. 9.5, which shows the spatial distribution of early lateral energy in a few differently shaped two-dimensional enclosures,[1] computed using eqn (7.17); the area of all halls was assumed to be 600 m². The position of the sound source is marked by a cross; the densities of shading of the various areas correspond to the following intervals of the early lateral energy fraction (LEF): 0–0.06, 0.06–0.12, 0.12–0.25, 0.25–0.5 and >0.5. In all examples the LEF is very low at locations next to the sound source, but it is highest in the vicinity of the side walls. Accordingly, the largest areas with high LEF and hence with satisfactory 'spaciousness' are to be expected in long and narrow rectangular halls. On the other hand, particular large areas with low early lateral energy fraction appear in fan-shaped halls, a fact which can easily be verified by a simple construction of the first-order image sources. These findings explain—at least partially—why so many concert halls with excellent acoustics have rectangular floor plans with relatively narrow side walls. It may be noted, by the way, that the requirement of strong lateral sound reflections favours room shapes which are different from those leading to strong direct sound (see Section 9.1).

In real, i.e. in three-dimensional halls, additional lateral energy is provided by the double reflection from the edges formed by a side wall and horizontal surfaces such as the ceiling or the soffits of side balconies (Fig. 9.6). These contributions are especially useful since they are less attenuated by the audience than reflections just from the side walls. If no balconies are planned

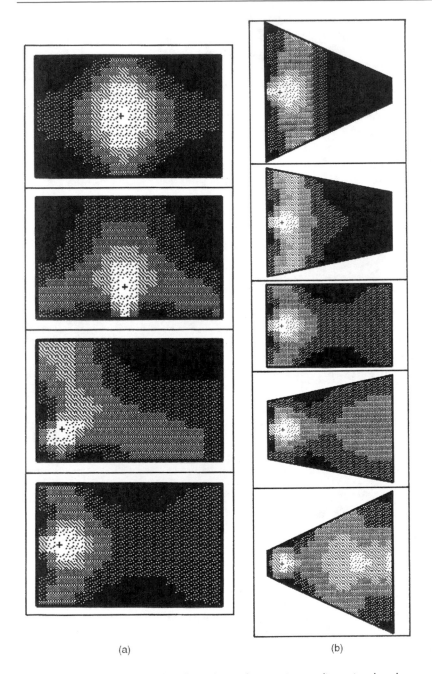

*Figure 9.5* Distribution of early reflected sound energy in two-dimensional enclosures with area 600 m$^2$: (a) rectangular enclosure, different source positions; (b) fan-shaped enclosures (after Ref. 1).

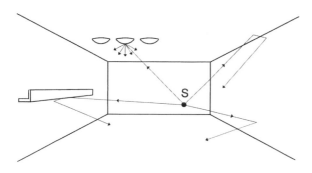

*Figure 9.6* Origin of lateral or partly lateral reflections (S = sound source).

the beneficial effect of soffits can be achieved as well by properly arranged surfaces or bodies protruding from the side walls.

With regard to the performance of orchestral music one should remember that various instruments have quite different directivity of sound radiation, which depends also on the frequency. Accordingly, sounds from certain instruments or groups of instruments are predominantly reflected by particular wall or ceiling portions. Since every concert hall is expected to house orchestras of varying composition and arrangement, only some general conclusions can be drawn from this fact, however. Thus the high-frequency components, especially from string instruments, which are responsible for the brilliance of the sound, are reflected mainly from the ceiling overhead and in front of the stage, whereas the side walls are very important for the reflection of components in the range of about 1000 Hz and hence for the volume and sonority of the orchestral sounds.[2]

Some further comments may be appropriate on the acoustical design of the stage or the orchestra's platform of concert halls. From the acoustical point of view, the stage enclosure of a concert hall has the purpose of collecting sounds produced by the musical instruments, to blend them and finally to project them towards the auditorium, but also to reflect part of the sound energy back to the performers. This is necessary to establish the mutual auditory contact they need to maintain ensemble, i.e. proper intonation and synchronism.

At first glance platforms arranged in a recess of the hall seem to serve these purposes better in that their walls can be designed in such a way as to direct the sound in the desired way. As a matter of fact, however, several famous concert halls have more exposed stages which form just one end of the hall. From this it may be concluded that the height and inclination of the ceiling over the platform deserve particular attention.

Marshall et al[3] have found by systematic experimental work that the surfaces surrounding the orchestra should be far enough away from the performers to delay the reflected sound by more than 15 ms but not more than 35 ms. This agrees roughly with the observation that the optimum height of the ceiling (or of overhead reflectors) in the stage area is somewhere between 5 and 10 m. With regard to the side walls, this guideline can be followed only for small performing groups, since a large orchestra will usually occupy the whole platform. If symphonic music is performed in an opera theatre it may be useful to install a carefully designed portable orchestra shell to provide for the necessary reflections. It goes without saying that this shell should be made of sufficiently heavy elements with non-porous surface.

Another important aspect of stage design is raking of the platform,[4] which is often achieved with adjustable or movable risers. It has, of course, the effect of improving the sightlines between listeners and performers. From the acoustical standpoint it increases the strength of the direct sound and reduces the obstruction of sound propagation by intervening players. It seems, however, that this kind of exposure can be carried too far; probably the optimum rake has to be determined by some experimentation.

The examination of room geometry can lead to the result that some wall areas, particularly if they are curved, will give rise to very delayed reflections with relatively high energy, which will neither support the direct sound nor be masked by other reflections, but will be heard as echoes. The simplest way of avoiding such effects is to cover these wall portions with highly absorbent material. If this precaution would cause an intolerable drop in reverberation time or is impossible for other reasons, a reorientation of those surfaces could be suggested, or they could be split up into irregularly shaped surfaces so that the sound is scattered in all directions. Of course the size of these irregularities must at least be comparable with the wavelength in order to be effective. If desired, any treatment of these walls can be concealed behind acoustically transparent screens, consisting of grids, nets or perforated panels, whose transmission properties were discussed in Section 6.3.

Another undesirable effect is flutter echo, as already described in Section 4.2, which is caused by repeated sound reflections between parallel walls with smooth surface. Whether such a periodic or nearly periodic train of reflections is audible at all depends on the strength of competing, non-periodic reflections (see also Section 8.4). If necessary, a flutter echo can be avoided by absorbing lining of the critical wall portions, or by providing for surface irregularities which scatter the impinging sound rather than reflect it specularly. Another way to avoid flutter echoes is reorientation of originally parallel walls by about 5°.

## 9.3 Reverberation time

Among all significant room acoustical parameters and indices the reverberation time is the only one which is related to room data by relatively reliable and tractable formulae, despite certain limitations which have been discussed in Chapter 5. Their application permits us to lay down practical procedures, provided we have certain ideas on the desired values of the reverberation times at various frequencies.

The room data needed for the application of these formulae are the volume of the room, the materials and the surface treatment of the walls and of the ceiling, and the number, arrangement and type of seats. Many of these details are not yet fixed in the early phases of planning. For this reason it makes no sense to carry out a detailed calculation of the reverberation time at this stage; instead, a rough estimate may be sufficient

An upper limit of the attainable reverberation time can be obtained from Sabine's formula (5.25) with $m = 0$ by attributing an absorption coefficient of 1 to the areas covered by audience, and an absorption coefficient of 0.05–0.1 to the remaining areas, which need not be known too exactly at this stage.

For halls with a full audience and without any additional sound absorbing materials, i.e. in particular for concert halls, a few rules of thumb for estimating the reverberation time are in use. The simplest one is

$$T = \frac{V}{4N} \tag{9.5}$$

with $N$ denoting the number of occupied seats. Another estimate is based on the 'effective seating area' $S_a$,

$$T = 0.15 \frac{V}{S_a} \tag{9.6}$$

Here $S_a$ is the area occupied by the audience, the orchestra and the chorus; furthermore, it includes a strip of 0.5 m around each block of seats. Aisles are added into $S_a$ if they are narrower than 1 m (see Section 6.7). In order to give an idea of how reliable these formula are, the mid-frequency reverberation times (500–1000 Hz) of many concert halls are plotted in Fig. 9.7 as a function of the 'specific volume' $V/N$ (Fig 9.7a) and of volume per square metre of audience $V/S_a$ (Fig 9.7b). Both diagrams are based on data from Ref. 5 of Chapter 6; each point corresponds to one hall. In both cases, the points show considerable scatter. Furthermore, it seems that eqns (9.5) and (9.6) overestimate the reverberation time.

More reliable is an estimate which involves two sorts of areas, the audience area $S_a$, as before, and the remaining area $S_r$ of the boundary:

$$T \approx 0.161 \frac{V}{S_a \alpha_a + S_r \alpha_r} \tag{9.7}$$

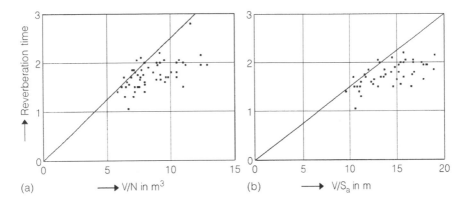

*Figure 9.7* Mid-frequency reverberation times of occupied concert halls: (a) as a function of the volume per seat; (b) as a function of the volume divided by the effective audience area (data from Ref 5 of Chapter 6). The straight lines represent eqns (9.5) and (9.6).

The absorption coefficients $\alpha_s$ and $\alpha_s$ of both types of areas may be found in Tables 6.4 and 6.5, which have been collected by Beranek and Hidaka (see Ref. 6 of Chapter 6).

In any case, a more detailed reverberation calculation should definitely be carried out at a more advanced phase of planning, during which it is still possible to make changes in the interior finish of the hall without incurring extra expense. The most critical aspect is the absorption of the audience. The factors on which it depends have already been discussed in Section 6.7. Regarding the uncertainties caused by audience absorption it is almost meaningless to try to decide whether Sabine's formula (5.25) would be sufficient or whether the more accurate Eyring equation (5.24) would be more adequate. Therefore the simpler Sabine formula is preferable, with the term $4mV$ accounting for the sound attenuation in air. If the auditorium is a concert hall which is to be equipped with pseudorandom diffusers, their absorption should be considered as well. The same holds for an organ (see Section 6.8).

As regards the absorption coefficients of the various materials and wall linings, use can be made of compilations which have been published by several authors (see, for instance, Ref. 22 of Chapter 8). It should be emphasised that the actual absorption, especially of highly absorptive materials, may vary considerably from one sample to the other and depends strongly on the particular way in which they are mounted. Likewise, the coefficients presented in Table 9.1 are to be considered as average values only. In case of doubt it is recommended to test actual materials and the influence of their mounting by measuring the absorption coefficient, either in the impedance tube or, more reliably, in a reverberation chamber (see Chapter 8). This applies particularly to chairs whose acoustical properties can vary considerably depending on the quantity and quality of the fabric used for the upholstery. If possible the empty chairs should have the about same absorption as the occupied ones. This has the favourable effect that the reverberation time of the hall will not depend too strongly on the degree of occupation. With tip-up chairs this can be accomplished by perforating the underside of the plywood or hardboard seats and backing them with rock wool. Likewise, an absorbent treatment of the rear of the backrests could be advantageous.

According to Table 9.1, light partitions such as suspended ceilings or wall linings have their maximum absorption at low frequencies. Therefore, such constructions can make up for the low absorptivity of an audience and thus reduce the frequency dependence of the reverberation time.

Very often a hall has to serve various types of purposes, including not only the performance of music of different styles or the presentation of the-atre plays but also conferences, fashion shows and many other events. Then, not only is a very flexible stage machinery needed, including a demount-able orchestra shell, but also provisions have to be made for adapting the acoustics of the auditorium to the varying uses. This concerns in the first place its reverberation time. When both speech and orchestral music are to be presented under optimum acoustical conditions, some variability is indispensable by which the conflicting requirements can be reconciled.

Variations of the reverberation time can be achieved by installing mov-able wall or ceiling elements, for instance, panels which can be turned back to front and thus exhibit reflecting surfaces in one position and absorbent

Table 9.1 Typical absorption coefficients of various types of wall materials

| Material | Centre frequency of octave band (Hz) | | | | | |
|---|---|---|---|---|---|---|
| | 125 | 250 | 500 | 1000 | 2000 | 4000 |
| Hard surfaces (brick walls, plaster, hard floors, etc.) | 0.02 | 0.02 | 0.03 | 0.03 | 0.04 | 0.05 |
| Slightly vibrating walls (suspended ceilings, etc.) | 0.10 | 0.07 | 0.05 | 0.04 | 0.04 | 0.05 |
| Strongly vibrating surfaces (wooden panelling over air space, etc.) | 0.40 | 0.20 | 0.12 | 0.07 | 0.05 | 0.05 |
| Carpet, 5 mm thick, on hard floor | 0.02 | 0.03 | 0.05 | 0.10 | 0.30 | 0.50 |
| Plush curtain, flow resistance 450 Ns/m$^3$, deeply folded, in front of a solid wall | 0.15 | 0.45 | 0.90 | 0.92 | 0.95 | 0.95 |
| Polyurethane foam, 27 kg/m$^3$, 15 mm thick on solid wall | 0.08 | 0.22 | 0.55 | 0.70 | 0.85 | 0.75 |
| Acoustic plaster, 10 mm thick, sprayed on solid wall | 0.08 | 0.15 | 0.30 | 0.50 | 0.60 | 0.70 |

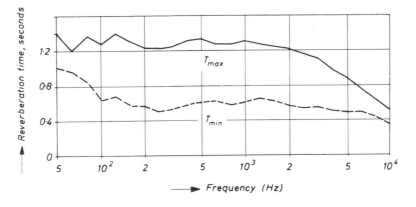

*Figure 9.8* Maximum and minimum reverberation times in a broadcasting studio with variable absorption.

ones in the other. The resulting difference in reverberation time depends on the fraction of area treated in this way and on the difference in absorption coefficients of those elements. Installations of this type are usually quite costly and sometimes give rise to considerable mechanical problems.

As an example, Fig. 9.8 shows the effect of variable wall absorption in a broadcasting studio with a volume of 726 m³. Its walls are fitted with strips of glass wool tissue which can be rolled up and unrolled electrically. Behind the fabric there is an air space with an average depth of 20 cm, subdivided laterally in 'boxes' of 0.5 m × 0.6 m. The reverberation times of the studio for the two extreme situations (glass wool rolled up and glass wool completely unrolled) can be changed from 0.6 to 1.25 s.

There are also multipurpose halls with variable volume. In this case the change of reverberation time is rather a side effect while the main goal is to adapt the seating capacity to different uses. One example of this kind is the Stadthalle in Braunschweig which has a maximum volume of 18000 m³ and accommodates nearly 2200 persons. Basically, its ground plan has the shape of a regular triangle with truncated corners. One of these corners contain the stage while the other ones are elevated parts of the seating area. They can be separated from the main auditorium by folding walls. This measure reduces the volume and the seating capacity of the auditorium by 1800 m³ and 650, respectively. At the same time, the reverberation time is increased from 1.3 to 1.6 s. Thus, the reduced configuration is more suited for concerts while for more popular events or for meetings the full capacity of the hall should be employed.

Quite a different way to vary the acoustics of a hall is by using special electroacoustic systems by which the reverberation time can be increased. Such methods will be described in Chapter 10.

In practice it is not uncommon to find that a room actually consists of several subspaces which are coupled to each other by some openings. Examples of coupled rooms are theatres with boxes which communicate with the main room through relatively small openings, or the stage (including the stage house which may by quite voluminous) of a theatre or opera house which is coupled to the auditorium by the proscenium, or churches with several naves or chapels. Cremer[5] was probably the first author to point out the necessity of considering coupling effects when calculating the reverberation time of such a room. This necessity arises if the area of the coupling aperture is substantially smaller than the equivalent absorption area of the partial rooms involved.

A general treatment of coupled rooms has been presented in Section 5.8. The following discussion is restricted to a more qualitative description of a system consisting of two coupled subspaces called Room 1 and Room 2. They are coupled to each other by a coupling area $S'$ (see Fig. 9.9), their damping constants and reverberation times are

$$\delta_i = \frac{c}{8V_i}(A_i + S'), \qquad T_i = \frac{6.91}{\delta_i} \qquad (i = 1, 2) \tag{9.8}$$

$A_i$ is the equivalent absorption area in room $i$. If the listener finds himself in the more reverberant room (Room 1), the only effect of the coupling to another space is a slight increase of the total absorption, i.e. a slight reduction of reverberation since the coupling area acts merely as an 'open window'. The listener will hardly hear any difference, even if there is a substantial difference between reverberation times $T_1$ and $T_2$. This situation is typical for an auditorium with large seating areas under balconies and with the receiver located in the main room. Nevertheless, whenever an auditorium has deep balcony overhangs, it is advisable to carry out an alternative calculation of its decay time by treating the 'mouths' of the overhangs as completely absorbing areas. Likewise, instead of including the whole stage of a concert hall with

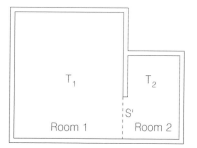

Figure 9.9 Coupled rooms.

all its uncertainties in the calculation, its opening can be treated as an area having an absorption coefficient rising from about 0.4 at 125 Hz to about 0.8 at 4 kHz.

Matters are different if the listener is in the less reverberant partial room, which we call Room 2. Whether he will become aware of the longer reverberation in Room 1 or not depends mainly on the way the system is excited and in the strength of coupling. If the sound source excites predominantly Room 2, the (logarithmic) sound decay in that room will be composed of two straight parts with a bend in between, similar to the decay curve 1 in Fig. 5.12. This holds for the impulse response of the system as well as for the decay process following its steady-state excitation. In the former case, the bent occurs at[5]

$$\Delta L = \frac{10}{1 - \delta_1/\delta_2} \cdot \log_{10}(1/C) - 3 \text{ dB} \qquad (9.9)$$

below the beginning of the curve. The constant $C$ is given by

$$C = \frac{k_{12}k_{21}}{4(\delta_1 - \delta_2)^2} = \frac{1}{V_1 V_2 (\delta_1 - \delta_2)^2} \cdot \left(\frac{cS'}{8}\right)^2 \qquad (9.9a)$$

If the level at which the bent occurs is high enough the listener will hear the reverberant 'tail', at least when the exciting signal contains impulsive components (loud cries, isolated musical chords, drumming, etc.). In no case will he or she experience this tail as natural because it is not a property of the listener's room but originates from the coupling aperture. However, if the location of the sound source is such that it excites both rooms, as may be the case with actors performing on the stage of a theatre, then the longer reverberation from Room 1 will be heard continually or it may even be the only reverberation to appear. In any event it is useful to calculate the reverberation times of both subspaces separately using eqn (9.8).

Coupling phenomena can also occur in enclosures lacking sound field diffusion. This is sometimes observed in rooms with regular geometry, with smooth wall surfaces and non-uniform distribution of the boundary absorption. Thus, a fully occupied hall with a relatively high ceiling and smooth and reflecting side walls will often build up a two-dimensional, highly reverberant sound field in its upper part. It is caused by horizontal or nearly horizontal sound paths and is influenced only slightly by the absorption of the audience. Whether a listener will perceive this particular kind of reverberation depends again on the strength of its excitation and other factors.

For lecture halls, theatre foyers, etc., such an extra reverberation is of course undesirable. In a concert hall, however, this particular lack of diffusion can sometimes lead to a badly needed increase in reverberation time beyond the Sabine or Eyring value, namely in those cases where the volume per seat is too small to yield a sufficiently long decay time under diffuse conditions. An example of this is the Stadthalle in Göttingen, which has a hexagonal ground plan and in fact has a reverberation time of about 2 s, although calculations had predicted a value of 1.6–1.7 s only (both for medium frequencies and for the fully occupied hall). Model experiments carried out afterwards demonstrated very clearly that a sound field of the described type was responsible for this unexpected increase in reverberation time.

## 9.4 Prediction of noise level

There are many spaces which are not intended for any acoustical presentations but where some acoustical treatment is nevertheless desirable or necessary. Although they show wide variations in character and structural details, they all fall into the category of rooms in which people are present and in which noise is produced: e.g. by noisy machinery or by the people themselves. Examples of this are staircases, concourses of railway stations and airports, and entrance halls and foyers of concert halls and theatres. Most important, however, are working spaces such as open-plan offices, workshops and factories. Here room acoustics has the relatively prosaic (however, important) task of reducing the noise level.

Traditionally, acoustics does not play any important role in the design of a factory or an open-plan office, to say the least; usually quite different aspects, as for instance those of efficient organisation, of the economical use of space or of safety, are predominant. Therefore the term 'acoustics' applied to such spaces does not have the meaning it has with respect to a lecture room or a theatre. Nevertheless, it is obvious that the way in which the noise from any kind of machinery propagates in such a room, and hence the noise level in it, depends highly on the acoustical properties of such a room.

A first idea of the steady-state sound pressure level that a sound source with power output $P$ produces in a room with the equivalent absorption area $A$ is obtained from eqn (5.37). Converting it into a logarithmic scale, with PL denoting the sound power level (see eqn (1.50)), yields

$$\text{SPL}^{\infty} = \text{PL} - 10 \log_{10}\left(\frac{A}{1\,m^2}\right) + 6 \text{ dB} \tag{9.10}$$

This relation holds for distances from the sound source which are significantly larger than the 'diffuse-field distance' $r_c$, as given by eqns (5.39) or (5.39a), i.e. it describes the sound pressure level of the reverberant field. For observation points at distances $r$ comparable to or smaller than $r_c$, the sound pressure level is, according to the more general eqn (5.41):

$$SPL = SPL^{\infty} + 10 \log_{10} \left(1 + \frac{r_c^2}{r^2}\right) \qquad (9.10a)$$

Both equations are valid under the assumption that the sound field is diffuse.

Numerous measurements in real spaces have shown, however, that the reverberant sound pressure level $SPL^{\infty}$ decreases more or less with increasing distance, in contrast to eqn (9.10). Obviously sound fields in such spaces are not completely diffuse, and this seems to affect the validity of eqn (9.10) much more than that of the common reverberation formulae. This lack of diffusion may have several reasons. Often one dimension of a working space is much larger (very long rooms) or smaller (very flat rooms) than the remaining ones. Another possible reason is non-uniform distribution of absorption. In all these cases a different approach is needed to calculate the sound pressure level.

For enclosures made up of plane walls the method of image sources could be employed, which has been discussed at some length in Section 4.1. It must be noted, however, that real working spaces are not empty but contain machines, piles of material, furniture, benches, etc.; in short, numerous obstacles which scatter the sound and may also partially absorb it.

One way to account for the scattering of sound in fitted working spaces is to replace sound propagation in the free space by that in an 'opaque' medium containing many randomly arranged scattering objects, as explained at the end of Section 5.1. We restrict the discussion to the steady state. Then, the energy density of the unscattered component, i.e. of the direct sound, is

$$w_0(r) = \frac{P}{4\pi c r^2} \exp(-r/\bar{r}_s) \qquad (9.11)$$

which is to be compared with eqn (5.38); $\bar{r}_s$ is the scattering mean free path length given by eqn (5.11). Now it is assumed that $\bar{r}_s$ is so small that virtually all sound particles will be scattered before reaching a wall of the enclosure. Then we need not consider any reflections of the direct sound. Instead, the scattered sound particles will uniformly fill the whole enclosure due to the equalising effect of multiple scattering. Since the scattered sound particles propagate in all directions, they constitute a diffuse sound field with its well-known properties. In particular, its energy density is $w_s = 4P/cA$. The steady-state level can be calculated from eqn (9.10) or eqn (9.10a). In the latter case, however, the 'diffuse-field distance' $r_c = (A/16\pi)^{1/2}$ from eqn (5.39) has to be replaced with a modified value $r_c'$ which is smaller than $r_c$. In fact, equating $w_s$ and $w_0$ from eqn (9.11) yields

$$r_c'/r_c = \exp(-r_c'/2\bar{r}_s) \qquad (9.12)$$

Solving this transcendental equation yields $r'_c/r_c = 0.7035$ if $r_c$ equals the scattering mean free path length $\bar{r}_s$ while this fraction becomes as small as 0.2653 for $\bar{r}_s/r_c = 0.1$, i.e. when a sound particle undergoes 10 collisions on average per distance $r_c$. However, there remains some uncertainty on the scattering cross-sections $Q_s$ of machinery or other pieces of equipment because there is no practical way to calculate them exactly from geometrical data. Several authors (see for instance Ref. 6) identify $Q_s$ with one-quarter of the scatterer's surface. This procedure agrees with the rule given at the end of Section 5.1.

The transient sound propagation in enclosures containing sound-scattering obstacles is much more complicated than the steady-state case. It has been treated successfully by several authors, for instance by Hodgson.[7]

In a different approach, the scatterers are imagined as being projected onto the walls, so to speak: i.e. it is assumed that the walls produce diffuse sound reflections rather than specular ones. Then the problem can be treated by application of the integral equation (4.24). This holds in particular for the calculation of the steady-state sound propagation in certain 'disproportionate' rooms for which the diffuse-field theory is not applicable. One of them is the infinite flat room, i.e. the space confined by two parallel planes. As already mentioned in Section 4.5, eqn (4.24) has a closed solution. This is of considerable practical interest since this kind of 'enclosure' may serve as a model for many working spaces in which the ceiling height $h$ is very small compared with the lateral dimensions (factories, or open-plan bureaus). Therefore, sound reflections from the ceiling are absolutely predominant over those from the side walls, and hence the latter can be neglected unless the source and/or the observation point are located next to them.

Equation (4.27) represents this solution for both planes having the same, constant absorption coefficient $\alpha$ or 'reflection coefficient' $\rho = 1 - \alpha$, and for both the sound source and the observation point being located in the middle between both planes. Fortunately, the awkward evaluation of eqn (4.27) can be circumvented by using the following approximation (see Ref. 6(1) of Chapter 4):

$$w(r) = \frac{P}{4\pi c}\left\{\frac{1}{r^2} + \frac{4\rho}{h^2}\left[\left(1 + \frac{r^2}{h^2}\right)^{-3/2} + \frac{b\rho}{1-\rho}\left(b^2 + \frac{r^2}{h^2}\right)^{-3/2}\right]\right\}$$
(9.13)

The constant $b$ depends on the absorption coefficient of the floor and the ceiling. Some of its values are listed in Table 9.2. Equation (9.13) may also be used if both boundaries have different absorption coefficients; in this case, $\alpha$ is the average absorption coefficient.

Figure 9.10b shows how the sound pressure level, calculated with this formula, depends on the distance from an omnidirectional sound source

Table 9.2 Values of the constant b in eqn (9.13)

| Absorption coefficient α | Reflection coefficient ρ | b |
|---|---|---|
| 0.7 | 0.3 | 1.806 |
| 0.6 | 0.4 | 1.840 |
| 0.5 | 0.5 | 1.903 |
| 0.4 | 0.6 | 2.002 |
| 0.3 | 0.7 | 2.154 |
| 0.2 | 0.8 | 2.425 |
| 0.1 | 0.9 | 3.052 |

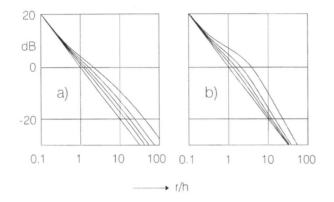

Figure 9.10 Sound pressure level in an infinite flat room as a function of distance r (h = room height). The absorption coefficient of both walls is (from bottom to top) 1, 0.7, 0.5, 0.3 and 0.1. (a) Smooth walls, calculated using eqn (4.3). (b) Diffusely reflecting walls.

for various values of the (average) absorption coefficient $\alpha = 1 - \rho$ of the walls. For comparison, the corresponding curves for specularly reflecting planes, computed using eqn (4.3), are presented in Fig. 9.10a. The plotted quantity is 10 times the logarithm of the energy density divided by $P/4\pi c h^2$. Both diagrams show characteristic differences: smooth boundaries direct all the reflected energy away from the source, this results in an increased level with the increment remaining constant at large distances. In contrast, diffusely reflecting boundaries reflect some energy back towards the source; accordingly, the level increment caused by the boundary—compared to that of free field propagation ($\alpha = 1$)—reaches a maximum at a certain distance and vanishes at large distances from the source. This behaviour is typical for enclosures containing scattering objects and was experimentally confirmed by numerous measurements carried out by Hodgson,[7,8] in models as well as in full-scale factories.

Both the aforementioned methods are well suited for predicting noise levels in working spaces and estimating the reduction which can be achieved by an absorbing treatment of the ceiling, for instance. Other possible methods are measurements in a scale model of the space under investigation or computer simulation as described in the following sections.

Another disproportionate room is the long room, i.e. a tube with cross-sectional dimensions that are large compared with the acoustical wavelength. Again, we simplify the problem by supposing that the tube is infinitely long. If it has smooth and rigid walls, and if its cross-section is rectangular, the sound propagation can be calculated by an obvious extension of eqn (4.3) in two dimensions, leading to a double sum over the full pattern of image sources. If the wall of the tube scatters the impinging sounds, a closed solution of eqn (4.24) is also available, as mentioned in Section 4.5, provided the tube is cylindrical, which is not a severe restriction (see Ref. 6(2) of Chapter 4). The outcomes of these calculations, both for smooth and scattering walls, are similar to those shown in Fig. 9.10; the deviations of the sound level from the $1/r^2$ law are even somewhat more pronounced than those in Fig. 9.10b.

Generally, absorbing treatment of the walls or the ceiling has a beneficial effect on the noise level, not only in working spaces such as factories or large offices but also in many other rooms where many people gather together, e.g. in staircases, or in theatre foyers. A noise level reduction of a few decibels only can increase the acoustical comfort to an amazing degree. If the sound level is too high due to insufficient absorption, people will talk more loudly than in a quieter environment. This in turn again increases the general noise level and so it continues until finally people must shout and still do not achieve satisfactory intelligibility. In contrast, an acoustically damped environment usually makes people behave in a 'damped' manner too—for reasons which are not primarily acoustical—and it makes them talk not louder than necessary.

There is still another psychologically favourable effect of an acoustically damped theatre or concert hall foyer: when a visitor leaves the foyer and enters the performance hall, he will suddenly find himself in a more reverberant environment, which gives him the impression of solemnity and raises his expectations.

The extensive use of absorbing materials in a room, however, is accompanied by an oppressive atmosphere, an effect which can be observed quite clearly when entering an anechoic room. Furthermore, since the level of the background noise is reduced too by the absorbing areas, a conversation held in a low voice can be understood at relatively large distances and can be irritating to unintentional listeners. Since this is more or less the opposite of what should be achieved in an open-plan office, the masking effect by background noise is sometimes increased in a controlled way by feeding loudspeakers with random white or 'coloured' noise, i.e. with a 'signal' without any temporal or spectral structure. The level of this noise should not exceed 50 dB(A). Even so it is still contested whether the advantages of such measures surpass their disadvantages.

## 9.5 Acoustical scale models

In Section 8.2 we described at length the methods of measuring impulse responses in rooms. However, for an acoustical consultant it would often be of equal interest to learn something about the sound transmission in a room when it is still on the drawing-board.

A well-tried method of gaining information on the acoustics of non-existing enclosures is to study the propagation of waves in a smaller model that is similar to the original room, at least geometrically. This method has the advantage that, once the model is at hand, many variations can be tried out with little expenditure: from the choice of various wall materials to changes in the shape of the room.

Since several properties of propagation are common to all sorts of waves, it is not absolutely necessary to use sound waves for the model tests. This was an important point, particularly in earlier times when acoustical measuring techniques were not yet at the advanced stage they have reached nowadays. So, waves on water surfaces were sometimes produced in 'ripple tanks'. The propagation of these waves can be studied visually with great ease. The use of them, however, is restricted to an examination of plane sections of the hall under test. More profitable is the use of light as a substitute for sound. In this case absorbent areas are painted black or covered with black paper or fabric, whereas reflecting areas are made of polished sheet metal. Likewise, diffusely reflecting areas can be quite well simulated by white matt paper. The energy distribution can be detected with photocells or by photography. However, because of the high speed of light, this method is restricted to the investigation of the steady-state energy distribution. Another limitation is the absence of realistic diffraction phenomena since the optical wavelengths are very small compared to the dimensions of all those objects which would diffract the sound waves in a real hall.

Nowadays, the techniques of electroacoustical transducers have reached a sufficiently high state of the art to generate and to receive sound waves under the condition of a model and hence to use sound waves for examining sound propagation in the model scale. For this purpose a few geometrical and acoustical modelling rules have to be observed. In the following, all quantities referring to the model are denoted by a prime. At first we define the scale factor $\sigma$ by

$$l' = \frac{l}{\sigma} \tag{9.14}$$

where $l$ and $l'$ are corresponding lengths. This leads us to the scale factor of times and time differences:

$$t' = \frac{l'}{c'} = \frac{l}{\sigma c'} = \frac{c}{c'} \cdot \frac{t}{\sigma} \tag{9.15}$$

where $c'$ is the sound velocity of the medium within the model. Since the rule (9.14) holds also for the wavelengths $\lambda$ and $\lambda'$, we obtain the following scaling rule for frequencies:

$$f' = \sigma \frac{c'}{c} \cdot f \qquad (9.16)$$

According to the last relationship, the sound frequencies applied in a scale model may well reach into the ultrasonic range. Suppose the model is filled with air ($c' = c$) and the model is scaled down by 1:10 ($\sigma = 10$), then a frequency range from 100 to 5000 Hz in the original room corresponds to 1–50 kHz in its model.

Since the model is supposed to be not just a geometrical replica of the original hall but an acoustical one as well, the wall absorption and its frequency dependence should be modelled with care. This means that any surface in the model should have the same absorption coefficient at frequency $f'$ as the corresponding surface in the original at frequency $f$:

$$\alpha_i'(f') = \alpha_i(f) \qquad (9.17)$$

Even more problematic is the modelling of the attenuation which the sound waves undergo in the medium. According to eqn (1.16a), $2/m$ is the travelling distance along which the sound pressure amplitude of a plane wave is attenuated by a factor e = 2.718.... From this we conclude that

$$m'(f') = \sigma \cdot m(f) \qquad (9.18)$$

These requirements can be fulfilled only approximately, and the required degree of acoustical similarity depends on the kind of information we wish to obtain from the model tests. If only the initial part of the impulse response or 'reflectogram' is to be studied (over, say, the first 100 or 200 ms in the original time scale), it may be sufficient to provide for only two different kinds of surfaces in the model, namely reflecting ones (made of metal, glass, gypsum, etc.) and absorbing ones (e.g. felt or plastic foam). The air absorption can be neglected in this case or its effect can be numerically compensated.

Matters are different if longer reflectograms are to be observed, for instance to create listening impressions from the auditorium as was first proposed by Spandöck.[9] According to this idea, tape-recorded music or speech signals are replayed in the model at a tape speed that is faster by a factor of $\sigma$. At some point within the model, the sound signal is picked up with a small microphone. After transforming the re-recorded signals back into the original time and frequency scale it can be presented by earphones to a listener who can judge subjectively the 'acoustics' of the hall and the effects of any modifications. Nowadays this technique is known as 'auralization'. More will be said about it in Section 9.7.

Concerning the instrumentation for measuring the impulse response in scale models, omnidirectional impulse excitation of the model is more difficult to achieve the higher the scale factor and hence the frequency range to be covered. Small spark gaps can be used as sound sources, but in any case it is advisable to check their directivity and frequency spectrum beforehand. Furthermore, electrostatic or piezoelectric transducers have been developed for this purpose; they have the advantage that they can be fed with any desired electrical signal and therefore allow the application of the more sophisticated methods described in Section 8.2. The microphone should be also have omnidirectional sensitivity. Sufficiently small condensor microphones are commercially available (1/4 inch or 1/8 inch microphones). Any further processing, including the evaluation of the various sound field parameters, as discussed in Chapters 7 and 8, is carried out, as with full-scale measurements, by means of a digital computer.

The construction of a scale model is relatively costly and requires special skills. Often the architect gets a model constructed on his own in order to study the visual impression of a hall. From this model the acoustical consultant can benefit since it is usually not too difficult to adapt it to simple acoustical tests such as the detection of echoes and the identification of critical wall portions. One should keep in mind that the sound field in a model includes all diffraction effects, and that it is hard to predict beforehand which of these effects are relevant and which are not. For this reason acoustical model measurements are still useful when the project under design is a prestigious and expansive hall.

## 9.6 Computer simulation

Despite the merits of acoustical model measurements these techniques are being superseded increasingly by cheaper, faster and more efficient methods, namely by digital simulation of sound propagation in enclosures. The introduction of the digital computer into room acoustics is due to Schroeder[10] and his co-workers. Since then this method has been employed by many authors to investigate various problems in room acoustics; examples have been presented in Sections 5.6 and 8.2 of this book. The first authors who applied digital simulation to concert hall acoustics were Krokstad et al,[11] who evaluated a variety of acoustical room parameters from impulse responses obtained by digital ray tracing techniques. Meanwhile digital computer simulation has been applied not only to all kinds of auditoria but also to factories and other working spaces.[6-8]

In this section only a brief description of the principles underlying computer simulation of sound fields in rooms can be given. A more detailed account is found in Vorländer's book on auralisation.[12] Basically, there are two methods of sound field simulation in use nowadays: namely, ray tracing and the method of image sources. Both are based on geometrical acoustics,

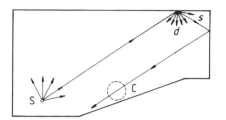

*Figure 9.11* Principle of digital ray tracing. S = sound source, C = counting sphere, s = specular reflection and d = diffuse reflection.

i.e. they rely on the validity and application of the laws of specular or diffuse reflection. So far no practical way has been found to include typical wave phenomena such as diffraction into these algorithms.

The principle of digital ray tracing is illustrated in Fig. 9.11. At some time $t = 0$, a sound source at a given position is imagined to release numerous sound particles in all directions. Each sound particle travels on a straight path until it hits a wall which we assume to be plane for the sake of simplicity. The point where this occurs is obtained by first calculating the intersections of the particle path with all wall planes, and then selecting the nearest of them in the forward direction. Provided this intersection is located within the polygon representing the actual wall, it is the point where the particle will be reflected, either specularly or diffusely. In the first case its new direction is calculated from the law of geometrical reflection. However, if perfectly diffuse reflection is supposed to occur, the computer generates two random numbers $z_1$ and $z_2$ from which the polar angle $\vartheta$ and the azimuth angle $\varphi$ of the new direction are calculated, both with respect to a local spherical coordinates system. Usually, it is assumed that the angle $\vartheta$ is distributed according to Lambert's law shown in eqn (4.22). This is the case if

$$\vartheta = \arccos \sqrt{z_1} \qquad \text{with } 0 \leq z_1 < 1 \tag{9.19}$$

$$\varphi = 2\pi z_2 \qquad \text{with } 0 \leq z_2 < 1 \tag{9.20}$$

Furthermore, we can assign a scattering coefficient $s$ to a boundary (see Section 8.8), which means that a fraction $s$ of all arriving sound particles undergo diffuse reflection while the remaining ones are specularly reflected.

After its reflection the particle continues on its way in the new direction towards the next wall, etc. The absorption coefficient $\alpha$ of a wall can be accounted for in two ways: either by reducing the energy of the particle by a factor of $1 - \alpha$ after each reflection or by interpreting $\alpha$ as 'absorption probability'. In this case another random number $z_3$ between 0 and 1 is generated; if $z_3$ exceeds $\alpha$, the particle will proceed, otherwise it is annihilated. In a similar way the attenuation of air can be taken into account.

Either the energy of a sound particle is reduced by a factor $\exp(-mct)$ if $t$ is the time after its release ($m$ = attenuation constant, see eqn (1.16a), or $\exp(-md_i)$ is considered the probability of survival within a section $d_i$ of the particle's zigzag path. As soon as the energy of the particle has fallen below a prescribed value or the particle has been absorbed, the process is truncated and the path of another particle will be 'traced'. This procedure is repeated until all the particles emitted by the sound source have been followed up.

The results of this procedure are collected by means of 'counters', i.e. of previously assigned counting areas or counting volumes. Whenever a particle crosses such a counter its energy and arrival time are stored, if needed also the direction from which it arrived. After the process has finished, i.e. the last particle has been followed up, the received energies are classified with respect to the arrival times of the particles; the result is a histogram (see Fig. 9.12), which can be considered as a short-time averaged energetic impulse response. The choice of the time intervals is not uncritical: if they are too long, the histogram will be only a crude approximation to the true impulse response since significant details are lost by averaging. Too short intervals, on the other hand, will afflict the results by strong random fluctuations superimposed on them. As a practical rule, the interval should be in the range 5–10 ms, since this figure is related to the time resolution of our hearing.

The problem of properly selecting the class width for classifying the arrival times is less critical if parameters listed in Table 8.1, for instance the

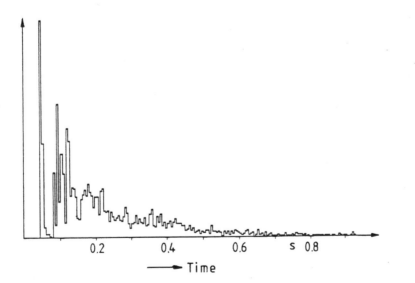

*Figure 9.12* Temporal distribution of received particle energy (energetic impulse response). The interval width is 5 ms.

'definition' $D$, the 'centre time' $t_s$ or the lateral energy parameters LEF and $LG_{80}^{\infty}$ are to be determined, since the calculation of these quantities involves integrations over the impulse response. The same holds for the steady-state energy or the 'strength factor' $G$ after eqn (7.15), which requires integrations over the whole energetic impulse response. As an example, Fig. 9.13 depicts the distribution of the stationary sound pressure level and of the definition obtained with ray tracing applied to a lecture hall with a volume of 3750 m³ and 775 seats.[13]

In any case, the achieved accuracy of the results depends on the number of sound particles counted with a particular counter. For this reason the counting area or volume must not be too small; furthermore, the total number of particles contained in one 'shot' of sound impulse must be sufficiently large. As a practical guideline, a total of $10^5$–$10^6$ sound particles will yield sufficiently precise results if the dimensions of the counters are of the order of 1 m.

The most tedious and time-consuming part of the whole process is the collection and input of room data such as the positions and orientations of the walls and their acoustic properties. If the boundary of the room contains curved portions these may be approximated by planes unless their shape is very simple, for instance spherical or cylindrical. The degree of approximation is left to the intuition and experience of the operator. It should be noted, however, that this approximation may cause systematic and sometimes intolerably large errors.[14] These are avoided by calculating

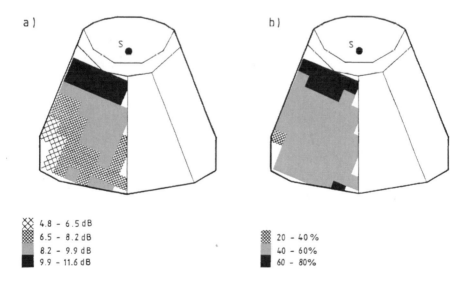

a )

b )

XX 4.8 – 6.5 dB
6.5 – 8.2 dB
8.2 – 9.9 dB
9.9 – 11.6 dB

20 – 40 %
40 – 60%
60 – 80%

*Figure 9.13* Spatial distribution (a) of the stationary sound level and (b) of 'definition' in a large lecture hall (after Vorländer[13]).

the path of the reflected particle directly from the curved wall by apply-ing eqn (4.1), which is relatively easy if the wall section is spherical or cylindrical.

The ray tracing process can be modified and refined in many ways. Thus, the sound radiation need not necessarily be omnidirectional; instead, the sound source can be given any desired directionality. Likewise, one can study the combined effect of more than one sound source, for instance of a real speaker and several loudspeakers with specified directional character-istics, amplifications and delays. This permits the designer to optimise the configuration of an electroacoustic system in a hall. Furthermore, any mix-ture of purely specular or ideally diffuse wall reflections can be taken into consideration; the same holds for the directional dependence of absorption coefficients.

The second method to be discussed here is based on the concept of image sources—also known as mirror sources—as has been described at some length in Section 4.1. In principle, this method is very old but its practical application started only with the advent of the digital computer by which constructing numerous image sources and collecting their contributions to the sound field has become very easy, at least in principle.

It should be noted that only reflections from the inside of a wall are rele-vant, as already mentioned in Section 4.1. Even more severe is the problem of 'invisible' or 'inaudible' image sources already addressed in Section 4.1. It is illustrated in Fig. 9.14, which shows two plane walls adjacent at an obtuse angle, along with a sound source A, both its first-order images $A_1$ and $A_2$ and the second-order images $A_{12}$ and $A_{21}$. It is easily seen that a path running

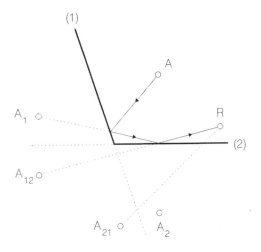

*Figure 9.14* Valid and invalid image sources. Image source $A_{21}$ is invalid, i.e. 'invisible' from receiver position R.

from the source A to the receiver R can be found which involves the images $A_1$ and $A_{12}$. But there is no path reaching R via the image $A_{21}$ since the intersection of the line $A_{21}$–R with the plane (1) is outside the physical wall; hence $A_{21}$ is inaudible.

Unfortunately, most higher-order source images are inaudible. Consider, for example, an enclosure made up of six plane walls with a total area of $3600\ m^2$ and a volume of $12\,000\ m^3$. According to eqn (4.8) a sound ray or sound particle would undergo 25.7 reflections per second on average. To compute only the first 400 ms of the impulse response, image sources of up to the 10th order (at least) must be considered. With this figure and $N = 6$, eqn (4.2) tells us that about $1.46 \times 10^7$(!) image sources must be constructed. However, if the considered enclosure were rectangular, there were only

$$N_r(i_0) = \frac{2}{3}\left(2i_0^3 + 3i_0^2 + 4i_0\right) \qquad (9.21)$$

image sources of order $\leq i_0$ (neglecting their multiplicity), and all of them are audible. For $i_0 = 10$ this formula yields $N_r(i_0) = 1560$. This consideration shows that the fraction of audible image sources is very small unless the room has a very special shape. And the set of audible image sources differs from one receiving point to another.

In principle an audibility check is carried out as explained before. However, several authors have developed algorithms by which these tests can be facilitated. One of them is due to Vorländer,[15] who performs an abbreviated ray-tracing process which precedes the actual simulation. Each sound path detected in this way is associated with a particular sequence of valid image sources, for instance $A \rightarrow A_1 \rightarrow A_{12} \rightarrow R$ in Fig. 9.14, which is identified by backtracing the path of the sound particle, starting from the receiver R. Hence $A_1$ and $A_{12}$ are certainly audible. The next particle which happens to hit the counting volume at the same time can be omitted since it would yield no new image sources. After running the ray tracing for a certain period one can be sure that all significant image sources—up to a certain maximum order—have been found, including their relative strengths, which depend on the absorption coefficients of the walls involved in the mirroring process. We want to emphasise that this method of simulation is still based on the image source model and that the only purpose of the ray tracing procedure is checking the audibility.

Another useful criterion of audibility is due to Mechel.[16] It employs the concept of the 'field angle' of an image source, i.e. the solid angle subtended by the physical wall polygon which a source irradiates. To create an audible image source, the reflecting wall polygon must be inside the field angle of the 'mother' image source. With each new generation of image sources the field angles are diminished. Besides a careful discussion of criteria for interrupting the process of image source construction, this publication offers many valuable computational details.

Having identified the relevant image sources we are ready to form the energy impulse response by adding their contributions. Suppose the original sound source produces a short power impulse at time $t = 0$, represented by a Dirac function $\delta(t)$, then the contribution of a particular image source of order $i$ to the energetic impulse response at a given point is

$$\frac{E_0}{4\pi cd^2_{m_1 m_2 \ldots m_i}} \cdot \rho_{m_1} \rho_{m_2} \ldots \rho_{m_i} \delta\left(t - \frac{d_{m_1 m_2 \ldots m_i}}{c}\right)$$

$E_0$ is the total energy released by the original sound source, $m_1, m_2, \ldots m_i$ indicate the walls involved in the particular sound path, $\rho_k = 1 - \alpha_k$ are their reflection coefficients, and $d_{m_1 m_2 \ldots m_i}$ denotes the distance of the considered image from the receiving point. Figure 9.15 depicts an impulse response obtained in this way. Note that no random processes whatsoever are involved in its generation. For frequency-dependent absorption coefficients this computation must be repeated for a sufficient number of frequency bands.

For the purpose of auralisation (see next section), an impulse response related to the sound pressure is more useful than an energetic impulse response. To obtain it, we have to replace in the equation above the reflection coefficients $\rho_{m_k}$ by the 'reflection responses' $r_{m_k}(t)$, as already introduced in Section 4.1, as the Fourier transform of the complex reflection factor $R_{m_k}(f)$. Suppose at free propagation the original sound source would produce a short

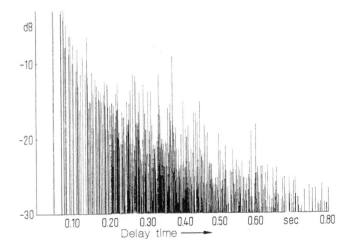

*Figure 9.15* Room impulse response computed from image sources.

pressure impulse $A\delta(t - d_0/c)$ at some distance $d_0$. Then the image source considered before contributes the component

$$(g_p)_{m_1 m_2 \ldots m_i} = A \frac{d_0}{d_{m_1 m_2 \ldots m_i}} \cdot r_{m_1} * r_{m_2} * \ldots * r_{m_i} * \delta \left( t - \frac{d_{m_1 m_2 \ldots m_i}}{c} \right)$$

$$(9.22)$$

to the pressure impulse response. Each asterisk symbolises one convolution operation. For the practical computation it is more convenient to deal with the Fourier transform of eqn (9.22), namely

$$(G_p)_{m_1 m_2 \ldots m_i} = A \frac{d_0}{d_{m_1 m_2 \ldots m_i}} \cdot R_{m_1} R_{m_2} \ldots R_{m_i} \cdot \exp \left( -2\pi i f \frac{d_{m_1 m_2 \ldots m_i}}{c} \right)$$

$$(9.23)$$

and to calculate $(g_p)_{m_1 m_2 \ldots m_i}$, if needed at all, as the inverse Fourier transform of $(G_p)_{m_1 m_2 \ldots m_i}$. The final impulse response is calculated by adding all contributions $(g_p)_{m_1 m_2 \ldots m_i}$, including that of the original sound source (the direct sound). In the frequency domain, this corresponds to calculating the transfer function as the sum of all components $(G_p)_{m_1 m_2 \ldots m_i}$.

In practical applications it turns out to be difficult to find the correct reflection factors $R_{m_k}$ which are complex functions of the frequency and of the angle of sound incidence. However, it may be permissible to determine the absolute values of the reflection factors from the absorption coefficients

$$\left| R_{m_k} \right| = \sqrt{\rho_{m_k}} = \sqrt{(1 - \alpha_{m_k})}$$

and to combine them with some arbitrary phase function which must be an uneven function of the frequency. To eliminate the angle dependence, the average according to Paris' formula (2.41) can be inserted into this formula.

Since the image model yields the complex transfer function of a room (or, more exactly, of a transmission path within the room) as well as its energetic impulse response, it is more 'powerful' in a way than ray tracing. Furthermore, its results are strictly deterministic. It fails, however, if the boundary of the room is not piecewise plane or if it is so complicated that it cannot be modelled in every detail. And its application is restricted to polyhedral enclosures. For small rooms and at low frequencies it may happen that the 'plane-wave condition' mentioned in Section 4.1 is violated. The ray-tracing method, on the other hand, is not subject to any limitations of room shape, and it permits us to treat heavily structured surfaces, for instance a coffered ceiling, by allowing for sound scattering. Therefore it suggests itself to combine both methods. In fact, several authors (see for instance

Refs 15 and 21) have developed hybrid procedures in order to combine the advantages of ray tracing and the image model. Typically, the latter is used to build up the early part of the impulse response, which can be considered as the acoustical fingerprint of a room, whereas the later parts of the response, which is not so characteristic are computed by ray tracing in one of its various versions. There are also simulation schemes which combine one of the aforementioned algorithms with the radiosity method described in Section 4.5.

The reader may have noted, that both the ray-tracing process and the method of image sources completely disregard all diffraction effects. This is a serious shortcoming. Since the wavelengths of all interesting sounds are of the same order of magnitude as many objects of everyday life, diffraction is a very common phenomenon. In a real hall, every protruding edge, every column or pillar, and every niche or balcony is the origin of scattered waves produced by diffraction. The same holds for every abrupt change in wall impedance of a seemingly smooth boundary. Many of these effects can be accounted for in ray tracing, but others cannot. The diffraction by the steps of a stair, for instance, can be taken into regard by treating the stair as an inclined plane and assigning a certain scattering coefficient to it. Obviously, the diffraction by a balcony face cannot be treated in this way.

Several authors have tried to fill this gap. The basic idea of all these attempts is to apply a simplified version of the diffraction pattern shown in Fig. 2.13 to every sound ray which passes a rigid edge within a certain range. It is obvious, however, that each diffraction process creates many secondary particles travelling in different directions.[17] This increases the complexity of the process and also the processing time drastically. To reduce the computational load Stephenson[18] has developed an algorithm which once in a while unifies energy portions occurring at neighbouring locations and about equal time in space-time cells of finite size. In this model, the carriers of sound energy are pyramids which include all rays connecting a particular (image) source with the points of a wall polygon. During each mirroring at a wall the wall's edges clip the pyramid, generating a narrower "daughter" pyramid. Hence the problem of invisible image sources cannot occur at all.

## 9.7 Auralisation

The term auralisation was coined to signify all techniques which intend to create audible impressions from enclosures not existing in reality but in the form of design data only. Its principles are outlined in Fig. 9.16. Music or speech signals originally recorded in an anechoic environment are fed to a transmission system which modifies the input signal in the same way as its propagation in a real room would modify it. This system is either a physical scale model of a room (see Section 9.5), equipped with a suitable sound source and receiver, or it is a digital filter which has the same impulse

*Figure 9.16* Principle of auralisation. The input signal is 'dry' music or speech.

response as the considered room. The impulse response may have been measured beforehand in a real room or in its scale model, or it is obtained by simulation as described in the preceding section. In order to permit direct comparisons of several impulse responses (i.e. rooms) it may be practical to realise the auralisation filter in the form of a digital real-time convolver.

In any case, the room simulator must produce a binaural output signal, otherwise no realistic, i.e. spatial, impressions can be conveyed to the listener. The output signal is presented to the listener by headphones or, preferably, by two loudspeakers in an anechoic room fed via a cross-talk cancellation system. The merits and shortcomings of both methods are discussed at the start of Chapter 7.

If an auralisation scheme is to convey a realistic impression of a room's acoustics it should cover a frequency range of at least eight octaves; hence the transducers used in a scale model have to meet very high standards. For the same reason the requirements concerning the acoustical similarity between an original room and its model are very stringent. The absorption of the various wall materials and of the audience must be modelled quite correctly, including their frequency dependence, a condition which must be carefully checked by separate measurements. It is even more difficult to control the attenuation by the medium according to eqn (9.18). Several research groups have tried to meet this requirement by filling the model either with air of very low humidity or with nitrogen. If at all, the frequency dependence of air attenuation can be modelled only approximately by such measures, and only for a limited frequency range and a particular scale factor $\sigma$. The first experiments of auralisation with scale models were started by Spandöck's group in Karlsruhe/ (Germany) in the early 1950s; a comprehensive report may be found in Ref. 19.

Auralisation based on a purely digital room model was first carried out by Allen and Berkley.[20] In this case most of the addressed problems do not exist since both the acoustical data of the medium and the enclosure's geometrical data are fed into the computer from its keyboard or from a database. Generally, digital models are much more flexible because the shape and the acoustical properties of the room under investigation can be changed quite easily. Therefore it is expected that in future nearly all auralisation experiments will be based on computer models.

Pressure-related impulse responses or transfer functions computed with the image source model according to eqn (9.22) or eqn (9.23) can be immediately used for auralisation, provided they are binaural. This is achieved by multiplying each sum term $(G_p)_{m_1 m_2 ... m_i}$ of eqn (9.23) with $L(f, \varphi, \vartheta)$ and $R(f, \varphi, \vartheta)$, the head-related transfer functions for the left and the right ear, respectively (see Section 1.6). The angles $\varphi$ and $\vartheta$ characterise the direction of incidence in a head-related coordinate system. Then the modified contributions to the binaural transfer function are

$$(G_{pl})_{m_1 m_2 ... m_i} = (G_p)_{m_1 m_2 ... m_i} \cdot L(f, \varphi, \vartheta)..) \tag{9.24a}$$

and

$$(G_{pr})_{m_1 m_2 ... m_i} = (G_p)_{m_1 m_2 ... m_i} \cdot R(f, \varphi, \vartheta)..) \tag{9.24b}$$

$L$ and $R$ are either a listener's individual head-related transfer functions or they are averages which can be regarded as representative of many individuals.

Very often, however, results obtained from room simulation are in the form of a set of energy impulse responses (see Fig. 9.12) i.e. one for each frequency band (octave band, for instance). This holds especially for ray-tracing results. Let us denote these results by $E(f_i, t_k)$, where $f_i$ is the mid-frequency of the $i$th frequency band, and $t_k$ is the $k$th time interval. Considered as a function of frequency, $E(f_i, t_k)$ approximates a short-time power spectrum valid for the time interval around $t_k$.

In the following, we describe a process by which these spectra can be converted into a pressure-related impulse, as developed by Heinz.[21] By properly smoothing, a steady and positive function of frequency $E_k(f)$ is produced. In a next step, a pressure-related transfer function $G_k(f)$ is derived by taking the square root of $E_k(f)$ and 'inventing' a suitable phase spectrum $\psi_k(f)$. Then

$$G_k(f) = \sqrt{E_k(f)} \cdot \exp\left[i\psi(f)\right] \tag{9.24}$$

As we know from Section 3.4 the phase spectrum is not critical at all since propagation in a room randomises all phases anyway. Therefore any odd phase function $\psi_k(f) = -\psi_k(-f)$ could be used for this purpose, provided it corresponds to a system with causal behaviour, i.e. to a system the impulse response of which vanishes for $t < 0$. A particular possibility is to derive $\psi_k(f)$ from $E_k(f)$ as the minimum phase function. This is achieved by applying the Hilbert transform, eqn (8.19), to the natural logarithm of $E_k(f)$:

$$\psi_k(f) = \frac{1}{\pi} \int_{-\infty}^{\infty} \frac{\ln\left[E_k(f - f')\right]}{f'} df' \tag{9.25}$$

In a second step, these transfer functions $G_k(f)$ are imposed on a random succession of Dirac impulses. The average number $\bar{n}$ of impulses per second is about 10 000. Finally, these impulses are attributed to properly selected groups of directions from which the sound is to arrive. The Fourier transforms of the spectra $G_k(f)$ computed in this way are 'short-time impulse responses', which are subsequently combined into the total impulse response of the auralisation filter in Fig. 9.16.

If all steps—including the simulation process—are carefully carried out, the listener to whom the ultimate result is presented will experience an excellent and quite realistic impression. To what extent this impression is identical with what the listener would have in the actual room is a question which is still to be investigated. In fact, there are many details which are open to improvement and further development. In any case, however, an old dream of the acoustician is going to come true through the techniques of auralisation. Many new insights into room acoustics are expected from its application. Furthermore, auralisation permits an acoustical consultant to convince the architect, the user of a hall and himself, on the efficiency of the measures he proposes in order to reach the original design goal.

## References

1 Vorländer M, Kuttruff H. Die Abhängigkeit des Seitenschallgrads von der Form und der Flächengestaltung eines Raumes. Acustica 1985; 58:118.
2 Meyer J. Der Einfluß der richtungsabhängigen Schallabstrahlung der Musikinstrumente auf die Wirksamkeit von Reflexions- und Absorptionsflächen in der Nähe des Orchesters. Acustica 1976; 36:147.
3 Marshall AH, Gottlob D, Alrutz H. Acoustical conditions preferred for ensemble. J Acoust Soc 1978; 64:1437.
4 Allen WA. Music stage design. J Sound Vib 1980; 69:143.
5 Cremer L. Die wissenschaftlichen Grundlagen der Raumakustik, Band II. Stuttgart: S Hirzel Verlag, 1961.
6 Ondet AM, Barbry JL. Modelling of sound propagation in fitted workshops using ray tracing. J Acoust Soc Am 1989; 85:787.
7 Hodgson M. Theoretical and physical models as tools for the study of factory sound fields. PhD thesis, University of Southampton, 1983.
8 Hodgson M. Measurements of the influence of fitting and roof pitch on the sound field in panel-roof factories. Appl Acoustics 1983; 16:369.
9 Spandöck F. Raumakustische Modellversuche. Ann d Physik V 1934; 20:345.
10 Schroeder MR. Digital computers in room acoustics. Proc Fourth Intern Congr on Acoustics, Copenhagen, 1962, Paper M21.
11 Krokstad A, Strøm S, Sørsdal S. Calculating the acoustical room response by the use of a ray-tracing technique. J Sound Vib 1968; 8:118.
12 Vorländer M. Auralization. Berlin: Springer-Verlag, 2008.
13 Vorländer M. Ein Strahlverfolgungsverfahren zur Berechnung von Schallfeldern in Räumen. Acustica 1988; 65:138.

14  Kuttruff H. Some remarks on the simulation of sound reflection from curved walls. Acustica 1992; 77:176.

15  Vorländer M. Simulation of the transient and the steady state sound propagation in rooms using a new combined sound particle-image source algorithm. J Acoust Soc Am 1989; 86:172.

16  Mechel F. Formulas of Acoustics. Berlin: Springer-Verlag, 2002.

17  Stephenson UM, Svensson P. Can also sound be handled as stream of particles?— An improved approach to diffraction-based uncertainty principle—from ray to beam tracing. Proc DAGA '07, Stuttgart.

18  Stephenson UM. Quantized beam tracing—a new algorithm for room acoustics and noise immission prognosis. Acustica/Acta Acustica 1996; 82:517.

19  Brebeck P, Bücklein R, Krauth E, Spandöck F. Akustisch ähnliche Modelle als Hilfsmittel für die Raumakustik. Acustica 1967; 18.

20  Allen JB, Berkley DA. Image method for efficiently simulating small room acoustics. J Acoust Soc Am 1979; 65:943.

21  Heinz R. Binaural room simulation based on the image source model with addition of statistical methods to include the diffuse sound scattering of walls and to predict the reverberant tail. Appl Acoust 1993; 38:145; Heinz R. Entwicklung und Beurteilung von computergestützten Methoden zur binauralen Raumsimulation. Dissertation, Aachen, Germany, 1994.

# Chapter 10

# Electroacoustical systems in rooms

Nowadays there are many points of contact between room acoustics and electroacoustics, even if we neglect the fact that modern measuring techniques in room acoustics could not exist without the aid of electroacoustics (loudspeakers, microphones, recorders). Thus we shall hardly ever find a meeting room of medium or large size which is not provided with a public address system for speech amplification; it matters not whether such a room is a church, a council chamber or a multipurpose hall. We could dispute whether such an acoustical 'prothesis' is really necessary for all these cases or whether sometimes they are more a misuse of technical aids; it is a fact that many speakers and singers are not only unable but also unwilling to exert themselves to such an extent and to articulate so distinctly that they can make themselves clearly heard even in a meeting room of moderate size. Instead they prefer to rely on the microphone which is readily offered to them. But the listeners are also demanding, to an increasing extent, a loudness which will make listening as effortless as it is in broadcasting, television or cinemas. Acousticians have to come to terms with this trend and they are well advised to try to make the best of it and to contribute to an optimum design of such installations.

But electroacoustical systems in rooms are by far more than a necessary evil. They open acoustical design possibilities which would be inconceivable with traditional means of room acoustical treatment. For one thing, there is a trend to build halls and performance spaces of increasing size, thus giving large audiences the opportunity to witness personally important cultural, entertainment or sports events. This would be impossible without electroacoustical sound reinforcement, since the human voice or a musical instrument alone would be unable to produce an adequate loudness at most listeners' ears. Furthermore, large halls are often used—largely for economic reasons—for very different kinds of presentations (see also Section 9.3).

In this situation it is a great advantage that electroacoustical systems permit the adaptation of the acoustical conditions in a hall to different kinds of presentations, at least within certain limits. Imagine a hall with a relative long reverberation time, well-suited for musical performances. Nevertheless, a carefully designed electrical sound system can provide for good speech intelligibility by directing the sounds towards the audience and hence avoiding the excitation of long reverberation.

The reverse way is even more versatile, but also more difficult technically: to render the natural reverberation of the hall short enough in order to match the needs of optimum speech transmission. For the performance of music, the reverberation is enhanced by electroacoustical means to a suitable and adjustable value. The particular circumstances will decide which of the two possibilities is more favourable.

Thus, electroacoustical systems for reverberation enhancement simulate acoustical conditions that are not encountered in the given hall as it is. At the same time, they can be considered as a first step towards producing new artificial effects which are not encountered in halls without an electroacoustical system. In the latter respect, we are just at the beginning of a development whose progress cannot yet be predicted.

Whatever the type and purpose of an electroacoustical system, there is a close interaction between the system and the room where it operates in that its performance depends to a high degree on the acoustical properties of the enclosure itself. Therefore, the installation and use of such a system does not dispense with careful acoustical planning. Furthermore, without the knowledge of the acoustical factors responsible for speech intelligibility and of the way in which these factors are influenced by sound reflections, reverberation and other acoustical effects, it would hardly be possible to plan, install and operate electroacoustical systems with optimal performance. It is the goal of this chapter to deal particularly with this interaction. More information on electroacoustical sound systems can be found in books on this subject—see Ref. 1, for instance.

## 10.1 Loudspeaker directivity

In the simplest case, a sound reinforcement system consists of a microphone, an amplifier and one or several loudspeakers. If the sounds of musical ensembles are to be reinforced, several microphones may be used, the output signals of which are electrically mixed.

The most critical component in this chain is the loudspeaker because it must generate high acoustical power in a very confined space (compare the size of a loudspeaker or loudspeaker cluster with that of a symphonic orchestra!) without producing intolerable non-linear distortions. At the same time it should distribute the sound it generates uniformly over the audience.

As a model of a loudspeaker we consider first a plane circular piston mounted flush in an infinite baffle and oscillating in the direction of its normal (see Fig. 10.1). Each of its surface elements contributes a spherical wave to the sound pressure at some point. At higher frequencies when the radius $a$ of

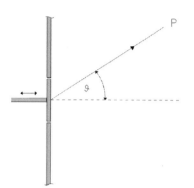

*Figure 10.1* Piston radiator, schematic.

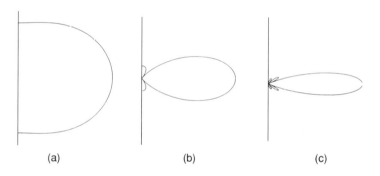

(a)                          (b)                          (c)

*Figure 10.2* Directional factor (magnitude $|\Gamma(\vartheta)|$) of the circular piston: (a) $ka = 2$; (b) $ka = 5$; (c) $ka = 10$.

the piston is not small compared with the wavelength of the radiated sound signal there may be noticeable phase differences between these contributions, resulting in an interference pattern. In the far field of the loudspeaker, i.e. when the distance from the source is considerably larger the $S/\lambda$ ($S$ = area of the piston), the directional structure of the sound field can be described by the directional factor (see eqn. (1.23)), which in the present case reads

$$\Gamma(\vartheta) = \frac{2J_1(ka \sin \vartheta)}{ka \sin \vartheta} \tag{10.1}$$

Here $\vartheta$ is the angle between the normal to the piston area and the direction of radiation, $J_1$ is the first-order Bessel function, and $ka = 2\pi a/\lambda$ is the so-called Helmholtz number, which equals the circumference of the piston divided by the acoustical wavelength. Figure 10.2 depicts a few directional patterns of the circular piston, i.e. polar representations of the directivity factor $|\Gamma|$ for various values of $ka$. (The three-dimensional directivity factors are obtained by rotating these diagrams around their horizontal axes.) For $ka = 2$ the radiation is nearly uniform, but with increasing $ka$ the sound is increasingly concentrated toward the middle axis of the piston. For $ka > 3.83$, smaller lobes appear additionally in the diagrams.

The shape of the main lobe of a radiator's directional diagram can be characterised by its half-width, i.e. the angular distance $2\Delta\vartheta$ of the points for which $|\Gamma|^2 = 0.5$, as shown in Fig. 10.5. For the circular piston with $ka \gg 1$, this figure is approximately

$$2\Delta\vartheta \approx \frac{\lambda}{a} \cdot 30° \tag{10.2}$$

Another important 'figure of merit' is the directivity or gain $g$, defined as the ratio of the maximum and the average intensity, both at the same distance from the source (see eqn. (5.40)). For the circular piston it is given by

$$g = \frac{(ka)^2}{2} \cdot \left(1 - \frac{2J_1(2ka)}{2ka}\right)^{-1} \tag{10.3}$$

The dependence of this function on $ka$ is shown in Fig. 10.3.

The membrane of a real loudspeaker is neither plane nor rigid, and in most cases it is not mounted flush in a plane infinite baffle but in the front side of a box. Hence its directivity differs more or less from that described by eqns (10.1)–(10.3). Nevertheless, these relationships yield at least a guideline for the directional properties of real loudspeakers.

Another type of loudspeaker which is commonly used in public address systems is the horn loudspeaker. It consists of a tube with a steadily

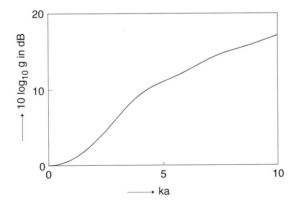

Figure 10.3 Directivity $g$ (in dB) of the circular piston as a function of $ka$.

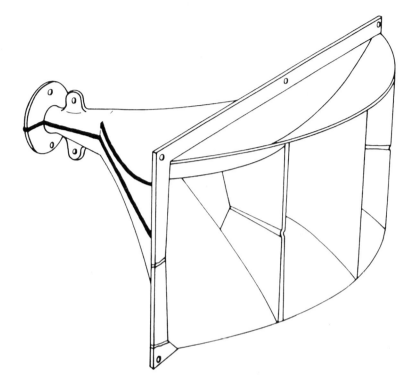

*Figure 10.4* Horn loudspeaker (multi-cellular horn).

increasing cross-sectional area, called a horn, and an electrodynamically driven diaphragm at its narrow end—the throat (see Fig. 10.4). Its main advantage is its high power efficiency because the horn improves the acoustical match between the diaphragm and the free field. Furthermore, by combining several horns a wide variety of directional patterns can be achieved. For these reasons, horn loudspeakers are often employed for large-scale sound reproduction.

Most horns are based on an exponential increase in cross-sectional area. Such horns have a marked cut-off frequency which depends on their flare and below which the horn cannot efficiently radiate sound. Other horn shapes are also in use. Unfortunately, no closed formulae are available for calculating or estimating the directional characteristics of a horn loudspeaker, which depend on the shape and the length of the horn as well as the size and the shape of its 'mouth'. Thus the directional pattern of a horn has to be determined experimentally or with numerical methods.

As mentioned, the directivity pattern of a horn loudspeaker can be shaped in a desired way by combining several or many horns. The most straightforward solution of this kind is the multi-cellular horn consisting of many single horns, the openings of which approximate a portion of a sphere and yield nearly uniform radiation into the solid angle subtended by the opening of the horn.

Another common loudspeaker arrangement is the linear loudspeaker array consisting of several loudspeakers of the same kind located along a straight line at equal distances. All loudspeakers are fed with the same electrical signals. As before, their contributions may interfere with each other. As a consequence, the loudspeaker array has a far-field directivity which is more pronounced the higher the sound frequency.

As a simple example we consider at first a linear array of $N$ point sources which are arranged along a straight line with equal spacing $d$. The sources are assumed to emit equal sine signals with the frequency $\omega = ck$. The directional factor of this array is given by

$$\Gamma_a(\vartheta) = \frac{\sin\left(\frac{1}{2}Nkd\sin\vartheta\right)}{N\sin\left(\frac{1}{2}kd\sin\vartheta\right)} \tag{10.4}$$

where $\vartheta$ is the angle which the considered direction includes with the normal of the array.

This is illustrated in Fig. 10.5, which plots $|\Gamma(\vartheta)|$ as a polar diagram for $kd = \pi/2$ and $N = 8$. The three-dimensional directivity pattern is obtained by rotating this diagram around the vertical axis of the array. It is noteworthy that the radiated sound is concentrated into the directions perpendicular to the array axis. As in the case of the circular piston, the directional pattern contains a main lobe which becomes narrower with increasing frequency. Furthermore, for $f > c/Nd$ it shows smaller satellite lobes, the number of which grows with increasing number of elements and with the frequency.

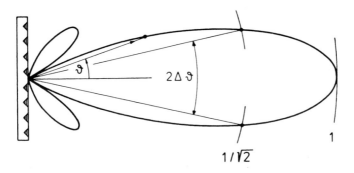

Figure 10.5 Directional factor (magnitude $|\Gamma(\vartheta)|$) of a linear loudspeaker array with $N = 8$ elements for $kd = \pi/2$.

For $N > 3$, the largest of these side lobes is at least 10 dB lower than the maximum of the main lobe. The angular half-width of the main lobe is

$$2\Delta\vartheta \approx \frac{\lambda}{Nd} \cdot 50° \tag{10.5}$$

This relationship, however, holds only if the resulting half-width $2\Delta\vartheta$ is less than 30°.

If the point sources in this array are replaced with loudspeakers fed with identical electrical signals, the total directivity function is simply obtained by multiplying the directional factor of the array $\Gamma_a$ with that of its elements $\Gamma_0$.

## 10.2 Reach of a loudspeaker

This section deals with some factors influencing the performance of public address systems, namely with the acoustical power supplied by the loudspeakers, with their directionality and with the reverberation time of the room. We are speaking here of loudspeakers, although the following considerations apply to any sound source operating in a room.

If speech intelligibility were a function merely of the loudness, i.e. of the energy density $w_r$ at a listener's position, we could use eqn (5.37) to estimate the necessary acoustical power of the loudspeaker, assuming a diffuse sound field:

$$P = \frac{c}{4}Aw_r = 13.8\frac{V}{T}w_r \tag{10.6}$$

with

$$A = \sum_i S_i \alpha_i \tag{10.6a}$$

denoting the total absorbing area in the room.

We know, however, from the discussions in Chapter 7, that the intelligibility of speech depends not only on the loudness of the signal but even more on the structure of the impulse response characterising the sound transmission from a sound source to a listener. In particular, it is of great importance how the total energy conveyed by the impulse response is distributed over the useful and the detrimental reflections.

To derive practical conclusions from this fact, we idealise the impulse responses of individual transmission paths by an exponential decay of sound energy with a decay constant $\delta = 6.91/T$:

$$E(t) = E_0 \exp(-2\delta t)$$

Now suppose a sound source supplies the constant sound power $P$ to the room. We regard as detrimental those contributions to the resulting energy that are conveyed by the reverberant 'tail' of the energetic impulse response, which is made up by reflections with delays exceeding 100 ms with respect to the direct sound. The energy they carry is

$$w_r' = \frac{P}{V} \int_{0.1s}^{\infty} \exp(-2\delta t)\,dt = \frac{P}{2\delta V} \exp(-0.2\delta) \tag{10.7}$$

On the other hand, the direct sound energy supplied by the loudspeaker is certainly useful. We assume that the loudspeaker or the loudspeaker array has some directivity and points with its main lobe towards the most remote parts of the audience. Then the density of the useful energy in that most critical region is obtained from eqn (5.40) by multiplying with the gain $g$:

$$w_d = \frac{gP}{4\pi c r^2} \tag{10.8}$$

Satisfactory speech intelligibility is achieved if

$$w_d \geq w_r'$$

in other words, we are searching for some kind of diffuse-field distance, as in Section 5.5. Inserting eqns (10.7) and (10.8) into this condition yields:

$$\frac{g}{2\pi r^2} \geq \frac{c}{V\delta} \exp(-0.2\delta) \tag{10.9}$$

Now we denote by $r_{max}$ the largest distance for which this condition can be fulfilled. By solving for $r_{max}$, we obtain an expression for the reach of the sound source, i.e. the range in which we can expect good intelligibility, provided the loudness of the sound signal is high enough. Introducing the reverberation time $T = 3 \cdot \ln 10/\delta$ and observing that $\exp(0.3 \cdot \ln 10) \approx 2$ yields

$$r_{max} \approx 0.057 \left(\frac{gV}{T}\right)^{1/2} \cdot 2^{1/T} \tag{10.10}$$

where $r_{max}$ is in metres and $V$ in cubic metres. This quantity differs from the critical distance $r_c$ in eqn (5.39a) by the factor $2^{1/T}$. In Fig. 10.6 it is plotted as a function of the reverberation time; the product $gV$ of the loudspeaker gain and the room volume is the parameter of the curves.

It should be noted that eqn (10.10) represents rather a rough estimate of the allowed $r_{max}$ than an exact limit. This is a consequence of the somewhat

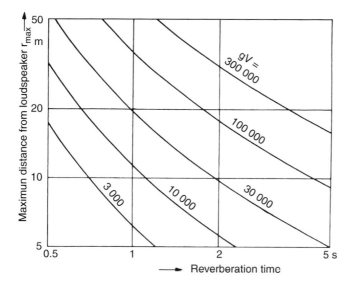

*Figure 10.6* Allowed maximum distance of listeners from loudspeaker as a function of the reverberation time for various values of $gV$ (*V* in m³).

crude assumptions on the derivation before. In particular, it underestimates the distance $r_{max}$ for two reasons:

(1) In most cases the loudspeaker will also produce early reflections which contribute to the left-hand side of eqn (10.9) but there is no general way to account for them.
(2) It is safe to assume that the main lobe of the loudspeaker's directional pattern will be directed towards the audience which is highly absorbing at mid and high frequencies. This reduces the power available for the excitation of the reverberant field roughly by a factor

$$\frac{1}{g'} = \frac{1 - \alpha_a}{1 - \overline{\alpha}} \tag{10.11}$$

with $\alpha_a$ denoting the absorption coefficient of the audience, while $\overline{\alpha} = A/S$ (see eqn (10.6a) is the mean absorption coefficient of the boundary. Consequently, the gain $g$ in eqn (10.10) and Fig. 10.6 can be replaced with $gg'$, which is larger than $g$.

The most important result of this consideration is that the reach of a sound source in a room is limited by the reverberant sound field and cannot be extended just by increasing the sound power. To illustrate this, let us consider a hall with a volume of 15 000 m$^3$ and a reverberation time of 2 s. If the product $gg'$ can be made as high as 16, the maximum distance which can be bridged acoustically will be $r_{max} \approx 28$ m.

At first glance it seems that the condition $r < r_{max}$ with $r_{max}$ from eqn (10.10) is not too stringent. This is indeed the case for medium and high frequencies. At low frequencies, however, $g$ as well as $g'$ is close to unity. As a consequence, most of the low-frequency energy supplied by the sound source will feed the reverberant part of the energy density where it is not of any use. This is the reason why so many halls equipped with a sound reinforcement system suffer from a low-frequency background which is unrelated to the transmitted signal and is perceived as a kind of noise. The simplest remedy against this evil is to suppress the low-frequency components of the signal which do not contribute to speech intelligibility by a suitable electrical filter.

Concerning the necessary power output $P$ of the loudspeaker, it should be clear that $P$ cannot be calculated from eqn (10.6). Instead, we require that the directly transmitted sound leads to a sufficiently high energy density $w_d$ at distances of up to $r_{max}$:

$$P = \frac{4\pi r_{max}^2}{g} \cdot w_d \approx \frac{12}{g} r_{max}^2 \cdot 10^{0.1 L_d - 12} \tag{10.12}$$

This is the same formula as would be applied to outdoor sound amplification since it does not include any room properties.

For speech, a level $L_d$ of 70–75 dB is adequate provided the environment is not very noisy. For music, the required sound level is considerably higher, depending on the type of music (95–105 dB). It should be noted, however, that there is usually some noise in a large hall, which is due to a restless audience, to the air-conditioning system or to insufficient insulation against exterior noise sources. If the noise level is $\leq 40$ dB it can be left out of consideration as far as the required loudspeaker power is concerned. Generally, the level produced by the loudspeaker(s) at the listener's place should surpass the noise level by about 10 dB. Since panic situations have to be accounted for, the level $L_d$ should cover the range from 75 to 105 dB. Of equal importance is a sufficiently high speech transmission index (STI), which must be at least 0.5. This indicates once more that good speech intelligibility is not just a matter of the acoustical power generated by the loudspeakers.

## 10.3 A few remarks on loudspeaker positions

In this section we consider a public address system which, in the simplest case, consists of a microphone, an amplifier, and one or several loudspeakers, as presented in Fig. 10.7. Additional components such as equalisers, delay

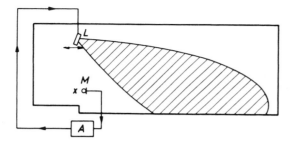

*Figure 10.7* Central loudspeaker system: $L$ = loudspeakers, $M$ = microphone, $A$ = amplifier and $x$ = Source.

units, etc., are not shown. The amplified microphone signal can be supplied to the room either by one central loudspeaker, or by several or many loud-speakers distributed throughout the room. (The term 'central loudspeaker' includes of course the possibility of combining several loudspeakers closely together in order to achieve a suitable directivity, for instance in a linear array.) This section will deal with several factors which should be considered when loudspeaker locations are to be selected in a room.

In any case the loudspeakers should ensure that all listeners are supplied with sufficient sound energy; furthermore, for speech installations, a satisfactory speech intelligibility should be attained. In addition a sound reinforcement system should yield a natural sound impression. In the ideal case (possibly with the exception of the presentation of electronic music) the listener would be unable to notice the electroacoustical aids at all. To achieve this, it would at least be necessary, apart from using high-quality microphones, loudspeakers and amplifiers, that the sounds produced by the loudspeakers reach the listener from about the same direction in which he sees the actual speaker or natural sound source.

Since the microphone and the loudspeakers are operated in the same room it is inevitable that the microphone will pick up not only sound produced by the natural source, for instance by a speaker's voice, but also sound arriving from the loudspeaker. This phenomenon, known as 'acoustical feedback', can result in instability of the whole equipment and lead to the well-known howling or whistling sounds. We shall discuss acoustical feedback in a more detailed manner in the next section.

With a central loudspeaker installation, the irradiation of the room is achieved by a single loudspeaker or loudspeaker combination, as shown in Fig. 10.7. The location of the loudspeaker, its directionality and its orientation have to be chosen in such a way that the audience is uniformly supplied with direct sound. At the same time the risk of acoustical feedback must be kept as low as possible.

In most cases, the loudspeaker will be mounted more or less above the natural sound source. This way of mounting has the advantage that the direct sound, coming from the loudspeaker, will always arrive from roughly the same direction (with regard to a horizontal plane) as the sound arriving directly from the sound source. The vertical deviation of directions is not very critical, since our ability to discriminate sound directions is not as sensitive in a vertical plane as in a horizontal one. The subjective impression is even more natural if care is taken that the loudspeaker sound reaches the listener simultaneously with the natural sound or a bit later. In the latter case, the listener benefits from the law of the first wavefront, which raises the illusion that all the sound he hears is produced by the natural sound source, i.e. that no electroacoustical system is in operation. This illusion can be maintained even if the level of the loudspeaker signal at the listener's position surpasses the level due to the natural source by 5–10 dB, provided the latter precedes the loudspeaker signal by about 10–15 ms (Haas effect, see Section 7.3).

The simultaneous or delayed arrival of the loudspeaker's signals at the listener's seat can be achieved by increasing the distance between the loudspeaker and the audience. The application of this simple measure is limited, however, by the increasing risk of acoustical feedback, since the microphone will lie more and more in the range of the main lobe of the loudspeaker's directional characteristics. Thus, a compromise must be found. Another way is to employ electrical methods for effecting the desired time delay. They are described below.

Very good results in sound amplification, even in large halls, have occasionally been obtained by using a speaker's desk which has loudspeakers built into the front facing panel; these loudspeakers were arranged in properly inclined, vertical columns with suitable directionality. With this arrangement, the sound from the loudspeakers will take almost the same direction as the sound from the speaker himself, which reduces problems due to feedback provided that the propagation of structure-born sound is prevented by resiliently mounting the loudspeakers and the microphone.

In very large or long halls, or in halls consisting of several sections, the supply of sound energy by one single loudspeaker only will usually not be possible, for one thing because condition (10.10) cannot be met without unreasonable expenditure. The use of several loudspeakers at different positions has the consequence that each loudspeaker supplies a smaller area, which makes it easier to satisfy eqn (10.10). Two simple examples are shown in Fig. 10.8. If all loudspeakers are fed with identical electrical signals, however, confusion zones will be created in which listeners are irritated by hearing sound from more than one source. In these areas it is not only the natural localisation of the sound source which is impaired but also the intelligibility is significantly diminished. This undesirable effect is avoided by electrically delaying the signals applied to the

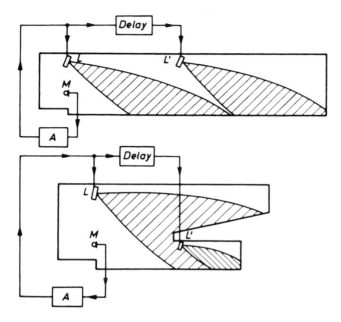

*Figure 10.8* Public address system with more that one loudspeaker. $L, L'$ = loudspeakers, $M$ = microphone, $A$ = amplifier.

auxiliary loudspeakers. The delay time should at least compensate the distances between the auxiliary loudspeaker(s) and the main loudspeaker. Furthermore, the power of the subsidiary loudspeakers must not be too high since this again would make the listener aware of it and hence destroy his illusion that all the sound he receives is arriving from the stage. The delay times needed in sound reinforcement systems are typically in the range 10–100 ms, sometimes even more. Nowadays, digital delay units are used for this purpose.

In halls where a high noise level is to be expected, but where nevertheless announcements or other information must be clearly understood by those present, the ideal of a natural-sounding sound transmission, which preserves or simulates the original direction of sound propagation, must be sacrificed. Accordingly, the amplified signals are reproduced by many loudspeakers which are distributed fairly uniformly and are fed with identical electrical signals. In this case it is important to ensure that all the loudspeakers which can be mounted on the ceiling or suspended from it are supplied with equally phased signals. The listeners are then, so to speak, in the near field of a vibrating piston. Sound signals of opposite phases would be noticed in the region of superposition in a very peculiar and unpleasant manner.

If the sound irradiation is effected by directional loudspeakers from the stage towards the back of the room, the main lobe of one loudspeaker will inevitably project sound towards the rear wall of the room, as we particularly wish to reach those listeners who are seated the furthest away immediately in front of the rear wall. Thus, a substantial fraction of the sound energy is reflected from the rear wall and can cause echoes in other parts of the room, which can irritate or disturb listeners as well as speakers. For this reason it is recommended that the remote portions of walls being irradiated by the loudspeakers are rendered highly absorbent. In principle, an echo could also be avoided by a diffusely reflecting wall treatment which scatters the sound in all possible directions. But then the scattered sound would excite the reverberation of the room, which, as was explained earlier, is not favoured for speech intelligibility.

The performance of an electroacoustical system can be tested by applying stationary test signals to the amplifier input. More detailed information can be gained with impulse or impulse-equivalent signals (maximum length sequences, sine sweeps, see Section 8.2). Such measurements permit us to discriminate the 'useful' signal received at a given place from the reverberant component, which is not of much use. In particular, the indicators of speech intelligibility, as discussed in Section 7.4, can be evaluated from the impulse response either of the public address system itself or of its combination with a loudspeaker which simulates a natural speaker. As an alternative, the STI can be determined for various typical listener locations.

The preceding discussions refer mainly to the transmission of speech. The electroacoustical amplification of music—apart from entertainment or dance music—is firmly rejected by many musicians and music lovers for reasons which are partly irrational. Obviously, many of these people have the suspicion that the music could be manipulated in an undue way which is outside the artist's influence. And finally, almost everybody has experienced the poor performance of technically imperfect reinforcement systems. If, in spite of objections, electroacoustical amplification is mandatory in a performance hall, the installation must be carefully designed and constructed with first-class components and it must preserve, under all circumstances, the natural direction of sound incidence. Care must be taken to avoid linear as well as non-linear distortions and the amplification should be kept at a moderate level only. A particular problem is to comply with the large dynamical range of symphonic music. For entertainment music, the requirements are not as stringent; in this case people have long been accustomed to the fact that a singer has a microphone in his or her hand and the audience will more readily accept that it will be conscious of the sound amplification.

These remarks have no significance whatsoever for the presentation of electronic music; here the acoustician can safely leave the arrangement of loudspeakers and the operation of the whole equipment to the performers.

## 10.4 Acoustical feedback and its suppression

Acoustical feedback of sound reinforcement systems in rooms has already been mentioned in the last section. In principle, feedback will occur whenever the loudspeaker is in the same room as the microphone, which inevitably will pick up a portion of the loudspeaker signal. Only if this portion is sufficiently small are the effects of feedback negligibly faint; with higher amplification of the feedback signal it may cause substantial linear distortions such as ringing effects. At still higher amplification the whole system will perform self-sustained oscillations at some frequency which makes the system useless.

Before discussing measures for the reduction or suppression of feedback effects, we shall deal with its mechanism in a somewhat more detailed manner.

We assume that the original sound source, for instance a speaker, will produce at the microphone a sound signal whose spectrum is denoted by $S(\omega)$ (see Fig. 10.9). Its output voltage is amplified with a frequency-independent amplifier gain $q$ and is fed to the loudspeaker. The loudspeaker signal will reach the listener by passing along a transmission path in the room with a complex transfer function $\overline{G}(\omega)$; at the same time, it will reach the microphone by a path with the transfer function $G(\omega)$. The latter path, together with the microphone, the amplifier and the loudspeaker, constitutes a closed loop which the signal passes through repeatedly.

The lower part of Fig. 10.9 shows the mechanism of acoustical feedback in a more schematic form. The complex amplitude spectrum of the output

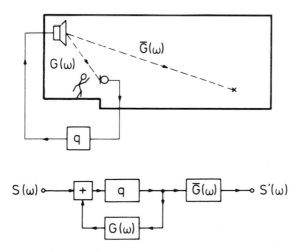

*Figure 10.9* Acoustical feedback in a room. In the lower part the transmission paths in the room including the loudspeakers are represented by 'black boxes'.

signal (i.e. of the signal at the listener's seat) is given by

$$S'(\omega) = q\overline{G}(\omega)\left[S(\omega) + G(\omega)\frac{S'(\omega)}{\overline{G}(\omega)}\right]$$

From this expression we calculate the transfer function of the whole system including the effects of acoustical feedback, $G'(\omega) = S'(\omega)/S(\omega)$:

$$G'(\omega) = \frac{q\overline{G}(\omega)}{1 - qG(\omega)} = q\overline{G}(\omega)\sum_{n=0}^{\infty}[qG(\omega)]^n \tag{10.13}$$

The latter expression clearly shows that acoustical feedback is brought about by the signal repeatedly passing through the same loop. The factor $qG(\omega)$, which is characteristic for the amount of feedback, is called the 'open loop gain' of the system. Depending on its magnitude, the spectrum $S'(\omega)$ of the received signal and hence the signal itself may be quite different from the original signal with the spectrum $S(\omega)$.

A general idea of the properties of the 'effective transfer function' $G'$ can easily be given by means of the Nyquist diagram, in which the locus of the complex open loop gain is represented in the complex plane (see Fig. 10.10). Each point of this curve corresponds to a particular frequency; abscissa and ordinate are the real part and the imaginary part of $qG$ respectively. The arrows point in the direction of increasing frequency. The whole system will remain stable if this curve does not include the point $+1$, which is certainly not the case if $|qG| < 1$ for all frequencies. Mathematically, this condition guarantees the convergence of the series at the right side of eqn (10.13).

Now let us suppose that we start with a very small amplifier gain, i.e. with $qG$ being small compared to unity. If we increase the gain gradually, the curve in Fig. 10.10 is inflated, keeping its shape. In the course of this process, the distance between the curve and the point $+1$, i.e. the quantity $|1 - qG|$, could become very small for certain frequencies. At these frequencies, the absolute value of the transfer function $G'$ will consequently become very large. Then the signal received by the listener will sound 'coloured' or, if the system is excited by an impulsive signal, ringing effects are heard. With a further increase of $q$, $|qG|$ will exceed unity and this will happen at a frequency close to that of the absolute maximum of $|G(\omega)|$. Then the system becomes unstable and performs self-excited oscillation at that frequency.

The effect of feedback on the performance of a public address system can also be illustrated by plotting $|G'(\omega)|$ on a logarithmic scale as a function of the frequency. This leads to 'frequency curves' similar to that shown in Fig. 3.7b. Figure 10.11 represents several such curves for various values of the open loop gain $qG$, obtained by simulation with a digital computer.[2] With increasing gain one particular maximum starts growing more rapidly than the other maxima and becomes more and more dominating. This is the

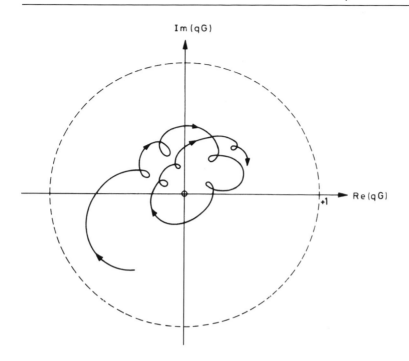

*Figure 10.10* Nyquist diagram illustrating stability of acoustical feedback.

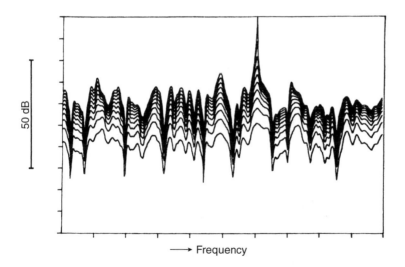

*Figure 10.11* Frequency curves of a room equipped with an electroacoustical sound system, simulated for different amplifier gains. The latter vary in steps of 2 dB from −20 dB to 0 dB relative to the critical gain $q_0$. The total frequency range is 90/$T$ hertz ($T$ = reverberation time, after Ref. 2).

condition of audible colouration. When a critical value $q_0$ of the amplifier gain is reached, this leading maximum becomes infinite, which means that the system will start to perform self-sustained oscillations. (In real systems, the amplitude of these oscillations remains finite because of inevitable nonlinearities of its components.)

A question of great practical importance concerns the amplifier gain $q$ which must not be exceeded if colouration is to be avoided or to be kept within tolerable limits. According to listening tests as well as to theoretical considerations, colouration remains imperceptible as long as

$$20 \log_{10}(q/q_0) \leq -12\,\text{dB} \tag{10.14}$$

For speech transmission, it is sufficient to keep the relative amplification 5 dB below the instability threshold to avoid audible colouration.

Another effect of acoustical feedback is the increase of reverberance which is again restricted to those frequencies for which $G'(\omega)$ is particularly high. To show this, we simplify eqn (10.13) by putting $\overline{G} = G$. Then, the second version of this equation reads

$$G'(\omega) = \sum_{n=1}^{\infty} [qG(\omega)]^n \tag{10.15}$$

The corresponding impulse response is obtained as the (inverse) Fourier transform of that expression:

$$g'(t) = \frac{1}{2\pi} \int_{-\infty}^{\infty} G'(\omega) \exp(i\omega t)d\omega = \sum_{n=1}^{\infty} q^n g^{(n)}(t) \tag{10.16}$$

In the latter formula $g^{(n)}$ denotes the $n$-fold convolution of the impulse response $g(t)$ with itself, defined by the recursion

$$g^{(n+1)}(t) = \int_0^t g^{(n)}(t')g(t-t')\,dt' \quad \text{and} \quad g^{(1)}(t) = g(t)$$

For our present purpose it is sufficient to use $g(t) = A \exp(-\delta t)$ (for $t \geq 0$, otherwise $g(t) = 0$) as a model response. Applying the above recursion to it yields

$$g^{(n)}(t) = A(At)^{n-1} \frac{\exp(-\delta t)}{(n-1)!}$$

If this expression is inserted into eqn (10.16), the sum turns out to be the series expansion of the exponential function, hence

$$g'(t) = Aq \exp[-(\delta - Aq)t] \tag{10.17}$$

Evidently, the decay constant of the feedback loop is $\delta' = \delta - Aq$, and the acoustical feedback increases the reverberation time by a factor

$$\frac{T'}{T} = \frac{\delta}{\delta'} = \frac{1}{1 - Aq/\delta} \tag{10.18}$$

Apparently, $Aq/\delta$ plays the role of the open loop gain. When this quantity approaches unity, the reverberation time $T'$ becomes infinite. On account of our oversimplified assumption concerning the impulse response $g(t)$, eqns (10.17) and (10.18) do not show that in reality the increase of reverberation time is limited to one or a few discrete frequencies, and that therefore the reverberation sounds coloured.

Acoustical feedback can be avoided by keeping the open loop gain $qG$ in eqn (10.13) small compared to unity. However, effecting this by reducing the amplifier gain $q$ may result in too low loudness of the loudspeaker signal at the listener's seat; then the system will become virtually useless. A better way is to make the mean absolute value of $G(\omega)$ in the interesting frequency range as small as possible without reducing that of $\overline{G}(\omega)$ (see Fig. 10.9). For this purpose the directivity of the loudspeaker must be carefully selected and adjusted; the main lobe of it should point towards the listeners, while the microphone is located in a direction of low radiation. Likewise, a microphone with some directivity may be used, for instance a cardioid microphone oriented in such a way that it favours the original sound source while suppressing the signal arriving from the loudspeaker. In any case it is advantageous to arrange the microphone close to the original source. With these rather simple measures, acoustical feedback cannot be completely eliminated to be sure, but often the point of instability can be raised high enough so that it will never be reached during normal operation.

In view of the irregular shape of room transfer functions, further increase of feedback stability should be attainable by smoothing these functions, either by frequency or by space averaging. Let us denote—as in Section 3.4— by $\Delta L_{max}$ the difference by which the level of the absolute maximum of a frequency curve exceeds that of the average. Then the stability of the system could be increased theoretically by this difference, which is given by eqn (3.38) or eqn (3.38a), and amounts typically to about 10 dB. Unfortunately, this averaging cannot be achieved by using several loudspeakers fed with the same signal, or with several microphones with their outputs signals just added. For the resultant transfer function would be the sum of several complex transfer functions; hence its general properties would be the same as those of a single room transfer function, which is itself the vector sum of numerous components superimposed with random phases (see Section 3.4). In particular, its squared absolute values are again exponentially distributed according to eqn (3.34).

The situation is different when we employ a variable transfer function G in eqn (10.13). In an early attempt to average the transfer functions of a room in this way the microphone was moved on a circular path during its operation (see Ref. 3). Since each point of the path has its own transmission characteristics, each harmonic component will undergo different amplifications and phase shifts during one round trip of the microphone. This results in some averaging of the open loop gain, provided the diameter of the circle is large compared with all sound wavelengths of interest. The use of a gradient microphone rotating around an axis which is perpendicular to the direction of its maximum sensitivity has also been proposed and this has an effect which is similar to that of a moving microphone. However, the mechanical movement makes it difficult to pick up a speaker's voice with such a microphone; furthermore, it produces amplitude and phase modulation of the signal.

A more practical method of virtually flattening the frequency characteristics of the open loop gain has been proposed and tested by Schroeder (see Refs. 8 and 9 of Chapter 3). It is effected by modifying the signal during its repeated roundtrips in the feedback loop instead of varying the transfer function. As we saw, acoustical feedback is brought about by particular spectral components which always experience the same 'favourable' amplitude and phase conditions when circulating in the closed loop in Fig. 10.9. If, however, the feedback loop contains a device which shifts the frequencies of all spectral components by a small amount $\Delta\omega$, then a particular component will experience favourable as well as unfavourable conditions, which, in effect, is tantamount to frequency averaging the transfer function. After $N$ trips around the feedback loop, the signal power is increased or decreased by a factor

$$K(\omega) = |qG(\omega + \Delta\omega)|^2 \ldots |qG(\omega + 2\Delta\omega)|^2 \ldots |qG(\omega + N\Delta\omega)|^2$$

Setting $L(\omega') = 10 \cdot \log_{10} |qG(\omega')|^2$, we obtain the change in level:

$$10 \cdot \log_{10}[K(\omega)] = L(\omega + \Delta\omega) + L(\omega + 2\Delta\omega) + \cdots + L(\omega + N\Delta\omega) \approx N\langle L \rangle$$
$$(10.19)$$

where $\langle L \rangle$ is the average of the logarithmic frequency curve from $\omega$ to $\omega + N\Delta\omega$. The system will remain stable if $N\langle L \rangle \to -\infty$ as $N$ approaches infinity, i.e. if $\langle L \rangle$ is negative.

Now it is no longer the absolute maximum of the frequency curve which determines the onset of instability, but a certain average value. As mentioned, the difference $\Delta L_{\mathrm{max}}$ between the absolute maximum and the mean value of a frequency curve is about 10 dB for most large rooms; it is this level difference by which the amplifier gain theoretically may be increased without the risk of instability, compared with the operation without frequency shifting. Note that the frequency shift $\Delta f = \Delta\omega/2\pi$ must be small enough to make the frequency shift inaudible. On the other hand, it must be high enough to yield effective averaging after a few round trips. The best compromise is met if $\Delta f$ corresponds to the mean spacing of frequency curve maxima according to eqn (3.37b):

$$\Delta f = \langle \Delta f_{\mathrm{max}} \rangle = \frac{4}{T} \tag{10.20}$$

This works quite well with speech; with music, however, even very small frequency shifts are audible, since they change the musical intervals. Therefore, this method is applicable to speech only. In practice the theoretical increase in amplification of about 10 dB cannot be reached without unacceptable reduction of sound quality; if the increase exceeds 5–6 dB, beating effects of the multiply shifted signals with each other become audible. Usually, the frequency shift is achieved by single side-band modulation of the signal.

Another method of reducing the risk of acoustical feedback by employing time-variable signals was proposed by Guelke and Broadhurst[4] who replaced the frequency-shifting device by a phase modulator. The effect of phase modulation is to add symmetrical side lines to each spectral line. By suitably choosing the width of phase variations, the centre line (i.e. the carrier) can be removed altogether. In this case, the authors were able to obtain an additional gain of 4 dB. They stated that the modulation is not noticeable even in music if the modulation frequency is as low as 1 Hz.

## 10.5 Reverberation enhancement with external reverberators

As mentioned earlier, many large halls have to accommodate quite different events such as meetings, lectures, performance of concerts, theatre and opera pieces, and sometimes even sports events, banquets and balls. It is obvious that the acoustical design of such a multipurpose hall cannot create optimum conditions for each type of presentation. At best, some compromise can be reached which necessarily will not satisfy all expectations.

A much better solution is to provide for a variable reverberation time that could be adapted to the different requirements. One way to achieve this is by changing the absorption within the hall by mechanical devices, as described in Section 9.3. Such devices, however, are costly and subject to mechanical wear. A more versatile and less expensive solution to this problem is offered by electroacoustical systems designed for the control of reverberation.

Electroacoustical enhancement of reverberance can be achieved princip-ially in two different ways. The first method employs the regenerative reverberation within the room brought about by acoustical feedback, as described in the preceding section. With the second method, artificial rever-beration is created by some external reverberator and is imposed on the sound signal which is originally 'dry'. First we shall describe the second method in more detail, while the discussion of the first one is postponed to the next section.

The principle of reverberation enhancement with separate reverberators is depicted in Fig. 10.12. The sounds produced by the orchestra are picked up by microphones which are close to the performers. The output signals of the microphones are fed into a reverberator. This is a linear system with an impulse response fairly similar to that of an enclosure and hence provides the signals with its inherent reverberance. The signals modified in this manner are re-radiated in the original room by loudspeakers distributed in a suitable way within the hall. In addition, delaying devices must usually be inserted into the electrical circuit in order to ensure that the reverberated loudspeaker signals will not reach any listener's place earlier than the direct sound signal from the natural sound source, at the same time taking into account the various sound paths in the room.

It should be noted that the selection of loudspeaker locations has a great influence on the effectiveness of the system and on the quality of the rever-berated sound. Another important point is that not all the loudspeakers are

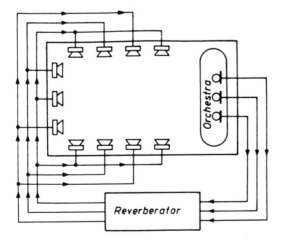

*Figure 10.12* Principle of electroacoustical reverberation enhancement employing an external reverberator.

fed with identical signals; instead, the loudspeaker signals must be incoherent since this is the condition for creating the impression of spaciousness
described in Section 7.7. Therefore, the reverberator should have several output terminals yielding mutually incoherent signals which are all derived from
the same input signal. In order to provide each listener with sound incident
from several substantially different, mainly lateral directions, it may be necessary to use far more loudspeakers than incoherent signals. It is quite obvious
that all the loudspeakers must be sufficiently distant from all the listeners
in order to prevent one particular loudspeaker being heard much louder
than the others. Finally, care must be taken to avoid significant acoustical
feedback.

Let us now discuss the various methods to reverberate the electrical signals coming from the stage microphone(s). The most natural way is to apply
them to one or several loudspeakers in a separate reverberation chamber
which has the desired reverberation time, including the proper frequency
dependence. The sound signal in the chamber is again picked up by microphones which are far apart from each other to guarantee the incoherence
of the output signals. The reverberation chamber should be free of flutter
echoes and its volume should not be less than 150–200 m$^3$, otherwise, the
density of eigenfrequencies would be too small at low frequencies.

A system of this type was installed in 1963 for permanent use with music
in the 'Jahrhunderthalle' of the Farbwerke Hoechst AG at Hoechst near
Frankfurt am Main.[5] This hall, the volume of which is 75 000 m$^3$, has a
cylindrical side wall with a diameter of 76 m; its roof is a spherical dome.
In order to avoid echoes, the dome as well as the side wall were treated
with highly absorbing materials. In this state the auditorium has a natural
reverberation time of about 1s. To increase the reverberation time, the sound
signals are picked up by several microphones on the stage, passed through a
reverberation chamber and finally fed to a total of 90 loudspeakers, which
are distributed in a suspended ceiling and along the cylindrical side and rear
wall. This system, which underwent several modifications in the course of
time, raised the reverberation time to about 2 s.

Other reverberators which have found wide application in the past
employed bending waves propagating in metal plates, or torsional waves
travelling along helical springs, excited and picked up with suitable electroacoustical transducers. The reverberation was brought about by repeated
reflections of the waves from the boundary or the terminations of these
waveguides.

The essential thing about these devices is the finite travelling time between
successive reflections. Therefore, in order to produce some kind of reverberation, we only need, in principle, a delaying device and a suitable feedback
path by which the delayed signal is transferred again and again from the
output to the input of the delay unit (see Fig. 10.13a). If $q$ denotes the
open loop gain in the feedback loop, which must be smaller than unity for

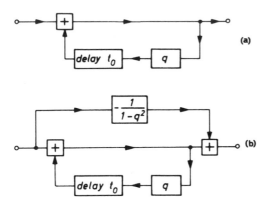

*Figure 10.13* Reverberators employing one delay unit: (a) comb filter type; (b) all-pass type.

stable conditions, and $t_0$ denotes the delay time, the impulse response of the circuit is given by eqn (7.4). With each round trip, the signal is attenuated by $-20 \log_{10}(q)$ dB, and hence after $-60/(20 \log 10(q))$ passages, the level has fallen by 60 dB. The associated total delay is the reverberation time of the reverberator and is given by

$$T = -\frac{3t_0}{\log_{10} q} \qquad (10.21)$$

It can be controlled by varying the open loop gain or the delay time $t_0$.

In order to reach a sufficiently long reverberation time, either $q$ must be fairly close to unity, which makes the adjustment of the open loop gain very critical, or $t_0$ must be relatively long. Suppose we aim at a reverberation time of 2 s. With a delay time of 10 ms, the gain must be set at 0.966 while with $t_0 = 100$ ms the required gain is still 0.708. In both cases, the reverberation has an undesirable tonal quality. In the first case the reverberator produces 'coloured' sounds due to the regularly spaced maxima and minima of its transfer function, as shown in Fig. 7.11. In the second case the regular succession of 'reflections' is heard as flutter .

The quality of such a reverberator can be improved to a certain degree, according to Schroeder and Logan,[6] by giving it all-pass characteristics. For this purpose, the fraction $1/(1 - q^2)$ of the input signal is subtracted from its output (see Fig. 10.13b). The impulse response of the modified reverberator is

$$g(t) = -\frac{1}{1 - q^2} \delta(t) + \sum_{n=0}^{\infty} q^n \delta(t - nt_0) \qquad (10.22)$$

Its Fourier transform, i.e. the transfer function of the reverberator, is given by

$$G(\omega) = -\frac{1}{1-q^2} + \frac{1}{1-q\exp(-i\omega t_0)}$$

$$= \frac{q\exp(-i\omega t_0)}{1-q^2} \cdot \frac{1-q\exp(i\omega t_0)}{1-q\exp(-i\omega t_0)} \qquad (10.23)$$

Since the second factor in eqn (10.23) has the absolute value 1, $G(\omega)$ has no maxima and minima, the system is an all-pass. Subjectively, however, the undesirable properties of the reverberation produced in this way have not completely disappeared, since our ear does not perform a Fourier analysis in the mathematical sense, but rather a 'short-time frequency analysis', thus also being sensitive to the temporal structure of a signal. A substantial improvement can be effected, however, by combining several reverberation units with and without all-pass characteristics and with different delay times. These units are connected partly in parallel, partly in series. An example is presented in Fig. 10.14. Of course it is important to avoid simple ratios between the various delay times; moreover, the impulse response of the reverberator should be free of long repetition periods which can be checked by performing an autocorrelation analysis on it (see Section 8.3). As far as the practical implementation is concerned, time delays are produced with digital circuits.

More recently, a sophisticated system has been developed by Berkhout et al.[7,8] which tries to modify the original signals in such a way that they

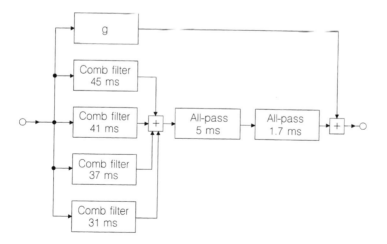

*Figure 10.14* Electrical reverberator consisting of four comb filter units and two all-pass units. The numbers indicate the delay of each unit. Optionally, the unreverberated signal attenuated by a factor g can be added to the output.

contain and hence transplant not only the reverberation but also the complete wave field from a fictive hall (preferably one with excellent acoustics) into the actual environment. This system, called the Acoustic Control System (ACS), is based on Huygens' principle, according to which each point hit by a wave may be considered as the origin of a secondary wave which effects the propagation of the wave to the next points. The ACS is intended to simulate this process by hardware components, i.e. by microphones, amplifiers, filters and loudspeakers. In the following explanation of 'wavefront synthesis', we describe all signals in the frequency domain, i.e. as functions of the angular frequency $\omega$.

Let us consider, as shown in Fig. 10.15, an auditorium in which a plane and regular array of N loudspeakers is installed. If properly fed these loudspeakers should synthesise the wave fronts originating from the sound sources. For this purpose, the sounds produced in the stage area are picked up by M microphones regularly arranged next to the stage (for example, in the ceiling above the stage). These microphones have some directional characteristics, each of them covering a subarea of the stage with one 'notional sound source' in its centre which is at $r_m$. If we denote the signals produced by these sound sources with $S(r_m, \omega)$, the signal received by the $m$th microphone at the location $r'_m$ is

$$M(r'_m, \omega) = W(r_m, r'_m) \cdot S(r_m, \omega) \qquad (m = 1, 2, \ldots M) \qquad (10.24)$$

Where W is a 'propagator' describing the propagation of a spherical wave from $r_m$ to $r'_m$:

$$W(r_m, r'_m) = \frac{\exp(-ik|r_m - r'_m|)}{|r_m - r'_m|} \qquad (10.25)$$

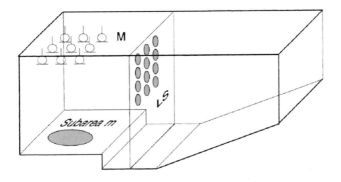

Figure 10.15 Principle of wavefront synthesis.

Each of these propagators involves an amplitude change and a delay $\tau$:

$$W(\mathbf{r}_m, \mathbf{r}'_m) = A \exp(-i\omega\tau) \tag{10.26}$$

with

$$\tau = |\mathbf{r}_m - \mathbf{r}'_m|/c$$

The microphone signals $M(\mathbf{r}'_m, \omega)$ are fed to the loudspeakers after processing them as if the source signals $S(\mathbf{r}_m, \omega)$ had reached the loudspeaker locations directly, i.e. as sound waves. Hence we have to undo the effect of the propagator $W(\mathbf{r}_m, \mathbf{r}'_m)$ and to replace it with another one $W(\mathbf{r}_m, \mathbf{r}_n)$ describing the direct propagation from the $m$th notional source to the $n$th loudspeaker. Finally the input signal of this loudspeaker is obtained by adding the contributions of all notional sources:

$$L(\mathbf{r}_n) = \sum_{m=1}^{M} W^{-1}(\mathbf{r}_m, \mathbf{r}'_m) \cdot W(\mathbf{r}_m, \mathbf{r}_n) \cdot M(\mathbf{r}'_m, \omega) \tag{10.27}$$

The loudspeakers will correctly synthesise the original wave fronts if their mutual distances are small enough and if they have dipole characteristics. (The latter follows from Kirchhoff's formula, which is the mathematical expression of Huygens' principle but will not be discussed here.) For a practical application it is sufficient to substitute the planar loudspeaker array by a linear one in horizontal orientation since our ability to localise sound sources in vertical directions is rather limited.

This relatively simple version of an ACS can be used not only for enhancing the sounds produced on stage but also for improving the balance between different sources, for instance between singers and an orchestra. It has the advantage that it preserves the natural localisation of the sound sources. Although the derivation presented above neglects all reflections from the boundary of the auditorium, the system works well if the reverberation time of the hall is not too long.

Sound reflections from the boundaries could be accounted for by constructing the mirror images of the notional sources at $\mathbf{r}_m$ and including their contributions into the loudspeaker input signals. At this point, however, it is much more interesting to construct image sources, not with respect to the actual auditorium but to a virtual hall with desired acoustical conditions, and hence to transplant these conditions into the actual hall. This process is illustrated in Fig. 10.16. It shows the actual auditorium (assumed as fanshaped) drawn in the system of mirror images of a virtual rectangular hall. Furthermore, it shows the images of just one notional source. Suppose the positions and the relative strengths of the image sources are numbered in some way,

$$\mathbf{r}_m^{(1)}, \mathbf{r}_m^{(2)}, \mathbf{r}_m^{(3)} \ldots \quad \text{and} \quad B_m^{(1)}, B_m^{(2)}, B_m^{(3)} \ldots$$

Then the propagator $W(\mathbf{r}_m, \mathbf{r}_n)$ in eqn (10.27) has to be replaced with

$$W(\mathbf{r}_m, \mathbf{r}_n) + B_m^{(1)} W(\mathbf{r}_m^{(1)}, \mathbf{r}_n) + B_m^{(2)} W(\mathbf{r}_m^{(2)}, \mathbf{r}_n) + \cdots \quad (m = 1, 2, \ldots M)$$

which leads to the following loudspeaker input signal

$$L(\mathbf{r}_n) = \sum_{m=1}^{M} W^{-1}(\mathbf{r}_m, \mathbf{r}'_m) \left[ W(\mathbf{r}_m, \mathbf{r}_n) + \sum_k B_m^{(k)} W(\mathbf{r}_m^{(k)}, \mathbf{r}_n) \right] \cdot S(\mathbf{r}_m, \omega)$$

$$(10.28)$$

In practical applications it is useful to arrange loudspeaker arrays along the side walls of the actual auditorium and to allocate each of them to the right-hand and the left-hand image sources, respectively, as indicated in Fig. 10.16.

Since the coefficients $B_m^{(k)}$ contain in a cumulative way the absorption coefficients of all walls involved in the formation of a particular image source (see Section 4.1), the reverberation time and the reverberation level, including their frequency dependence, are easily controlled by varying these coefficients. Similarly, the shape and the volume of the virtual hall can be changed. Thus, an ACS permits the simulation of a great variety number of different environments in a given hall.

Since the number of image sources increases rapidly with the order of reflection, a vast number of amplitude-delay units, as in eqn (10.26), would be required to synthesise the whole impulse response of the virtual room. Therefore this treatment is restricted to the early part of the impulse response. The later parts, i.e. those corresponding to reverberation, can be synthesised in a more statistical way because the auditive impression conveyed by them does not depend on individual reflections. More can be found on this matter in Ref. 9.

Systems of this kind have been installed in many halls and theatres and are successfully used for sound reinforcement and for reverberation enhancement. If carefully installed and adjusted, even experienced listeners will not be aware that any electroacoustical system is being operated during the performance.

## 10.6 Reverberation enhancement by controlled feedback

This last section deals with another method of reverberation enhancement, namely that mentioned first in the preceding section. Again, reverberation is created by time delays, but in contrast to the earlier method these

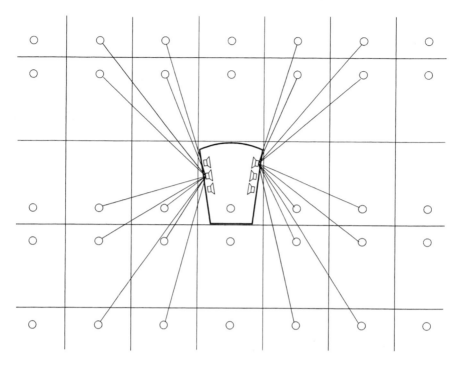

*Figure 10.16* Actual hall and image sources of a virtual rectangular hall (after Berkhout et al[9]).

delays are brought about not by external devices but by sound paths inside the room.

As discussed in Section 10.3, the occurrence of acoustical feedback in a room is accompanied by an increase in reverberation time, which depends on the open loop gain. In a usual public address system this effect is restricted to one frequency. In order to achieve wide-band enhancement of reverberation with satisfactory tonal quality, the system has to comprise numerous acoustical transmission channels operated simultaneously in the same enclosure.

Figure 10.17 depicts the multi-channel system invented by Franssen.[3] It consists of $N(\gg 1)$ microphones, amplifiers and loudspeakers, each of the latter being electrically connected with one particular microphone. All microphone–loudspeaker distances are significantly larger than the critical distance (see eqn (5.39) or eqn (5.39a)). Thus there are $N^2$ transmission paths which are interconnected by the sound field. Accordingly, the output voltage of the $k$th microphone contains the contribution $S_0(\omega)$ made by the sound source SS as well as the contributions of all loudspeakers. Therefore,

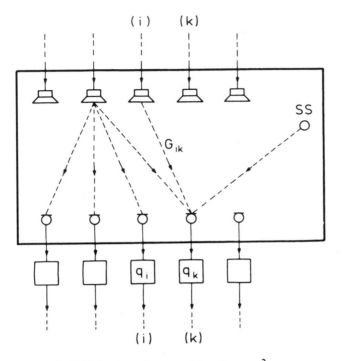

*Figure 10.17* Multi-channel system (after Franssen[3]).

its amplitude spectrum is given by

$$S_k(\omega) = S_0(\omega) + \sum_{i=1}^{N} q_i G_{ik}(\omega) S_i(\omega) \qquad (k = 1, 2, \ldots N) \tag{10.29}$$

where $q_i$ denotes the gain of the $i$th amplifier and $G_{ik}(\omega)$ is the transfer function of the acoustic transmission path from the $i$th loudspeaker to the $k$th microphone, including the properties of both transducers.

The expression (10.29) represents a system of $N$ linear equations from which the unknown signal spectra $S_k(\omega)$ can be determined, at least in principle. To get a basic idea of what the solution of this system is like we neglect all phase relations and hence replace all complex quantities by their squared magnitudes averaged over a small frequency range. Performing these operations on eqns (10.29) we obtain $k$ equations relating the real quantities $s_k$ (derived from $S_k$), $s_0$ (derived from $S_0$) and $g_{ik}$ (derived from $G_{ik}$):

$$s_k = s_0 + \sum_{i=1}^{N} q_i^2 g_{ik} s_i \qquad (k = 1, 2, \ldots N) \tag{10.29a}$$

This is tantamount to superimposing energies instead of complex amplitudes and may be justified if the number $N$ of channels is sufficiently high. In a second step we assume as a further simplification that all amplifier gains and energetic transfer functions are equal, $q_i \approx q$, and $g_{ik} \approx g$ for all $i$ and $k$; furthermore, we suppose that $s_i \approx s$. Then we obtain immediately from eqn (10.29a):

$$s = \frac{s_0}{1 - Nq^2 g} \tag{10.30}$$

The ratio $s/s_0$ characterises the increase of the energy density caused by the electroacoustical system. On the other hand, the simple diffuse-field theory presented in Section 5.1 tells us that the steady-state energy density in a reverberant space is inversely proportional to the equivalent absorption area and hence inversely proportional to the reverberation time (see eqns (5.6) and (5.9)). Therefore the ratio of reverberation times with and without the system is

$$\frac{T'}{T} = \frac{1}{1 - Nq^2 g} \tag{10.31}$$

This formula is similar to eqn (10.18), but in the present case one can afford to keep the (energetic) open loop gain of each channel low enough to exclude the risk of sound colouration by feedback, due to the large number $N$ of channels. Franssen recommended making $q^2 g$ as low as 0.01; then 50 independent channels would be needed to double the reverberation time.

However, more recent investigations by Behler[10] and by Ohsmann[11] into the properties of such multi-channel systems have shown that eqn (10.31) is too optimistic in that the actual gain of reverberation time is lower. According to the latter author, a system consisting of 100 amplifier channels will increase the reverberation time by slightly more than 50% if all channels are operated with gains 3 dB below instability.

For the performance of a multi-channel system of this type it is of crucial importance that all open loop gains are virtually frequency independent within a wide frequency range. To a certain degree, this can be achieved by carefully adjusted equalisers which are inserted into the electrical paths. In any case there remains the problem that such a system comprises $N^2$ feedback channels, but only $N$ amplifiers gains and equalisers to influence them.

Nevertheless, systems of this kind have been successfully installed and operated at several places, for instance in the Concert House at Stockholm.[12] This hall has a volume of 16 000 m$^3$ and seats 2000 listeners. The electroacoustical system consists of 54 dynamic microphones and 104 loudspeakers. That means there are microphones which are connected to more than one loudspeaker. It increases the reverberation time from 2.1 s (without audience) to about 2.9 s. The tonal quality is reportedly so good that unbiased listeners are not aware that an electroacoustical system is in operation.

An electroacoustical multi-channel system of quite a different kind, but to be used for the same purpose, has been developed by Parkin and Morgan[13] and has become known as 'assisted resonance system'. Unlike Franssen's system, each channel has to handle only a very narrow frequency band. Since the amplification and the phase shift occurring in each channel can be adjusted independently (or almost independently), all unpleasant colouration effects can be avoided. Furthermore, electroacoustical components, i.e. the microphones and loudspeakers, need not meet high fidelity standards.

The 'assisted resonance system' was originally developed for the Royal Festival Hall in London. This hall, which was designed and constructed to be used solely as a concert hall, has a volume of $22\,000$ m$^3$ and a seating capacity of 3000 persons. It has been felt, since its opening in 1951, that the reverberation time is not as long as it should have been for optimum conditions, especially at low frequencies. For this reason, an electroacoustical system for increasing the reverberation time was installed in 1964; at first this was on an experimental basis, but in the ensuing years several aspects of the installation have been improved and it has been made a permanent fixture.

In the final state of the system, each channel consists of a condenser microphone, tuned by an acoustical resonator to a certain narrow frequency band, a phase shifter, a very stable 20 W amplifier, a broadband frequency filter and a 10- or 12-inch loudspeaker, which is tuned by a quarter wavelength tube to its particular operating frequency at frequencies lower than 100 Hz. (For higher frequencies, each loudspeaker must be used for two different frequency bands in order to save space and therefore has to be left untuned.) The feedback loop is completed by the acoustical path between the loudspeaker and the microphone. For tuning the microphone, Helmholtz resonators with a $Q$ factor of 30 are used for frequencies up to 300 Hz; at higher frequencies they are replaced by quarter wave tubes. The loudspeaker and the microphone of each channel are positioned in the ceiling in such a way that they are situated at the antinodes of a particular room mode.

There are 172 channels altogether, covering a frequency range 58–700 Hz. The spacing of operating frequencies is 2 Hz from 58 Hz to 150 Hz, 3 Hz for the range 150–180 Hz, 4 Hz up to 300 Hz, and 5 Hz for all higher frequencies.

In Fig. 10.18 the reverberation time of the occupied hall is plotted as a function of frequency with both the system on and the system off. These results were obtained by evaluating recordings of suitable pieces of music which were taken in the hall. The difference in reverberation time below 700 Hz is quite obvious. Apart from this, the system has the very desirable effect of increasing the overall loudness of the sounds perceived by the listeners and of increasing the variety of directions from which sound reaches the

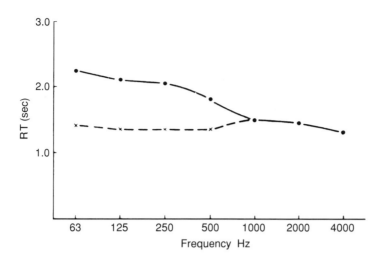

*Figure 10.18* Reverberation time of occupied Royal Festival Hall, London, both with (•————•) and without (×−−−−×) 'assisted resonance system'.

listeners' ears. In fact, from a subjective point of view, the acoustics of the hall seem to be greatly improved by the system and well-known performers have commented enthusiastically on the achievements, particularly on a more resonant and warmer sound.[14]

During the past years, assisted resonance systems have been installed successfully in several other places. These more recent experiences seem to indicate that the number of independent channels need not be as high as was chosen for the Royal Festival Hall.

The foregoing discussions should have made clear that there is a great potential in sophisticated electroacoustical systems for creating acoustical environments which can be adapted to nearly any type of performance. Their widespread and successful application depends, of course, on the technical perfection of their components and on further technical progresses, and equally on the skill and experience of the persons who operate them.

In the future, however, the 'human factor' will certainly be reduced by more sophisticated systems, allowing application also in places where no specially trained personnel are available.

## References

1 Ahnert, W, Steffen F. Sound Reinforcement Engineering. London: E & FN Spon, 1999.
2 Kuttruff H, Hesselmann N. Zur Klangfärbung durch akustische Rückkopplung bei Lautsprecheranlagen. Acustica 1976; 36:105.

3 Franssen NV. Sur l'amplification des champs acoustiques. Acustica 1968; 20:315.
4 Guelke RW, Broadhurst AD. Reverberation time control by direct feedback. Acustica 1971; 24:33.
5 Meyer E, Kuttruff H. Zur Raumakustik einer großen Festhalle. Acustica 1964; 14:138.
6 Schroeder MR, Logan BF. Colorless artificial reverberation. J Audio Eng Soc 1961; 9:192.
7 Berkhout AJ. A holographic approach to acoustic control. J Audio Eng Soc 1988; 36:977.
8 Berkhout AJ, de VriesD, Vogel P. Acoustical control by wave field synthesis. J Acoust Soc Am 1993; 93:2764.
9 Berkhout AJ. de Vries D, Boone MM. Application of wave field synthesis in enclosed spaces: new developments. Proc 15th Intern Congr on Acoustics, Vol. II, Trondheim, 1995, p 377.
10 Behler G. [Investigation of multichannel loudspeaker installations for increasing the reverberation of enclosures] Acustica 1989; 69:95 [in German].
11 Ohsmann M. [Analytic treatment of multiple-channel reverberation systems]. Acustica 1990; 70:233 [in German].
12 Dahlstedt S. Electronic reverberation equipment in the Stockholm Concert Hall. J Audio Eng Soc 1974; 22:626.
13 Parkin PH, Morgan KJ. Sound Vibr 1965; 2:74.
14 Barron M. Auditorium Acoustics and Architectural Design. London: E & FN Spon, 1993.

# List of symbols

## Latin capital letters

| | |
|---|---|
| $A$ | constant, equivalent absorption area, absorption cross-section |
| $B$ | constant, frequency bandwidth, irradiation density |
| $C$ | constant, clarity index, circumference |
| $D$ | diameter, thickness, definition, diffusion constant |
| $E$ | energy |
| $F$ | force, arbitrary function |
| $G$ | strength factor |
| $G(\omega), G(f)$ | transfer function, arbitrary function |
| $H$ | transfer function, distribution of damping constants |
| $J_n(z)$ | Bessel function of order $n$ |
| $L$ | length, sound level |
| $M$ | mass |
| $M'$ | specific mass ( = mass per unit area) |
| $N$ | integer |
| $P$ | power, probability |
| $Q$ | volume velocity, quality factor |
| $Q_s$ | scattering cross-section |
| $R$ | reflection factor, radius of curvature |
| $R_r$ | radiation resistance |
| $S$ | area |
| $S(\omega), S(f)$ | spectral density or spectral function |
| $T$ | period of an oscillation, transmission factor, reverberation or decay time |
| $V$ | volume |
| $W$ | probability |
| $W(\omega), W(f)$ | power spectrum |
| $Y$ | admittance |
| $Z$ | impedance |
| $Z_r$ | radiation impedance |

## Latin lower case letters

| | |
|---|---|
| $a$ | radius, constant |
| $a(t)$ | weighting function |
| $b$ | thickness |
| $b(t)$ | weighting function |
| $c$ | sound velocity |
| $d$ | thickness, distance |
| $f$ | frequency |
| $g$ | directivity or gain |
| $g(t)$ | impulse response |
| $h$ | height |
| $h(t)$ | decaying sound pressure |
| $i$ | imaginary unit $(= \sqrt{-1})$ |
| $k$ | angular wavenumber |
| $l$ | integer, length |
| $m$ | integer, attenuation constant, modulation transfer function |
| $n$ | integer, normal direction |
| $p$ | sound pressure |
| $q$ | amplifier gain, volume velocity per unit volume |
| $r$ | radius, distance, resistance |
| $r_{\mathrm{c}}$ | critical distance or diffuse-field distance |
| $r_{\mathrm{s}}$ | flow resistance |
| $r(t)$ | white noise (time function) |
| $s$ | scattering coefficient |
| $s(t)$ | time function of a signal |
| $t$ | time or duration |
| $t_{\mathrm{s}}$ | centre time or centre-of-gravity time |
| $v$ | velocity, particle velocity |
| $w$ | energy density |
| $x$ | coordinate |
| $y$ | coordinate |
| $z$ | coordinate |

## Greek capital letters

| | |
|---|---|
| $\Gamma(\vartheta, \varphi)$ | directional factor |
| $\Delta$ | difference, Laplacian operator |
| $\theta$ | angle, temperature in centigrade |
| $\Xi$ | specific flow resistance |
| $\phi$ | angle, autocorrelation function |
| $\Psi$ | correlation coefficient |
| $\Omega$ | solid angle |

## Greek lower case letters

$\alpha$      angle, absorption coefficient
$\beta$      angle, specific admittance
$\beta'$     phase constant
$\gamma$      angle
$\gamma'$     attenuation constant
$\gamma^2$    relative variance of path length distribution
$\delta$      decay constant, difference
$\delta(t)$   Dirac or delta function
$\varepsilon$      angle
$\zeta$      specific impedance
$\eta$      imaginary part of the specific impedance
$\vartheta$      angle
$\kappa$      adiabatic or isentropic exponent
$\lambda$      wavelength
$\xi$      real part of the specific impedance
$\rho$      density
$\sigma$      porosity, standard deviation
$\tau$      transit time or delay time
$\varphi$      phase angle
$\chi$      phase angle of reflection factor
$\psi$      phase angle
$\omega$      angular frequency

# Index